Jerusalem Studies in Philosophy and History of Science

Jerusalem Studies in Philosophy and History of Science sets out to present state of the art research in a variety of thematic issues related to the fields of Philosophy of Science, History of Science, and Philosophy of Language and Linguistics in their relation to science, stemming from research activities in Israel and the near region and especially the fruits of collaborations between Israeli, regional and visiting scholars.

Yemima Ben-Menahem
Editor

Rethinking the Concept of Law of Nature

Natural Order in the Light of Contemporary Science

Springer

Editor
Yemima Ben-Menahem
Department of Philosophy
The Hebrew University of Jerusalem
Jerusalem, Israel

ISSN 2524-4248 ISSN 2524-4256 (electronic)
Jerusalem Studies in Philosophy and History of Science
ISBN 978-3-030-96777-2 ISBN 978-3-030-96775-8 (eBook)
https://doi.org/10.1007/978-3-030-96775-8

This Springer imprint is published by the registered company Springer Nature Switzerland AG
The registered company address is: Gewerbestrasse 11, 6330 Cham, Switzerland

Introduction

There are two kinds of reasons that invite a fresh look at the concept of law of nature: one deriving from theoretical and practical changes in science, the other motivated by philosophical concerns over the traditional concept of natural law. The papers in this volume, all of which are written with contemporary science in view, address these types of concerns either separately or in tandem. Despite this common focus, the papers vary considerably in the positions they advocate and the arguments they adduce in their support. Ideally, all papers should have been distributed to contributors so that the numerous links between them could have been made more explicit, but for obvious reasons, this ideal has not been realized.

Laws of nature are commonly (stereotypically?) perceived as universal and eternal, that is, they apply everywhere and at all times to all natural events and processes. From the epistemic point of view, laws, together with initial conditions, are also thought to enable, at least in principle, the prediction and explanation of events and processes falling under their purview. Laws of nature thus organize the immense multitude of diverse phenomena into a unified structure that achieves explanatory, aesthetic, and practical goals. It is usually also part of the common notion of natural law that some laws are more fundamental than others, the latter being derivable from the former or reducible to them. The structure of science is thus believed to be hierarchic with the most fundamental laws serving as guidelines for the construction of new theories, applicable to new territories, but still abiding by the old regime of law. The merits of this picture and its contribution to the advance of science during the scientific revolution and in its wake can hardly be overestimated. And yet, as the papers collected here show, contemporary science makes it necessary to reconsider some of these characteristics and draw a more realistic picture of the status and function of laws.

Here are some of the science-based considerations that encourage revision of the traditional account. First, quantum mechanics (QM), on all of its various interpretations, strains the traditional concept of law. The most radical deviation from the classical concept, perhaps, is brought about by recent epistemic interpretations that construe quantum laws not as referring to events and processes taking place in the world, but as cataloguing the probabilities that should be assigned by rational observers to measurement results.[1] Secondly, the status of laws as truths "out there" on the one hand, and their predictive utility, on the other, have parted ways in theories such as chaos theory, where, due to the complexity of the equations, predictability may fail even when the processes themselves are taken to obey strict laws. Thirdly, the distinction between the physical content of a theory and by-products of its mathematical representation proves to be a thorny problem, aggravated in particular by gauge theories. Fourth, the hierarchic structure of science might be called into question due to the possibility of structuring the same system of laws in various ways; what is deemed fundamental may therefore vary from one systematization to another. Fifth, even sanctified laws, such as the principle of conservation of energy and momentum, have turned out to be difficult to extend to all areas of physics; in fact, the status of this conservation principle in the general theory of relativity (GTR) has been a matter of ongoing concern. Sixth, the traditional conception respects two distinctions: laws are distinguished from initial conditions and dynamical laws, applicable to specific kinds of interaction, are distinguished from the kinematical framework in which dynamical laws operate. It turns out, however, that neither of these distinctions should be seen as clear cut as it used to be. Finally, science has reached the point of confronting the baffling question regarding the origin of laws. This final question leads from scientific concerns to philosophical ones, the latter setting off not from any particular development in science, but from the very meaning of the concept of law.

Metaphors can be perplexing: How do we ascertain that an expression is used literally rather than figuratively? After all, linguistic expressions like metaphors do not wear their semantic categories on their sleeves. We are mostly confident that we understand which way an expression is being used, but cases of doubt are also familiar. Man was created in the image of God, the Bible tells us, but whether this means God has a body of which the human body is a kind of replica, or whether this is, instead, a powerful metaphor, has puzzled scholars and laypersons alike. The law of nature idiom provides another example of an expression that has received a wide spectrum of interpretations from the more literal to the metaphorical. Even though laws of nature as we know them only became central to science during the scientific revolution, the idea that nature is regular and subject to law had been voiced much earlier. Aware of the difference between legal and scientific contexts (both of which invoke the concept of law), philosophers were often puzzled about the application of the term "law" to nature. Thomas Aquinas, for example (Thomas, 1948, *Summa Theologica* I-II, Q 93), reckoned that laws could only be administered to rational beings and was therefore hard pressed to explain how nature could obey laws. While his solution—that God "imprints" the laws on nature—was rather ambiguous, Francisco Suarez was more direct.

[1] Since there is another volume in this series devoted entirely to interpretations of QM, highlighting in particular the epistemic interpretation, this issue does not figure significantly in the present volume; See *Quantum Probability, Logic* (Hemmo & Shenker, 2020).

> For the present we refer by the expression "natural law" not to that law which dwells in mankind ... but rather to that which befits all things, in accordance with the inclination imparted to them by the Author of nature .... This ... acceptation of law is therefore metaphorical, since things which lack reason are not, strictly speaking, susceptible to law, just as they are not capable of obedience. Accordingly, the efficacy of divine power, and the natural necessity resulting therefrom in this connection, are metaphorically given the name of law (Suarez, 1944, *De Legibus* chapter 1).

The metaphorical construal of the term "law" (repeated by others, for instance Spinoza and Boyle) does not actually solve the problem for it does not answer the question of what the term *does* mean. (This is of course true of many metaphors, "the image of God" included). One would have expected that with secularization the question would be considered more pertinent: Could one retain the concept of law of nature while abandoning faith in God as the author of these laws? In fact, interest in the question has waned. The more closely science became associated with the endeavor of discovering laws, it seems, the less its practitioners stopped to worry over the precise meaning of the concept of law. In moving the problem to center stage, the best system account offered by David Lewis generated a significant change, at least among philosophers. Lewis urged "Humean supervenience," "the doctrine that all there is to the world is a vast mosaic of local matters of fact, just one little thing and then another"(1986, x).[2] Laws, accordingly, have no ontological standing beyond actual facts. Rather than governing nature, laws are *our* way of organizing the facts into useful and simple structures. This deflationary account of laws has been critiqued by philosophers such as Tim Maudlin (2007), who seek to retain the fundamentality of laws.

Whether by way of agreement, or by way of critique, the best system account provides the springboard for several of the papers in this volume. In his witty Galilean dialogue, Gordon Belot wrestles with the worry that the best system conception of laws places too much weight on "soft" human criteria and too little on the input of objective reality. Hence the likelihood of different and incompatible "best" systems and consequently the threat of an idealist or relativist account of laws. Belot makes clear that Lewis was aware of this threat, but finds his hopeful response—nature's "kindness" being sufficient to secure the convergence of alternative best systems to a single one—wanting. This deficiency of the Lewisian account, we should note, is not represented by Belot as motivating the polar position of laws as more fundamental than facts. A defense of the latter, however, is put forward by Eddy Keming Chen and Shelly Goldstein, whose paper argues for a version of the "governing" law position, which they call "minimal primitivism about laws." Minimal primitivism unpacks the "governing" metaphor as meaning that laws are *constraints* on physical possibilities. It is minimal in the sense that it is committed neither to the dynamic character of laws, nor to a fundamental arrow of time along which the dynamic laws operate, generating, as it were, later states of the world on the basis of earlier ones. Surrendering these commitments

[2] As the terms "Humean mosaic" and "Humean supervenience" indicate, there were predecessors to this view, but contemporary discussions generally take off from Lewis.

distinguishes minimal primitivism from Maudlin's position (which involves both of them) and allows non dynamical laws such as conservation principles to be considered fundamental.

Two papers, one by Marc Lange, the other by Meir Hemmo and Orly Shenker, provide a critical examination of The Mentaculus—the candidate for a best system, proposed by Barry Loewer and David Albert.[3] Lange focusses on the autonomy of the special sciences—the independence of their laws from those of fundamental physics. Whereas Loewer and Albert's position is that The Mentaculus casts doubt on this autonomy (by leading, for example, to overdetermination of explanations of the phenomena that the special sciences purport to explain), Lange argues that with a little bit of "tweaking," The Mentaculus is perfectly compatible with the special sciences' autonomy and independence. In other words, The Mentaculus, according to Lange, is not in conflict with non-reductive physicalism (as Loewer and Albert have it). Lange's argument is based on an observation that has further implications than those bearing on The Mentaculus. Laws, he reminds us, are distinguished from accidental generalizations by supporting counterfactuals, that is, laws are taken to hold true under a variety of conditions that differ from the actual ones. Lange now observes that one could introduce an analogous distinction between different kinds of laws. Some laws are more robust than others in the sense that they support "counterlegals" (as Lange calls them), that is, they are taken to hold true even if the fundamental micro-level of physics were to differ from the actual one in the kinds of particles and interactions that inhabit it, and consequently differ in the laws that the fundamental level obeys. Lange cites the principle of conservation of energy as an example of a law that supports such counterlegals. Precisely its *insensitivity* to the details of the micro-level endows the principle with this stronger kind of robustness and independence.[4]

Hemmo and Shenker are also intrigued by independence and reduction, but unlike Lange, they aim at an even firmer kind of reductionism than Loewer and Albert. Indeed, Hemmo and Shenker seek to improve on The Mentaculus in this respect. Their new candidate for a "best system," "flat physicalism," is claimed to be superior to The Mentaculus in strength and simplicity: According to them, it enables the *derivation* of the past hypothesis (the assumption of a low-entropy macrostate at the beginning of the universe) and the probabilistic hypothesis (a uniform probability density distribution over microstates belonging to this macrostate, measured by the Lebesgue measure), both of which Loewer and Albert take to be *postulates*.[5] Hemmo and Shenker argue further that flat

[3] See for example Albert (2014) and Loewer (2008). I set aside small differences between these works as well as a slight difference between Lange's reading of them and that of Hemmo and Shenker.

[4] Interestingly, Wheeler apparently saw this insensitivity as a weakness rather than strength of the most general principles. See, for example Deutsch (1986).

[5] Flat physicalism pays for this advantage by accepting the existence of certain brute facts, contingent and inexplicable.

physicalism allows for the reduction of the special sciences to fundamental physics, a reduction that The Mentaculus (on their view) cannot guarantee.

John Norton and Richard Healey address the problem of the meaning and status of laws from two different points of view, both orthogonal to the best system account discussed thus far, though related to it via the persisting concern over the concept of law and the metaphysical aura surrounding it. Focusing on the notions of necessity and possibility, Norton takes issue with a prevailing unpacking of these modalities in terms of possible world semantics. For an empiricist, Norton claims, there are no metaphysical possibilities and necessities, only logical and empirical ones. Empirical possibility and necessity are anchored in evidence—what the evidence allows is possible, what it compels is necessary. Laws and theories provide concise and systematic formulations of such possibilities and necessities; they are hypothetical and contingent, though their scope is much wider than that of singular contingent truths. The novelty of this characterization is manifest in its implications: For example, depending on the strength of the evidence, both possibility and necessity come in degrees. Moreover, a proposition can be empirically necessary though its negation is not impossible.

Sharing Norton's distaste for the kind of necessity traditionally associated with laws, Healey goes a step further and forsakes not only the necessity of laws, but also their truth. His position is pragmatic: Laws are "inference tickets;" as such they are immensely useful to science even if they have no representational function and are not universally true (let alone necessarily true). He analyses a number of theories that support this pragmatic account and attest to the utility of laws that are not considered to be true in predicting phenomena, constructing models, and calibrating instruments. In parting with truth as a desideratum, this pragmatist construal of laws is clearly distinguished from both deflationary and metaphysical accounts and has affinities with instrumentalism. The analogy with inference rules, however, suggests that while truth values need not be ascribed to laws, they should be ascribed to the conclusions derived from the laws. The departure from realism is therefore specific to laws and does not involve nonrealist semantics in general.

Mathias Frisch offers another way of escaping the dichotomy between the fundamental interpretation of laws and its best system rival. Inspired by Feynman's Babylonian conception of laws (Feynman, 1967), Frisch contends that while the truth values and scope of laws are objective matters of fact, the *structure* of science is up to us. Typically, there are various ways of organizing a system of laws into a logical structure that leads from premises to conclusions. Though it is quite common to view the former as more fundamental than the latter, in the light of the existence of alternative ways of structuring, fundamentality is not an attribute of the laws as such but of particular systematizations. On the basis of detailed examples from contemporary science, Frisch goes on to argue that this plasticity of structure calls into question not only any rigid hierarchy of laws, but also the distinction between dynamical laws and kinematical frameworks. Lorentz invariance, for example, could be construed as a consequence of dynamical laws or as a constitutive element of the kinematic framework. Similarly, GTR allows for both a dynamic and a kinematic construal of gravity. The same plasticity leads Frisch to question the objectivity of the causal nexus and the distinction between matters of law and matters of fact as reflected in the distinction between laws and initial conditions. The latter distinction is also undermined by the best system account, but Frisch sees no reason to expect the existence of such a best system—a unique system superior to all others.

Initial conditions are also taken up by Matt Farr, who examines their explanatory role and the link between this role and the direction of time. Since, in general, laws are formulated as conditional statements, their explanatory and predictive uses require (statements describing) initial conditions as premises. The initial conditions of the universe as a whole confront us with a problem, for being initial, these conditions do not seem to be explicable by other conditions and are therefore just postulated or accepted as contingent brute facts. Moreover, the special status afforded to them as explanatory but inexplicable raises the question of whether, by endowing them with this special explanatory role, we are *ipso facto* committing ourselves to an objective arrow of time. Farr answers the latter question in the negative. His argument adduces the conventionality of the direction of time first proposed by Reichenbach in his discussion of the entropy gradient in statistical mechanics. This conventionality also bears on how we construe causation: Causal asymmetry, on this view, could in principle be reversed, resulting in teleological explanations that take earlier events to be explained by later ones.[6]

Fundamentality comes into sharp focus in Shelly Goldstein and Nino Zanghi's paper. One of the most challenging problems of physics is to distinguish between facts and modes of description. The challenge, in other words, is not to mistake artifacts of the description we happen to use (a particular system of coordinates, say) for objective facts. Can we transcend descriptions and touch base with reality? Aware of the connection between this question and Kant's philosophical project, Goldstein and Zanghi profess to invest physical theory with "a somewhat Kantian aspect." The historical debate between relationist and substantivalist theories of space and time clearly illustrates the difficulty of separating representation and reality, a difficulty that is also at the heart of recent discussion of gauge symmetry. In their paper, Goldstein and Zanghi put forward a radical form of relationism, which is in line with Barbour's path breaking work on relationism and shape dynamics, and which they apply to both space and time. On this version of relationism, in addition to the usual translational and rotational symmetries that leave the physical content of a theory invariant, expansion and contraction are also considered as symmetries that yield equivalent descriptions of the same physical reality. Metrical parameterization thus disappears from what we take to be real and reinterpreted as gauge. Reality, on this view, consists in free motion in shape space while physical theories as we know them are emergent, owing their structure in large degree to modes of representation that the human mind finds manageable. Goldstein and Zanghi take

[6] A similar point about causation comes up in Frisch's paper. Note also the relation between Farr's conclusion about the objectivity of the direction of time and the Chen-Goldstein argument for minimal primitivism about laws.

Bohmian quantum theory to be particularly amenable to this vision and therefore devote a substantial part of their paper to a detailed examination of this theory.

A number of papers remind us that the laws we know are not actually universal. J. Brian Pitts addresses the controversial case of the principle of conservation of energy. He brings together two controversies centered on this principle: one pertaining to the conservation of energy and momentum in GTR, the other to the compatibility of the principle with mental causation (as understood in many versions of dualism/interactionism). Both of these problems, Pitts argues, have stumbled over confusions, in particular imprecision in applying Noether's first theorem to the issues at hand. Surveying the history of these controversies, he comes to the following interrelated conclusions. First, the difficulties in extending energy conservation to GTR notwithstanding, Noether's theorem does apply to the theory, yielding symmetries and conservation laws that are all too often neglected. Second, in view of the former, the attempt to draw support for dualism from the failure of energy conservation in GTR is unsuccessful.

We now come to the tough question regarding the origin of laws and John Archibald Wheeler's ambitious attempt to answer it.[7] In their captivating paper, Alex Blum and Stefano Furlan tell the story of how Wheeler, who was at first strongly committed to the universality of the laws of fundamental physics, gradually moved to a position that underlines their limits, and how recognition of these limits eventually led him to his celebrated "law without law" position. The paper charts the contributions of developments in QM and GTR to Wheeler's voyage and distinguishes between different versions of the "law without law" program. Specifically, Wheeler came to the conclusion that absolute lawlessness does not constitute a feasible explanatory option and modified his position accordingly. Lawlessness in a more restricted sense is also the subject of my paper. I discuss three examples: 1. Feynman's derivation of the principle of least action from his path integral version of quantum theory; 2. An amended version of nonreductive physicalism; 3. Lawlessness as linked to randomness and its compatibility with determinism.

To conclude this volume, Eli Lichtenstein, provides a thought-provoking inquiry into the complex history of the concept of law. Whereas the other papers in this collection (whether leaning toward a fundamentalist conception of laws or toward a Humean-Lewisian account), take it for granted that the connection of laws of nature with a divine lawgiver has been completely severed, Lichtenstein shows that the history of the concept of law is much more nuanced. He detects traces of divine attributes that found their way into secular conceptions of nature and humanity and explores the metaphysical and ethical implications of these conceptions.

[7] Proponents of the best system account presume to obliterate the question, but in my paper, I express doubts about this presumption.

References

Albert, D. Z. (2014). *After physics.* Harvard University Press.

Deutsch, D. (1986) On Wheeler's notion of "law without law" in physics. *Foundations of Physics, 16*(6), 565–572.

Feynman, R. (1967). *The character of physical law.* MIT Press.

Hemmo, M., & Shenker, O. (2020). *Quantum probability, logic.* Springer.

Lewis, D. (1986). *Philosophical papers* I. Oxford University Press.

Loewer, B. (2008). Why there is anything except physics. In: J. Hohwy & J. Kallestrup (Eds.), *Being reduced: New essays on reduction, explanation, and causation* (pp. 149–163). Oxford University Press.

Maudlin, T. (2007). *The metaphysics within physics.* Oxford University Press.

Suarez, F. (1944). In B. Scott (Ed.), *Selections from three works of Francisco Suarez S.J. Theclassics of international law*. The Clarendon Press.

Thomas Aquinas. (1948). *The Summa Theologica*, translation of the Fathers of the English Dominican Province, Burns. Oates and Washbourne.

Department of Philosophy,
The Hebrew University of Jerusalem,
Jerusalem, Israel

Yemima Ben-Menahem

Contents

Contributors

Gordon Belot University of Michigan, Ann Arbor, MI, USA

Yemima Ben-Menahem Department of Philosophy, The Hebrew University of Jerusalem, Jerusalem, Israel

Alexander Blum Max Planck Institute for the History of Science, Berlin, GermanyAlbert Einstein Institute, Potsdam, Germany

Eddy Keming Chen Department of Philosophy, University of California, San Diego, CA, USA

Matt Farr Department of History and Philosophy of Science, University of Cambridge, Cambridge, UK

Mathias Frisch Leibniz University Hannover, Hannover, Germany

Stefano Furlan Max Planck Institute for the History of Science, Berlin, GermanyUniversity of Geneva, Geneva, Switzerland

Sheldon Goldstein The State University of New Jersey, Rutgers, New Brunswick, NJ, USA

Richard Healey University of Arizona, Tucson, AZ, USA

Meir Hemmo University of Haifa, Haifa, Israel

Marc Lange University of North Carolina, Chapel Hill, NC, USA

Eli Lichtenstein University of Michigan, Ann Arbor, MI, USA

John Norton University of Pittsburgh, Pittsburgh, PA, USA

J. Brian Pitts University of Lincoln, Lincoln, UKUniversity of Cambridge, Cambridge, UKUniversity of South Carolina, Columbia, SC, USA

Orly Shenker The Hebrew University of Jerusalem, Jerusalem, Israel

Nino Zanghi Dipartimento di Fisica dell'Universita di Genova and INFN sezione di Genova, Genova, Italy

Ratbag Idealism

Gordon Belot

Abstract A discussion of the sense in which reality is mind-dependent for Kant and for David Lewis. Plus a lot about space-aliens (and a bit about pimple-worms).

Keywords Best-system analysis · David Lewis · Immanuel Kant · Transcendental idealism

> Well, I grant that there are indeed these ineffable, unanswerable questions—but who ever promised me that there wouldn't be such questions? The worst part of being a quidditist, I guess, is that I am in danger of agreeing with Kant about something (Kant as told by Rae Langton, anyway) and I'd never want *that* to happen!
>
> David Lewis

First Day

Sagredo I spent a lot of time working on an anthology of philosophical horror that I am editing, *The Human Serpent is Over Everything: An Anthology of Philosophical Horror.* I am not sure that's the right title, though. I would prefer something longer and more obscurely allusive.

Salviati I know what you mean. All of my manuscripts are entitled "Miscellaneous Tracts on Some Curious, and Very Interesting Subjects in Mechanics, Physical-Astronomy, and Speculative Mathematics," until some well-meaning editor gets involved. Your project sounds very odd.

G. Belot (✉)
University of Michigan, Ann Arbor, MI, USA
e-mail: belot@umich.edu

Y. Ben-Menahem (ed.), *Rethinking the Concept of Law of Nature*,
Jerusalem Studies in Philosophy and History of Science,
https://doi.org/10.1007/978-3-030-96775-8_1

Sagredo Yes! So far as I can tell, what really creeps out most scientists and philosophers of science is the idea that the world as described by science is at least partially constituted by the mental. I want to publish a collection of quaint philosophical writing in this vein—I reckon that horror of any sort will be a huge money maker again next time we have a plague year.

Salviati Unquestionably. But why 'quaint'?

Sagredo Well, these days it is sociologists that come out with the really scary stuff—you know, the stuff that gets Nobel Prize winners to complain about philosophers. But back in the day, it was philosophers who provoked scientists to complain about philosophers. Think of the conventionalist-pragmatist-Bergsonian positivism of Le Roy, according to which scientific facts are made by the scientists who observe them, rather than being externally imposed. It sounds even better in French, especially if you leave out most of the words: *les faits sont faits.*

Salviati Most things do. Is this the same Le Roy that drove Poincaré to fantasize about strange alien beings and to argue that their science would coincide with our own?

Sagredo Yes, that's the one. It seems that an exposure to a certain sort of horror drives a certain sort of person to a certain sort of science fiction, as a way of affirming their unshakable confidence in the objectivity of science. Mach's phenomenalist positivism drove Planck to pronouncements on the inevitable congruence between human and Martian science.

Simplicio Planck's student Schlick followed a similar path.

Salviati Okay. But I am still puzzled by the 'quant' bit. Some of the most influential, interesting, and widely-read philosophers seem like good candidates for your anthology, if I understand what it is all about.

Sagredo Well, I am certainly going to include Peirce. Perhaps that is who you are thinking of? He famously tells us that:

> The opinion which is fated to be ultimately agreed to by all who investigate, is what we mean by truth, and the object represented in this opinion is the real.

This is pretty spooky! For one thing, it seems to imply that there can be no buried secrets, facts about the past that it is beyond our power to discover. But then to counter the spookiness, he elsewhere turns to science fiction. He tells us that it is at least possible that some facts that we might think would be buried forever could nonetheless come to light: perhaps beings of a distant planet have recorded historical events on Earth and the recordings will come into our possession. In general, he seems to have really enjoyed a good science fiction tale:

> if we ever succeed in interpreting the communications of the inhabitants of Mars, I will risk a nickel that their first dispatch will read: 'O yes, we know all about your attempts to learn from us. Well learn this: you are the most stupid of all organized beings. You even allow creatures to have a voice or a vote in your government who hardly can solve a partial differential equation in its generality and have not mastered the higher reaches of the calculus of variations. Bring a diploma from a colony of bees before you presume to seek instruction from Martians!'

Salviati I am glad you are including Peirce! But I had others in mind.

Sagredo Pray tell.

Salviati Let me introduce a technical notion: a *ratbag idealist* is someone who takes some pretheoretically fundamental aspect of the world—its causal structure, or its laws, or its spatiotemporal geometry—to depend on human cognitive constitution.

Simplicio Of course, this notion is due to Lewis—although he only mentions the possibility of ratbag idealism concerning laws—and he has a decisive argument that the correct account of laws saves us from ratbag idealism.

Salviati Let us come back to Lewis. First ...

Sagredo What is a 'ratbag,' exactly?

Simplicio No one knows, except Australians. Sometimes they say a ratbag is a lunatic, sometimes a loveable rascal, sometimes a troublemaker. It is hard to know exactly what Lewis had in mind when he introduced ratbag idealists about laws:

> Now, some ratbag idealist might say that if we don't like the misfortunes that the laws of nature visit upon us, we can change the laws—in fact, we can make them always have been different-just by changing the way we think! (Talk about the power of positive thinking.) It would be very bad if my analysis endorsed such lunacy.

Salviati First, Sagredo, are my ratbag idealists the sort of characters that you are interested in?

Sagredo I am unsure because I don't know how to think of the notion of dependence that figures in your characterization of ratbaggery.

Salviati Let us agree to think of it this way: a good test of whether someone might be a ratbag idealist is whether they think that, for all we know, there may be other beings differently constituted, for whom geometry, or laws, or causal facts might be different.

Sagredo Agreed. Then your ratbag idealists are exactly the sort of people I have in mind. Doesn't it give you goosebumps just thinking about people like that?

Salviati Thinking about Kant gives you goosebumps? I think that might be unusual.

Sagredo I guess I wasn't really thinking of Kant. I have my doubts that he was a ratbag idealist.

Salviati Hmm. Perhaps we can continue our discussion on that point tomorrow.

Second Day

Salviati Consider terms in which Kant explains the transcendental ideality of space in the *Critique of Pure Reason.*

> We can accordingly speak of space, extended beings, and so on, only from the human standpoint. If we depart from the subjective condition under which alone we can acquire

> outer intuition, namely that through which we may be affected by objects, then the representation of space signifies nothing at all. (A26/B42)

> Since we cannot make the special conditions of sensibility into conditions of the possibility of things, but only of their appearances, we can well say that space comprehends all things that may appear to us externally, but not all things in themselves, whether they be intuited or not, or by whatever subject they may be intuited. For we cannot judge at all whether the intuitions of other thinking beings are bound to the same conditions that limit our intuition and that are universally valid for us. (A27/B43)

This is a form of ratbag idealism.

Sagredo Very nice! But I hardly think that Kant, in the eighteenth century, would have had space aliens in mind.

Salviati As someone in the habit of flipping to final pages of long books to find out how they end, I can assure you that Kant *did* like to think about space aliens. Having observed that asking someone to back up a pronouncement of belief with a wager is a good way to help them to clarify how strongly held their belief really is, Kant tells us that

> If it were possible to settle by any sort of experience whether there are inhabitants of at least some of the planets that we see, I might well bet everything that I have on it. Hence I say that it is not merely an opinion but a strong belief (on the correctness of which I would wager many advantages in life) that there are also inhabitants of other worlds. (A825/B853)

Sagredo I don't know whether to be more surprised that Kant was so sure that we had company or that he *didn't* disapprove of betting.

Simplicio Kant's chum Heilsberg recalled that as undergraduates, they and another chap, Wlömer, made enough money at billiards that eventually they could no longer find anyone to play against—at which point they switched to the card game *l'hombre.* After Kant's death, his house became a student tavern, *Au Billard Royal,* complete with a bowling-alley and a billiards-room.

Salviati So far as extra-terrestrials go, there is more. His early essay, "Universal Natural History and Theory of the Heavens, or, Essay on the Constitution and the Mechanical Origin of the Whole Universe according to Newtonian Principles," shows that Kant liked to think about space aliens *a lot.*

There Kant takes space to be infinite and Euclidean, of course, but he also takes the material world to have a centre of attraction at which the density of matter achieves its maximum, with average matter density falling off as one travels away from this point (1: 311 f.). He thinks that inhabited planets can be found throughout infinite space. Further, since there is a lower bound to the possible degree of intellectual perfection of a spiritual being, but no upper bound on the intellectual perfection of such a being,

> if a law is to be in place according to which the domiciles of intelligent creatures are distributed in the order of their relation to the common centre point, we shall have to place the lowest and least complete type that constitutes, as it were, the beginning of the type of the spiritual world, at that region that can be called the beginning of the entire universe in order to fill simultaneously with this and in equal progression all infinity of time and spaces

> with increasing degrees of perfection of the capacity to think and as it were gradually to approach the goal of the highest excellence, namely the divinity without, however, ever being able to attain it. (1: 331).

Kant sees a related hierarchy within our solar system.

> The material of which the inhabitants of different planets, indeed even the animals and plants on them, are formed must altogether be of a lighter and finer type and the elasticity of the fibres together with the advantageous arrangement of their build be more perfect the further away they are away from the Sun. (1: 358)

And the gradations in material fineness and elasticity are reflected in intellectual and moral gradations. For Kant, there are of course only six planets in the solar system: Mercury, Venus, Earth, Mars, Jupiter, and Saturn.

> Human nature, which occupies as it were the middle rung on the ladder of beings, sees itself as being between the two extreme limits of perfection, equally distant from both ends. If the idea of the most sublime classes of rational creatures that inhabit Jupiter or Saturn arouses their jealousy and humiliates them by the knowledge of their own baseness, then they can be satisfied again and comforted by the sight of the low stages on the planets Venus and Mercury, which are lowered far below the perfection of human nature. (1: 359)

Simplicio Elsewhere, Kant suggests that if only our sense organs were more fine, we could perceive magnetic fields (A226/B273).

Sagredo Very well—as far as the stuff about space aliens goes, I am happy to concede that Kant positively relished the idea that we are not alone. But I simply do not believe that the possibility of aliens whose space and time are different from ours is what Kant was alluding to in the passages you started us with today. Indeed, I recall once sitting next to a Kant scholar at dinner who assured me that in passages like this Kant has in mind God, for whom there is no space and there is no time.

For consider Kant's distinction (B144–B150) between sensible and intellectual intuition and his account of their respective relation to the understanding. In beings like ourselves, the understanding operates on material given to it via sensible intuition, and in this way cognizes objects in space and time. We can distinguish between two types of sensible intuition: pure intuition (space and time, the forms of sensible intuition) and empirical intuition (the contribution of sensation). For a being like God, however, understanding is itself intuited and to cognize an object is to produce that object. For such a being, nothing is externally given. There is no sensation, no sensible intuition, no form of sensible intuition—there is no space and no time. Kant later tells us that even the possibility of intellectual intuition is something that we cannot understand (B307).

This is what is going on when he tells us that "we cannot judge at all whether the intuitions of other thinking beings are bound to the same conditions that limit our intuition and that are universally valid for us." He has in mind not beings whose forms of sensible intuition are different from our own, but a Being whose intuition is not sensible at all.

Salviati I have heard people say that. But I think that there are some texts that are difficult to square with such a reading.

One is a note that Kant inserted in his copy of the first edition of the *Critique,* immediately following the first passage that we started with today.

> Perhaps all created beings are bound to it, that we do not know. This much one can know, that it is a merely sensible form. The most important thing is that it yields a determinate concept *a priori,* and through inner intuition we would not have sensations, thus no empirical representations and no science of objects *a priori.*

The question raised here is not whether space and time are forms of sensible intuition for God (the one being with intellectual rather then sensible intuition), but rather whether space and time are forms of sensible intuition for all ordinary cognizing beings, or just for humans.

The second example, new to the second edition, is General Remark IV of the Transcendental Aesthetic (B71 f.).

In the first half of this Remark, Kant observes that in natural theology, one does not regard space and time as conditions of the intuition of God. He then notes that this means that they therefore cannot be regarded as objective forms of all things—since if they were such, the intuitions of God would need to be subject to them. The natural option, then, is to regard them as forms of sensible intuition—that type of intuition which is dependent on the separate existence of objects, to be distinguished from that type of intuition (here called *original intuition*), which would appear to pertain only to God (the *original being*), through which objects are given.

In the second half of the Remark, Kant tells us, as clearly as he can, that for all we know, we may share our world with beings for whom space and time are not the same as they are for us—and also distinguishes, as clearly as he can, between this thought and the thought that for God there is no space and time.

> It is also not necessary for us to limit the kind of intuition in space and time to the sensibility of human beings; it may well be that all finite thinking beings must necessarily agree with human beings in this regard (though we cannot decide this), yet even given such universal validity this kind of intuition would not cease to be sensibility, for the very reason that it is derived (*intuitus derivativus*), not original (*intuitius originarius*), thus not intellectual intuition, which for the ground already adduced seems to pertain only to the original being, never to one that is dependent as regards both its existence and its intuition (which determines its existence in relation to given objects); although the last remark must be counted only as an illustration of our aesthetic theory and not as a ground of its proof.

Sagredo This is disappointingly convincing. It seems that Kant is more correct, more trivial, and more Anglo-Saxon than I took him to be.

Salviati But still deeply weird! In the second edition version of the Transcendental Deduction, he makes it clear that his ratbaggery extends even to the categories—and so to the causal structure of the world of experience.

> But for the peculiarity of our understanding, that it is able to bring about the unity of apperception *a priori* only by means of the categories and only through precisely this kind and number of them, a further ground may be offered just as little as one can be offered for why we have precisely these and no other functions for judgment or for why space and time are the sole forms of our possible intuition. (B145 f.)

Sagredo Now you have gone too far! It sounds like you are attributing to Kant a view on which the categories are merely psychological apparatus with which we happen to find ourselves equipped. But Kant explicitly tells us that

> the concept of cause, which asserts the necessity of a consequent under a presupposed condition, would be false if it rested only on a subjective necessity, arbitrarily implanted in us, of combining certain empirical representations according to such a rule of relation I would not be able to say that the effect is combined with the cause in the object (i.e., necessarily), but only that I am so constituted that I cannot think of this representation otherwise than as so connected; which is precisely what the skeptic wishes most.... (B168)

Salviati Ah! Perhaps it will help to zoom out a bit from the passage you quote. In the final section of the revised version of the Transcendental Deduction, Kant considers three accounts of the origins of our concepts and he links these with three accounts of the generation of living beings, which he calls *generatio aequivoca,* epigenesis, and preformation.

Simplicio Kant lived at a time of radical transformation in the scientific understanding of embryology. Kant himself shifted his views on this topic over time. In this he was certainly influenced, to some extent, by his younger contemporary Blumenbach. In the first edition of his famous work on *Bildungstrieb* Blumenbach defends the preformation account (and reports that boys in the Middle East are born pre-circumcised). In the second edition, admired by Kant, Blumenbach defends epigenesis (and discusses jackelopes).

Sagredo You seem very well-informed, Simplicio. Perhaps you could tell me a bit more about what Kant might have had in mind in mentioning these accounts of the growth and development of organisms?

Simplicio On the epigenetic account, an individual animal or plant develops out of material that is initially formless, with the form of the individual emerging gradually over time. On the preformation account, each animal or plant is fully formed from the moment it comes into being (which on, some versions of this theory, is when the first member of its species comes into being—on one such version, each human is preformed in a seed in a seed in a seed ... in Adam, on another, each is preformed in an egg in an egg in an egg ... in Eve). The term *generatio aequivoca* has a tangled history of use. But in a footnote to §80 of the *Critique of Judgement* (5: 419 f.), Kant distinguishes between *generatio homonyma,* in which the being produced is of the same kind as those which produce it, and *generatio heteronyma,* in which the being produced is of a different kind. Kant recognizes two types of *generatio heteronyma*: *generatio univoca,* in which organic things gives rise to organic things of different types (as a far-fetched possible example, he suggests a scenario in which, over the course of generations, aquatic animals become terrestrial animals), and *generatio aequivoca* in which a living being arises out of non-living matter. By way of illustration, consider an interesting case discussed by Blumenbach. The pimple-worm (*Hydatis finna*) is found only in domestic varieties of swine, never in wild varieties. Now, we can be sure that domestic swine are descendants of wild swine. So it is reasonable to assume that the first pimple-worms originated some time after the first domestication of swine. So here we have a case *generatio heteronyma*: either these first pimple-worms had no parents (*generatio aequivoca*) or had parents of some other species (*generatio univoca*).

Salviati That *is* an interesting case. But let us leave the pimple-worms alone for the time being. Naturally, Kant compares the empiricist account of concept formation with the notion of *generatio aequivoca*: the categories and the pure forms of sensible intuition are *a priori* and cannot be the result of experience. His own account is compared with epigenesis: the categories "contain the grounds of the possibility of all experience in general from the side of the understanding." A third approach is compared with preformation, an approach on which "subjective predispositions for thinking, implanted in us along with our existence by our author in such a way that their use would agree exactly with the laws of nature along which experience runs...." It is this third approach that is subject to the objection levelled in the passage that you quoted to us, Sagredo.

Sagredo Wait—I am having a little trouble picturing how this preformationist view is supposed to work. Does Kant have somebody in mind here?

Salviati It appears that he does: in the corresponding passage in the *Prolegomena,* he mentions Crusius (4: 320 n.). I suspect that Simplicio could tell us a thing or two about him?

Simplicio The semi-empiricist, semi-rationalist system of Crusius is extremely interesting. In his *Sketch of the Necessary Truths of Reason,* he identifies the *principle of the inseparable,* according to which it is genuinely impossible to separate things which cannot be separated in our thought, and the *principle of the uncombinable,* according to which it is genuinely impossible to combine things that cannot be combined in our thought. These principles go beyond mere logic: indeed, it is not just that their denial entails no contradiction, we can even imagine beings whose thought is not governed by them. It is his view that each thing that comes into existence has a cause, that this follows from the principles of human reason, and that it does not follow from logic alone. Special cases aside, "the essence of our understanding is the criterion of truth."

Salviati Excellent. It is that sort of view that Kant is complaining about in the passage that you pointed us towards, Sagredo. Recall that there he says that on a preformation view "I would not be able to say that the effect is combined with the cause in the object: (i.e., necessarily), but only that I am so constituted that I cannot think of this representation otherwise than as so connected...." Of course, we are to understand that on Kant's own account, we *can* say that the effect is combined with the cause in the object, because under transcendental idealism, the categories are constitutive of experience. Things are quite different on the preformation view: on this view, our thought is required to mirror the causal relation between things in themselves rather than the structure of appearances: but as Kant tells us, "there are only two ways in which a *necessary* agreement of experience with the concepts of its objects can be thought: either the experience makes these concepts possible or these concepts make the experience possible" (of course, closer examination shows the former to be unacceptable). The preformation view doesn't fall under either of those conditions—and it is for this reason that it renders the concept of cause false. So his point is not that there is something wrong with views on which it is at least notionally possible for distinct types of beings to be equipped with different systems of categories—it is, rather, that there is something wrong with any view, empiricist or preformationist, on which in order for us to know that the same cause always produces the same effect, we must know that the structure of a mind-independent world obeys the causal principle(s) implicit in the structure of our minds.

This point is, arguably, more clearly expressed in the corresponding passage in the first edition (A128 f.). If the objects our cognition were things in themselves, could we have an a priori concept of them? No. For either these concepts would derive from experience or from ourselves. Clearly, the first of these options would not lead to a priori concepts (*generatio aequivoca* again). Neither would the second:

> If we take them from ourselves, then that which is merely in us cannot determine the constitution of an object distinct from our representations, i.e., be a ground why there should be a thing that corresponds to something we have in our thoughts, and why all this representation should not instead be empty.

The solution, of course, is to take the objects of experience to be mere appearances rather than things in themselves.

Sagredo Hummph. That is enough for me for one day, I think.

Third Day

Simplicio Can we finally talk about Lewis?

Salviati Yes! Perhaps we can begin by agreeing that Lewis is a ratbag idealist about laws of nature?

Simplicio That is just what I deny—because Lewis denies it.

Salviati Let us begin at the beginning, then. In his earliest published discussion of laws of nature, Lewis suggests that

> a contingent generalization is a *law of nature* if and only if it appears as a theorem (or axiom) of each of the true deductive systems that achieves a best combination of simplicity and strength. (73)

Sagredo This is a very interesting idea! But what has been done in the intervening years to substantiate Lewis's bold conjecture that the required intersubjective notions of simplicity and balance between simplicity and strength exist?

Salviati Hmm. I suspect that it will not be fruitful to pursue that line of questioning here.

Simplicio I disagree.

Sagredo Good. Rather than considering laws of nature, suppose that we consider a simpler case that is under better conceptual control: geometry. Suppose that someone were to conjecture that there is a natural intersubjective simplicity ordering on geometries. That would be very interesting! But a problem would immediately arise: it is not hard to find experts disagreeing about which of the classical geometries is simplest—Euclidean geometry, hyperbolic geometry, and elliptic geometry each has its fans.

Simplicio Granted. But I am confident that we will not find the same thing in the more complicated setting of laws of nature. Also, many fans of the best-system approach take talk of best balance between simplicity and strength to be a place-holder for some more subtle story about how scientists use data to choose between theories—and I think you will agree that there is little disagreement among expert scientists about which hypotheses are best-supported by data. Also, even if our program does rely on an unsubstantiated empirical conjecture, it is still a better philosophical account than its extant competitors. So it is the best choice among our available options—and after all, each of us *does* need to choose a favourite account of laws of nature, or where would we be?

Sagredo I see.

Salviati Yes. Let us return to Lewis. After the passage quoted above, he goes on to make the obvious explicit when he suggests that we think of the system of truths that best balances simplicity and strength in the following terms:

> Imagine that God had decided to provide mankind with a *Concise Encyclopedia of Unified Science,* chosen according to His standards of truthfulness and our standards of simplicity and strength. (74)

The notions of simplicity and of balance between simplicity and strength are at best species-relative. So, on the best-system approach, the notion of a law of nature is likewise species-relative. Lewis is a ratbag idealist about laws of nature.

Simplicio I feel compelled to point out that Lewis has a response:

> it does not follow that lawhood depends on us in the most straightforward way: namely, that if our standards were suitably different, then the laws would be different. For we can take our actual standards as fixed, and apply them in asking what the laws would be in various counterfactual situations, including in counterfactual situations in which people have different standards—or in which there are no people at all. (123)

Sagredo I don't see how rigidification will help Lewis here. Suppose that elsewhere in the galaxy there is a species of brainiacs who stand to us as we stand to dogs, intellectually. We can consider three sets of true generalizations at our world: the d-laws that are consequences of the set of truths that best balance simplicity and strength according to dogs; the h-laws that are consequences of the set of truths that best balance simplicity and strength according to humans; and the b-laws that are consequences of the set of truths that best balance simplicity and strength according to brainiacs. We can, if we like, insist that the word 'law' in our mouths always picks out the h-laws, even when we are entertaining counterfactual scenarios in which we come to have the intellectual capacities of dogs or brainiacs—but this is not to say that brainiac science would or should be concerned with h-laws rather than b-laws.

Salviati Yes. Lewis himself seems to have come around to this assessment.

> I used to think that rigidification came to the rescue: in talking about what the laws would be if we changed our thinking, we use not our hypothetical new standards of simplicity and strength and balance, but our actual and present standards. But now I think this is a cosmetic remedy only. It doesn't make the problem go away, it only makes it harder to state. (232)

Simplicio I know all that, of course—but there is nothing I can do in the face of my compulsion to mention rigidification every time the spectre of ratbag idealism is raised. But now we can enjoy Lewis's decisive rebuttal:

> Maybe some of the exchange rates between aspects of simplicity, etc., are a psychological matter, but not just anything goes. If nature is kind, the best system will be *robustly* best—so far ahead of its rivals that it will come out first under any standards of simplicity and strength and balance. We have no guarantee that nature is kind in this way, but no evidence that it isn't. It's a reasonable hope. Perhaps we presuppose it in our thinking about law. I can admit that *if* nature were unkind, and *if* disagreeing rival systems were running neck-and-neck, then lawhood might be a psychological matter, and that would be very peculiar. I can even, concede that in that case the theorems of the barely-best system would not very well deserve the name of laws. But I'd blame the trouble on unkind nature, not on the analysis; and I suggest we not cross these bridges unless we come to them. (232 f.)

I think you will agree that settles the matter. Lewis is not a ratbag idealist about laws.

Sagredo I am a little puzzled by this, frankly. Let us return to something you said a little while ago, Simplicio. You remarked that there is little disagreement between experts about how to choose theories in light of evidence. I found that claim surprising. Dirac famously thought that one ought to prefer the more mathematically beautiful theory whether or not it was simpler than its competitors. In this, he self-consciously departed from the standards of other physicists. Or again, today there vigorous debate over the question whether the mathematical fecundity of string theory enhances its plausibility as physics.

But now I see that perhaps you are hoping that the scope of such disagreement is narrow enough that we may, with luck, live at a world at which each of the sets of standards accepted by various experts will determine the same set of laws. The laws in *Dirac's Concise Encyclopedia of Science* would be exactly the same as the laws of *Bohr's Concise Encyclopedia of Science,* and so on.

Simplicio Yes, exactly. Lewis long emphasized (74, 124) that with luck, any vagueness in our standards of simplicity and of balance will be irrelevant because all reasonable standards will determine the same laws at our world. Only later did he note that, with luck, the dependence of laws on us (via their dependence on our standards) would likewise evaporate.

Sagredo I see. Even the intra-species version of this idea is very interesting. I would love to hear more about why you expect the sort of variation in the standards between scientists that we see at present to become irrelevant in the limit of complete knowledge of the distribution of (non-nomic) properties at a world. Perhaps you have some reason to think that the range of this variation is decreasing over time?

Or perhaps you have some reason to think that although this range remains constant in time, it tends to make less difference to theory choice, as time goes on?

Salviati I suspect that it will not be fruitful to pursue that line of questioning here.

Sagredo I see. Well, let us grant, then, that the range of standards used by human scientists is narrow enough that we can indeed reasonably hope that, with luck, they will underwrite a unique set of best-system laws. Now, think what Kant and Peirce would say at this point: a lot more luck will be required if our laws are to coincide with those of Martians, given their massive cognitive superiority to us—just as, very plausibly, the d-laws and the h-laws at our world do not coincide. And Kant will add: it seems reasonable to assume that there are or could be types of beings whose cognitive powers outstrip ours to an arbitrary degree; so no amount of luck could ensure that our laws coincide with those of each type of space alien.

Simplicio Perhaps, though, we are lucky, and there *are* no beings whose cognitive powers outstrip ours to the required extent?

Salviati Perhaps. But that is not relevant to the question of whether Lewis is a ratbag idealist about laws—we agreed that it sufficed to be a ratbag idealist if you agreed that, for all you know, there might be other beings for whom certain (pretheoretically) fundamental facts about our world differed. And that sufficient condition is met at present.

Simplicio Well, have it your way then: I suppose that Lewis *is* a ratbag idealist about laws. It seems that you think that is an objection to his account. But it is really more of a matter of taste.

Salviati I think it is more like an observation. It is not like I think it is an objection to Kant to label him a ratbag idealist—it just highlights a feature of transcendental idealism of which he was well aware. Nor do I think it is any objection to the best-system account of laws to observe that it renders laws mind-dependent. Lewis himself always seemed to hope to be able to finesse this feature away. But today, there are a number of relativist, indexical, or perspectivalist variants of the best-system account of laws that, to some extent, own up to their ratbaggery. For me, the point of making the observation is to goad fans of the best-system approach to take its ratbag idealism fully seriously.

Sagredo I would be interested to hear more about what you mean by that.

Simplicio First, I think it is only fair to let you know that you have walked into a trap! Let me reveal my *real* response (inspired by Ned Hall) to your worries about ratbag idealism. It is that all along you should have been saying: *This is going to hurt me more than it hurts you.*

Lewis is a reductionist about laws. A set s of standards of simplicity and balance determine the s-laws at each possible world, where a possible world is specified by specifying a Humean mosaic, a spatiotemporal pattern of instantiation of perfectly natural properties (themselves non-modal). So for each s, the s-laws at each world supervene on non-modal facts at that world.

Lewis's opponents are non-reductionists: they think that the laws at a world do not supervene on the pattern of instantiation of non-nomic properties at that world. Typically, they also buy into a picture on which the laws *govern* the pattern of instantiation of fundamental properties. They think that the laws of a world play a role in determining the shape of things at that world—whereas reductionists know that it is the other way around.

Now, all hands should admit that different sorts of beings will use different sets of standards of simplicity and balance in doing science.

For reductionists about law, this leads to the conclusion that at a given world, the laws may be species-relative. But that is nothing to worry about. After all, what is the point of the distinction between mere true generalizations and laws, by the reductionist's deflationary lights? Roughly, the distinction between mere facts and facts that are central and interesting enough to be targets of scientific investigation. It is no surprise that different sorts of beings, with different cognitive capacities and interests, will draw the border between the two sorts of facts in different places. So it is to the credit of Lewis's reductionism that it implies that the laws are to some extent "up to us."

How do things look to non-reductionists? Given the metaphysical centrality of laws on their picture, it would be hard for them to deny that science ought to strive to discover the laws. But now we have to worry whether our standards of simplicity and balance are the right ones—among all the kinds of cognizing beings at our world, most will be using standards that are too lax (leading them to count too many regularities as laws) or too stringent (leading them to count too few regularities as laws). Why think that natural selection has equipped us with the right standards for the goals of science?

This is a pressing challenge for non-reductionists—but for Lewisians, it can be met trivially, since the scientific goal of a type of being (the laws-for-them) covaries with their abilities.

Sagredo That is very interesting! I would like to hear what Salviati has to say in response—tomorrow.

Fourth Day

Sagredo I hope that today we will see whether or not Salviati is able to escape from Simplicio's trap.

Salviati I would like to try! Along the way, maybe I can also enlarge on something I said yesterday: that to me the point of raising the threat of ratbag idealism is to push fans of Lewis's best-system account of laws to take the account fully seriously. Consider scientific realism.

Simplicio Gladly—most best-systems enthusiasts are scientific realists, you know.

Salviati Are they, though? It is common to identify scientific realism as involving a semantic component, a metaphysical component, and an epistemic component: (i) the sentences expressing a scientific theory are true or false; (ii) very special cases aside, the relevant truth conditions are mind-independent; and (iii) the empirical success of our theories gives us (defeasible) reason to think them true.

Simplicio Humph. Yes. But the sort of mind-dependence involved in our account of law is not the bad kind.

Sagredo What is the bad kind?

Simplicio *You* should know—the spooky kind. The sort of thing Putnam went in for when he *wasn't* a scientific realist: "the mind and the world jointly make up the mind and the world." He went even further, you know: "the Universe makes up the Universe—with minds—collectively—playing a special role in making it up." Even he admitted that last bit was Hegelian. Scandalous. I am glad to say that he eventually repented.

Salviati You do concede, though, that your view about laws brings you into conflict with the metaphysical clause of the characterization given above?

Simplicio Yes. But it isn't important. We just need to tinker with that clause a bit.

Salviati Good—I genuinely look forward to hearing more about that. In the meantime, of course, you can continue to make common cause with classic and contemporary scientific realists against those anti-realists who question the epistemic credentials of science. But you will have to part ways with a realist like Planck, who was deeply concerned with mind-independence. As you will no doubt recall, the second sentence of his "Scientific Autobiography" reads:

> In this connection, it is of paramount importance that the outside world is something independent from man, something absolute, and the quest for the laws which apply to this absolute appeared to me as the most sublime scientific pursuit in life.

Simplicio I *do* recall that. *And* I recall that his first sentence reads:

> My original decision to devote myself to science was a direct result of the discovery which has never ceased to fill me with enthusiasm since my early youth—the comprehension of the far from obvious fact that the laws of human reasoning coincide with the laws governing the sequences of impressions we receive from the world about us; that, therefore, pure reasoning can enable man to gain an insight into the mechanism of the latter.

I am afraid this brings us right back to the trap that I laid for you, Salviati. Planck is in effect supposing some sort of pre-established harmony between the principles of human reasoning and the nomic structure of the world. And it is hard to see how non-reductionists about laws can do otherwise in the face of the gap that their view sets up between how we think and how the world is. The beauty of the best-system account of laws is that it erases that gap, as I was saying yesterday.

Sagredo We might almost say that where Planck adopts a preformationist account of human reason, Lewis adopts an epigeneticist one?

Simplicio Ha, ha. Very clever—perhaps on this point, Lewis *does* stand to Planck as Kant stands to Crusius. I am still waiting to hear how you are going to get out of the trap that *I* laid, however.

Salviati Good. Let us bring causation into the picture. For Lewis, the causal facts at a world depend on which counterfactual conditionals are true at that world, and which counterfactual conditionals are true at a world depends in part on the laws at that world. So Lewis should be a ratbag idealist about causation as well as about laws. That is an appealingly weird idea. But not, so far as I can tell, one that Lewis's followers have had much to say about.

Simplicio Cohen and Callender mention that if one adopts their relativized best-system account laws, it is natural to likewise accept relativized accounts of causation and chance. They say a bit about how this looks for a theory of chance.

Sagredo So you and your friends hold that the world as it is in itself, is not the sort of thing that is governed by scientific laws or that is structured by relations of cause and effect, but that the constitution of the human mind is such that, relative to it, the world has a rich nomic and causal structure?

Simplicio You could put it that way. Some of us add that the structure of space and time also have this status. But it is important to keep in mind that there is nothing remotely German about any of this.

Salviati Let us think a bit about all of this looks in the case of causation. Let us follow Peirce and Kant in assuming that Martians are far more intelligent than we are. And let us suppose that we are unlucky enough to find ourselves in a world in which this difference in intelligence, via the difference it makes to our respective standards of simplicity and strength, also makes a difference to laws. So there are some regularities at our world that count as laws for Martians, but not for us. Now, if you are Lewis—or indeed, anyone who accepts that there is a tight connection between laws and effective strategies—then you will expect that there will also be strategies for intervening in the world that Martians regard as effective strategies but which we do not.

Simplicio Yes. Different beings with different capacities and interests will and should draw the line between effective and ineffective strategies in different places.

Salviati I am not sure that they always will or that they always should. Suppose that we come into possession of a piece of Martian technology that exploits some of these strategies that are Martian-effective but not human-effective. Now, it does not follow from the difference in our cognitive capacities that we will be incapable of figuring out the principles by which this gizmo works—it is just that in working this out, we will eventually realize that the Martians take as lawful and counterfactual-supporting some regularity that we consider to be a mere regularity, one that is excluded from our best-system because adding it would not yield enough further strength, given its complexity and our standard of balance between strength and complexity. What will and should we do if we are interested in having devices that do what the gizmo does? Change our engineering practices so that we treat the relevant regularity as counterfactual-supporting, even though according to our pre-gizmo standards, that would appear most unwise.

Let us consider Simplicio's trap, then. It is true that anyone who takes laws to be mind-independent faces a substantive challenge: Why think that human science is well-adapted to discover the laws of our world? And it is true that views that take laws to be mind-dependent in the right sort of way have an easy answer to this question: laws-for-us are *automatically* the sort of thing that our science is good at discovering. But it would be better to say that all accounts of laws of nature face a double-barrelled challenge: Why think that laws of nature are the sort of things that we would want our science to discover—and why think that our science can discover them? Anti-reductionists who assign laws an exalted metaphysical status *may* feel that they have a good answer to the first of these challenges—but as we have just been saying, they have a hard time with the second. With Lewisian reductionism, it is the other way around. On this view, the second part of the challenge is more or less trivial. But the first part is not: Lewisian reductionists face the substantive challenge of showing that the notions of simplicity and balance that we have been equipped with by natural selection align with our scientific and technological goals. There are real challenges here for everyone.

Sagredo Ah, I see. Does it follow that whether or not we have gotten our hands on Martian technology, we should be open to the idea that our built-in standards are leading us to systematically under-shoot—that we may be ignoring lots of effective strategies that are in principle available to us, because we are being too conservative in determining which regularities are candidate laws?

Salviati Yes. Indeed, a natural consequence of Lewis's view is that there are regularities that are not good candidates to be laws by our standards—but which would count as laws for smarter beings. Why shouldn't we aim for super-human science and technology—even if it means using an exchange rate between simplicity and strength that feels unnatural to us?

Simplicio The question you raise seems very like an aspect of the problem of induction to me—you are asking us to justify our choice of inductive methods. I am really more interested in philosophy of science, you know.

Salviati I think that anyone interested in philosophy of science should be interested, at the very least, in the practical aspects of the problem of induction. The New Riddle of Induction is nothing but a sharp way of putting the point that you need some way of selecting which correlations in your data set you expect to hold up as more data come in (since it is incoherent to expect every correlation to continue to hold). This is a practical problem for anyone interested in designing autonomous learning systems. Similarly, the choice between different inductive methods is a practical problem faced by anyone intending to use machines to perform super-human scientific tasks—such as identifying patterns in huge data sets. When we talk about 'our' standards of simplicity and balance, should we be talking about the standards appropriate for unaided human scientists or for humans with super-computing prosthetic extensions?

Simplicio I think I have had enough horror-science fiction for now, thank you very much.

Salviati I expect you are not the only one.

Sagredo Let us leave the pimple-worms for another time, then.

References

For the epigraph, see Beebee and Fisher (2021, Letter 217). For Lewis, quidditism is the doctrine that two worlds can differ only in which property plays which role—for example, according to quidditists, corresponding to any Newtonian world in which R_1 is the relation of temporal betweenness and R_2 is the relation of spatial betweenness, there is another which differs only in that the roles played by these two relations are reversed. For discussion, see Lewis (2009) and Beebee and Fisher (2021, Letters 217, 405, and 510).

First Day See Le Roy (1901, p. 145) for a good statement of his positivism. For Poincaré's reaction, see his (1907, Part III). For others fantasizing about non-human science in response to perceived threats to scientific objectivity, see Glashow (1992), Planck (1909/1970), Schlick (1979), and Weinberg (1996). The passages from Peirce that Sagredo mentions can be found at de Waal (2014, pp. 98, 180, and 208). For the debut of the ratbag idealists, see Lewis (1999, p. 232). For further Lewisian uses of 'ratbag,' see Lewis (2001, p. 396), Lewis (2015, p. 208) and Beebee and Fisher (2021, Letters 168 and 475).

Second Day Quotations from Kant are from the translations in Guyer and Wood (2015) and Watkins (2012), with page numbers given using the usual conventions (first edition/second edition page numbers for the *Critique of Pure Reason*, Akademie edition volume:page number for other works). Note that the editorial apparatus of the Guyer and Wood edition of the *Critique* includes Kant's marginalia. For Kant as pool-shark, see, e.g., Stuckenberg (1882, 51) or Kuehn (2001, p. 64). On *Au Billard Royal,* see Stuckenberg (1882, p. 159) and Smith (1827, p. 87). For helpful discussions of B168 see Zöller (1989), Sloan (2002), and Callanan (2013). For a helpful discussion of the development of Kant's and Blumenbach's views on development, see again Sloan (2002). See Blumenbach (1781, p. 69) on circumcision and Blumenbach (1789/1792, p. 101/p. 80) on jackelopes. On the origin of pimple-worms, see Blumenbach (1806/1865, §I.5; 1810, §39); for further discussion and context, see McLaughlin (2005). For the views of Crusius summarized by Simplicio, see §§14, 15, 31, and 50 of his *Sketch* in Watkins (2009); the snippet quoted is from §50. For background on Crusius, see, e.g., Beck (1993, pp. 15–19). Following the last-quoted passage from the "Universal Natural History," Kant gives one of his appalling racist illustrations, the gist of which is that among the Venusians, the least impressive human would be as a Newton, whereas the Jovians would consider Newton himself even less impressive than Kant considers the least impressive humans. Despite what he says about the role of the Sun in shaping intellectual capacities, Kant does not consign the residents of the hemiboreal climate of Kaliningrad to the middle rungs of the ladder of human beings—instead, he singles out as inferior some inhabitants of milder climates and some inhabitants of harsher climates. On Kant's racial theories (and Blumenbach's), see, e.g., Smith (2015, Chapter 9).

Third Day For the canonical presentations of the best-system account of laws, see Lewis (1973, pp. 72–77; 1986, pp. *xi* f. and 122–131; 1999, pp. 39–43 and 231–236)—these are cited in the main text by page number alone. For an earlier version in Lewis's correspondence, see Beebee and Fisher (2021, Letter 427). For other discussions of Lewis and transcendental idealism, see Belot (2010, §1; 2011, §III.4), Langton (2015), Lewis (2009), and Lichtenstein (2022). For a best-system account of geometry, see Huggett (2006). For Sagredo's criticism of this approach, see again Belot (2011, §III.4). For a reply, see Callender (2017, p. 151 fn. 6). For approaches that take strongest-simplest to be a placeholder for the standards of scientists, see, e.g., Cohen and Callender (2009), Dorst (2019), Hall (2015), Hicks (2018), Jaag and Loew (2020), and Loewer (2021). On simplicity vs. beauty, see Dirac (1940) and the references in Bangu (2006). For a critical overview of the debate about the probative force of the mathematical fecundity of string theory, see Penrose (2004, §§31.18, 34.2, and 34.9). For indexical, relativist, or perspectivalist variants of the best-system account, see Cohen and Callender (2009), Halpin (2003), Massimi (2017), and Roberts (1999). For a classic discussion of the governing and non-governing conceptions of laws, see Beebee (2000). For Simplicio's trap, see Hall (2015, §15.5.1).

Fourth Day The characterization of scientific realism given is intended to be non-tendentious, being in broad conformity with: the "general recipe for realism" of Chakravartty (2017, §1.2); the characterization of Psillos (1999, pp. *xix* f.); and the "naïve statement of the position" offered by van Fraassen (1980, pp. 6 f.). See Putnam (1981, p. *xi*) for his Hegelian mottos, and Auxier et al. (2015, pp. 87, 481, and 771) for Putnam's repentance. See Planck (1949, p. 13) for the passage from his "Scientific Autobiography." For Lewis on the relationship between causal facts, counterfactual conditionals, and laws, see Lewis (1986, Essays 17 and 21). See Cohen and Callender (2009, §4.2) for their discussion of the relativity of causation and chance. For the classic case for the linking causal laws with effective strategies, see Cartwright (1979).

Acknowledgments Many thanks to Yemima Ben-Menahem for her editorial kindness, patience, and open-mindedness. Thanks, also, to Eli Lichtenstein, Laura Ruetsche, and Janum Sethi for their helpful comments and suggestions. This paper descends from a talk at conference at Columbia in 2018 and from conversations over the years with my teacher Carlo Rovelli, some of whose inimitable turns of phrase have found their way into the present version. I hope that my admiration for David Lewis shows through it all.

References

Auxier, R., Anderson, D., & Hahn, L. (Eds.). (2015). *The philosophy of Hilary Putnam.* Open Court.

Bangu, S. (2006). Pythagorean heuristic in physics. *Perspectives in Science, 14*, 387–416.

Beck, L.W. (1993). From Leibniz to Kant. In R. Solomon & K. Higgins (Eds.), *The age of German idealism* (pp. 5–39). Routledge.

Beebee, H. (2000). The non-governing conception of laws of Nature. *Philosophy and Phenomenological Research, 61*, 571–594.
Beebee, H., & Fisher, A. (Eds.). (2021). *Philosophical letters of David K. Lewis*. Oxford University Press.
Belot, G. (2010). Transcendental Idealism among the Jersey Metaphysicians. *Philosophical Studies, 150*, 429–438.
Belot, G. (2011). *Geometric possibility*. Oxford University Press.
Blumenbach, J. (1781). *Über den Bildungstrieb* (1st ed.). Johann Christian Dieterich.
Blumenbach, J. (1792). *An essay on generation*. T. Cadell. (Originally Blumenbach, J. (1789). *Über den Bildungstrieb* (2nd ed.). Johann Christian Dieterich).
Blumenbach, J. (1810). *Abbildungen naturhistorischer Gegenstände*. Heinrich Dieterich.
Blumenbach, J. (1865). Contributions to natural history. In T. Bendshye (Ed.), *The anthropological treatises of Johann Friedrich Blumenbach* (pp. 277–340). Longman, Green, Longman, Roberts, & Green. (Originally Blumenbach, J. (1806). *Beyträge zur Naturgeschichte*. Heinrich Dieterich.)
Callanan, J. (2013). Kant on nativism, scepticism and necessity. *Kantian Review, 18*, 1–27.
Callender, C. (2017). *What makes time special?* Oxford University Press.
Cartwright, N. (1979). Causal laws and effective strategies. *Noûs, 13*, 419–437.
Cohen, J., & Callender, C. (2009). A better best system account of lawhood. *Philosophical Studies, 145*, 1–34.
Chakravartty, A. (2017). Scientific realism. In E. Zalta (Ed.), *Stanford encyclopedia of philosophy*.
de Waal, C. (Ed.). (2014). *Charles S. Peirce: Illustrations of the logic of science*. Open Court.
Dirac, P. (1940). The relation between mathematics and physics. *Proceedings of the Royal Society of Edinburgh, 59*, 122–129.
Dorst, C. (2019). Towards a best predictive system account of laws of nature. *British Journal for the Philosophy of Science, 70*, 877–900.
Glashow, S. (1992). The death of science!? In R. Elvee (Ed.), *Nobel Conference XXV: The End of Science?* (pp. 23–32). University Press of America.
Guyer, P., & Wood, A. (Eds.). (2015). *Immanuel Kant: The critique of pure reason*. Cambridge University Press.
Hall, N. (2015). Humean reduction about laws of nature. In B. Loewer & J. Schaffer (Eds.), *A companion to David Lewis* (pp. 262–277). Wiley Blackwell.
Halpin, J. (2003). Scientific law. *Erkenntnis, 58*, 137–168.
Hicks, M. (2018). Dynamic Humeanism. *British Journal for the Philosophy of Science, 69*, 983–1007.
Huggett, N. (2006). The regularity account of relational spacetime. *Mind, 115*, 41–73.
Jaag, S., & Loew, C. (2020). Making best systems *best for us*. *Synthese, 197*, 2525–2550.
Kuehn, M. (2001). *Kant: A life*. Cambridge University Press.
Langton, R. (2015). Humility and coexistence in Kant and Lewis. In B. Loewer & J. Schaffer (Eds.), *A companion to David Lewis* (pp. 491–503). Wiley Blackwell.
Le Roy, É. (1901). Un positivisme nouveau. *Revue de Métaphysique et de Morale, 9*, 138–153.
Lewis, D. (1973). *Counterfactuals* Blackwell.
Lewis, D. (1986). *Philosophical papers* (Vol. II). Oxford University Press.
Lewis, D. (1999). *Papers in metaphysics and epistemology*. Cambridge University Press.
Lewis, D. (2001). Redefining 'intrinsic.' *Philosophy and Phenomenological Research, LXIII*, 381–398.
Lewis, D. (2009). Ramseyan humility. In R. Nola & D. Braddon–Mitchell (Eds.), *Conceptual analysis and philosophical naturalism* (pp. 203–222). MIT Press.
Lewis, S. (2015). Where (in logical space) is God? In B. Loewer & J. Schaffer (Eds.), *A companion to David Lewis* (pp. 206–219). Wiley Blackwell.
Lichtenstein, E. (2022). Revaluing Laws of Nature in Secularized Science. In Y. Ben-Menahem (Ed.), *Rethinking the concept of laws of nature*. (pp. xx-yy) Springer.
Loewer, B. (2021). The package deal account of laws and properties. *Synthese, 199*, 1065–1089.

Massimi, M. (2017). Laws of nature, natural properties, and the *robustly* best system. *The Monist, 100*, 406–421.
McLaughlin, P. (2005). Spontaneous and equivocal generation in early modern science. *Annals of the History and Philosophy of Biology, 10*, 79–88.
Penrose, R. (2004). *The road to reality.* Vintage.
Planck, M. (1949). *Scientific autobiography and other papers.* Philosophical Library.
Planck, M. (1970). The unity of the physical world-picture. In S. Toulmin (Ed.), *Physical reality: Philosophical essays on twentieth century physics* (pp. 1–27). Harper & Row.
Poincaré, H. (1907). *The value of science.* The Science Press.
Psillos, S. (1999). *Scientific realism: How science tracks truth.* Routledge.
Putnam, H. (1981). *Reason, truth, and history.* Cambridge University Press.
Roberts, J. (1999). 'Laws of nature' as an indexical term. *Philosophy of Science, 66*, S502–S511.
Schlick, M. (1979). The universe and the human mind. In H. Mulder & B. van de Velde–Schlick (Eds.), *Moritz Schlick: Philosophical papers* (Vol. 2, pp. 499–513). Reidel.
Sloan, P. (2002). Preforming the categories: Eighteenth century generation theory and the biological roots of Kant's *a priori*. *Journal of the History of Philosophy, 40*, 229–253.
Smith, J. (2015). *Nature, human nature, and human difference: Race in early modern philosophy.* Princeton University Press.
Smith, R. (1827). *Notes made during a tour in Denmark, Holstein, Mecklenburg–Schwerin, Pomerania, the Isle of Rugen, Prussia, Poland, Saxony, Brunswick, Hanover, the Hanseatic Territories, Oldenburg, Friesland, Holland, Brabant, the Rhine Country, and France. Interspersed with Some Observations on the Foreign Corn Trade.* C. & J. Rivington.
Stuckenberg, J. (1882). *The life of Immanuel Kant.* MacMillan and Co.
van Fraassen, B. (1980). *The scientific image.* Oxford University Press.
Watkins, E. (Ed.). (2009). *Kant's critique of pure reason: Background and source materials.* Cambridge University Press.
Watkins, E. (Ed.). (2012). *Immanuel Kant: Natural science.* Cambridge University Press.
Weinberg, S. (1996). Sokal's Hoax. *The New York review of books,* August 8, 11–15.
Zöller, G. (1989) From innate to *a priori*: Kant's radical transformation of a Cartesian–Leibnizean legacy. *The Monist, 72*, 222–235.

Governing Without a Fundamental Direction of Time: Minimal Primitivism About Laws of Nature

Eddy Keming Chen and Sheldon Goldstein

Abstract The Great Divide in metaphysical debates about laws of nature is between Humeans, who think that laws merely *describe* the distribution of matter, and non-Humeans, who think that laws *govern* it. The metaphysics can place demands on the proper formulations of physical theories. It is sometimes assumed that the governing view requires a fundamental/intrinsic direction of time: to govern, laws must be *dynamical*, producing later states of the world from earlier ones, in accord with the fundamental direction of time in the universe. In this paper, we propose a *minimal primitivism about laws of nature* (MinP) according to which there is no such requirement. On our view, laws govern by constraining the physical possibilities. Our view captures the essence of the governing view without taking on extraneous commitments about the direction of time or dynamic production. Moreover, as a version of primitivism, our view requires no reduction/analysis of laws in terms of universals, powers, or dispositions. Our view accommodates several potential candidates for fundamental laws, including the principle of least action, the Past Hypothesis, the Einstein equation of general relativity, and even controversial examples found in the Wheeler-Feynman theory of electrodynamics and retrocausal theories of quantum mechanics. By understanding governing as constraining, non-Humeans who accept MinP have the same freedom to contemplate a wide variety of candidate fundamental laws as Humeans do.

Keywords Laws of nature · Fundamentality · Explanation · Simplicity · Non-Humeanism · Humeanism · Governing · Dynamics · Constraint · Production · Direction of time · Causation · Primitivism · Objective Probability · Typicality

---

E. K. Chen (✉)
Department of Philosophy, University of California San Diego, La Jolla, CA, USA
e-mail: eddykemingchen@ucsd.edu

S. Goldstein
Departments of Mathematics, Physics, and Philosophy, Rutgers University, Piscataway, NJ, USA
e-mail: oldstein@math.rutgers.edu

Y. Ben-Menahem (ed.), *Rethinking the Concept of Law of Nature*,
Jerusalem Studies in Philosophy and History of Science,
https://doi.org/10.1007/978-3-030-96775-8_2

1 The Great Divide

The goal of this paper is to articulate *minimal primitivism about laws of nature* (MinP), a minimalist and primitivist view about laws, and to contrast it with some leading alternatives. MinP captures our conviction that the universe is governed by laws of nature in a way that does not presuppose a fundamental direction of time. Here we focus on laws of physics, particularly those suitable for being fundamental laws.

To begin, let us list a few paradigm examples of candidate laws of physics:

- Newton's laws of motion
- The Schrödinger equation
- The Dirac equation

They are all dynamical laws concerning how physical systems evolve in time. Here are some other equations or principles that, for one reason or another, may be controversial examples as candidate laws of physics:

- The Einstein equation (of general relativity)
- The Wheeler-DeWitt equation
- Conservation laws
- Symmetry principles
- The principle of least action
- The Past Hypothesis (of a low-entropy boundary condition of the universe)
- Equations of motion in Wheeler-Feynman electrodynamics

In physics, a significant amount of work has been devoted to the discovery of its true fundamental laws: the basic principles that govern the world.[1] The collection of all of these laws may be called the axioms of the final theory of physics or the Theory of Everything (TOE). The fundamental laws cannot be explained in terms of deeper principles (Weinberg 1992, p. 18); from them we can derive theorems of great importance and explain all significant observable regularities. Some of the equations and principles in the above lists, with suitable adaptation, may be included in such a collection. In this paper, we assume that there are fundamental laws and that they play important roles in scientific explanations. But what kind of things are fundamental laws? Most people believe that laws are different from material entities such as particles and fields, because, for one thing, laws seem to *govern* the material entities. But what is this governing relation? What makes material entities respect such laws? What is the role of laws in scientific explanations? Such questions do not have straightforward answers, and they cannot be directly tested in experiments. They fall in the domain of metaphysics.

The Great Divide in metaphysical debates about laws of nature is between Humeans, who think that laws are merely descriptions, and non-Humeans, who

[1] In this paper, we use "fundamental laws" and "laws" interchangeably unless noted otherwise.

think that laws govern.[2] Humeans maintain that laws merely describe how matter is distributed in the universe. In Lewis's version, laws are just certain efficient summaries of the distribution of matter in the universe, also known as the Humean mosaic. Nothing really enforces the patterns in the mosaic. A common theme in non-Humean views is that laws govern the distribution of matter. By appealing to the governing laws, the patterns are explained. How laws perform such a role is a matter of debate, and there are differences of opinion between reductionist non-Humeans such as Armstrong (1983) and primitivist non-Humeans such as Maudlin (2007).

One's metaphysical position can shape one's expectations about what physical laws should look like. It is sometimes assumed that the governing view of laws requires a fundamental direction of time: to govern, laws must be *dynamical* laws that *produce* later states of the world from earlier ones, in accord with the direction of time that makes a fundamental distinction between past and future. Call this conception of governing *dynamic production*. It is suggested by Maudlin (2007) and discussed at length by Loewer (2012). For Maudlin, primitivism about laws and primitivism about the direction of time should be postulated together, with this package supporting a particular kind of explanation associated with dynamic production. This emphasis on dynamic production is not unique to Maudlin and is important to some other types of non-Humeans. Although we subscribe to the governing view and the primitivist view about laws of nature, we do not share the view that a fundamental direction of time is essential to either.

Reflecting upon the variety of kinds of laws that physicists present as fundamental, we find many that do not fit in the form of dynamical laws. Moreover, even when physicists postulate dynamical laws, dynamic production in accord with a fundamental direction of time does not seem essential to how these laws govern the world or explain the observed phenomena.

In this paper, we propose a minimal primitivist view (MinP) about laws of nature that disentangles the governing conception from dynamic production. On our view, fundamental laws govern by constraining the physical possibilities of the entire spacetime and its contents.[3] They need not exclusively be dynamical laws, and their governance does not presuppose a fundamental direction of time. For example, they can take the form of global constraints or boundary-condition constraints for spacetime as a whole; they can govern even in an atemporal world; they may permit the existence of temporal loops. Our minimal view captures the essence of the governing view without taking on extraneous commitments about the direction

[2] This is an oversimplification as there are some non-Humeans, such as Aristotelian Reductionists, who do not think that laws govern. See Sect. 2.3.

[3] Throughout this paper, for simplicity, we assume that spacetime is fundamental. This assumption is not essential to MinP. One can consider non-spatio-temporal worlds governed by minimal primitivist laws. For those worlds, one can understand MinP as suggesting that laws constrain the physical possibilities of the world, whatever non-spatio-temporal structure it may have. Indeed, if one regards time itself as emergent, one may find it natural to understand governing in an atemporal and direction-less sense.

of time. Moreover, as a version of primitivism, our view requires no reduction or analysis of laws into universals, powers, or dispositions. Because of the minimalism and the primitivism, our view accommodates several candidate fundamental laws, such as the principle of least action, the Past Hypothesis, and the Einstein equation (which in its usual presentation is non-dynamical). It is also compatible with more controversial examples of fundamental laws in the Wheeler-Feynman theory of electrodynamics and retrocausal theories of quantum mechanics. The flexibility of MinP is, we believe, a virtue. From our viewpoint, it is an empirical matter what forms the fundamental laws take on; one's metaphysical theory of laws should be open to accommodating the diverse kinds of laws entertained by physicists. It may turn out that nature employs laws beyond those expressible in the form of differential equations that admit (Cauchy) initial value formulations or can be given a dynamic productive interpretation. The metaphysics of laws should not stand in the way of scientific investigations. Our view encourages openness.

The idea that fundamental laws produce later states of the world from earlier ones is related to causal fundamentalism, the idea that causation, or something like causation (such as dynamic production), is fundamental in the world.[4] If causation is fundamental and asymmetric, then either it defines a direction of time or the direction of time itself is metaphysically fundamental (such that it has no deeper explanation). On MinP, temporally asymmetric relations such as causation and dynamic production are not constitutive of how laws govern. We do not think causal fundamentalism is true. Neither do we think there must be a fundamental direction of time. However, rejecting them is not part of our view about laws; agnosticism about them is sufficient for our purposes. On MinP, there can be but need not be fundamental causal relations, and there can be but need not be a fundamental direction of time. Even if they existed, they would not be essential to how laws explain. Hence, MinP carves out conceptual space for non-Humeans such as ourselves who believe that laws govern but do not demand causal fundamentalism, a fundamental direction of time, or dynamic production.

This paper has been written by a philosopher of physics and a mathematical physicist. It is written for mathematicians, physicists, and philosophers who are interested in the nature of physical laws. We believe that MinP is close to what many mathematicians, physicists, and philosophers have in mind when they think about laws.

We start with a review of four leading approaches to laws of nature: Humean reductionism, Platonic reductionism, Aristotelian reductionism, and Maudlinian primitivism. Readers familiar with the philosophical literature on laws may skim the review. Next, we state two central theses of MinP and suggest how minimal primitivist laws can explain natural phenomena without presupposing a fundamental

[4] Causal fundamentalism does not imply that everyday causality is metaphysically fundamental. For example, Maudlin's notion of dynamic production is different from everyday causality (Maudlin 2007, ch.5). For some recent works on causal fundamentalism, physics, and everyday causality, see Blanchard (2016) and Weaver (2019).

direction of time. We illustrate MinP by providing interpretations of several types of candidate physical laws: dynamical laws, non-dynamical constraint laws, and probabilistic laws. Finally, we illustrate the key differences between MinP and the alternatives. Many working physicists, mathematicians, and philosophers of science may appreciate our view precisely because of its minimalism and primitivism. We also list some open questions for future work.

2 Some Existing Approaches

In this section, we survey some existing approaches to laws of nature.[5] We highlight the key motivations that underpin such approaches, the explanatory principles they employ, and the kinds of laws that they accommodate.

2.1 Humean Reductionism

A popular approach to laws in contemporary philosophical literature is that of Humean Reductionism. On this view, laws do not govern but merely describe by summarizing what actually happens in the world. Inspired by writings of Hume, Mill, and Ramsey, David Lewis pioneered the contemporary versions of this view. On Humean Reductionism, the fundamental ontology is that of a *Humean mosaic*, a concrete example of which is a 4-dimensional spacetime occupied by particles and fields. At the fundamental level, laws of nature do not exist and do not move particles and fields around. There are no "necessary connections" forged by governing laws. Laws of nature are derivative of and ontologically dependent on the actual Humean mosaic. The laws are the way they are *because of* what the actual trajectories of particles and histories of fields are, not the other way around, in contrast to the governing picture of laws. Laws are reducible to the Humean mosaic.

Lewis (1986) calls this view *Humean supervenience*.[6] Following Ramsey, Lewis proposes a "best-system" analysis of laws that shows how laws can be recovered from the Humean mosaic. The basic idea is that laws are certain regularities of the

[5] This survey is by no means exhaustive of the rich literature on laws. For example, against the view that there are fundamental laws that are universally true, Cartwright (1994a) advocates a patchwork view of laws where they are, at most, true *ceteris paribus*. Van Fraassen (1989) advocates a view where there are no laws of nature. See Carroll (2020), Hildebrand (2020), Bhogal (2020) for more detailed surveys.

[6] Whether contemporary Humean position in the metaphysics of science represents the historical Hume has been debated. See for example Strawson (2015).

Humean mosaic. However, not any regularity is a law, since some are accidental.[7] Hence, one needs to be selective about which regularities to count as laws. Lewis suggests we pick those regularities in the best system of true sentences about the Humean mosaic. The strategy is to consider various systems (collections) of true sentences about the Humean mosaic and pick the system that strikes the best balance among various theoretical virtues, such as simplicity and informativeness.

To get an intuitive grasp of this balancing act, consider an example. Let the Humean mosaic (the fundamental ontology) be a Minkowski spacetime occupied by massive, charged particles and an electromagnetic field. The locations and properties of those particles and the strengths and directions of the field at different points in spacetime is the matter distribution, which corresponds to the local matters of particular fact. Suppose the matter distribution is a solution to Maxwell's equations. Consider three systems of true statements (characterized below using the axioms of the systems) about this mosaic:

- System 1: {Spacetime point (x_1, y_1, z_1, t_1) has field strengths E_1 and B_1 with directions $\vec{v}_1$ and $\vec{v}_{1'}$ and is occupied by a particle of charge q_1; spacetime point (x_2, y_2, z_2, t_2) has field strengths E_2 and B_2 with directions $\vec{v}_2$ and $\vec{v}_{2'}$ and is not occupied by a charged particle; ......}
- System 2: {"Things exist."}
- System 3: {Maxwell's equations and the Lorentz force law}

System 1 lists all the facts about spacetime points one by one. It has much informational content but it is complicated. System 2 is just one sentence that says there are things but does not tell us what they are and how they are distributed. It is extremely simple but has little informational content. System 3 lists just five equations of Maxwellian electrodynamics. It has less information about the world than System 1 but has much more than System 2. It is more complicated than System 2 but much less so than System 1. System 1 and System 2 are two extremes; they have one virtue too much at the complete expense of the other. In contrast, System 3 strikes a good balance between simplicity and informativeness. System 3 is the best system of the mosaic. Therefore, according to the best-system analysis, Maxwell's equations are the fundamental laws of this world.

To emphasize, on Humean Reductionism, laws are descriptive of the Humean mosaic. Laws are not among the fundamental entities that push or pull things, enforce behaviors, or produce the patterns. Laws are just winners of the competition among systematic summaries of the mosaic. Beebee (2000) calls it the "non-governing conception of laws of nature." Laws are merely those generalizations which figure in the most economical true axiomatization of all the particular matters of fact that obtain.

[7] For example, the regularity that all uranium spheres are less than one mile in diameter may be a law or a consequence of some law, but the regularity that all gold spheres are less than one mile in diameter is not a law or a consequence of a law.

Despite the simplicity and appeal of Lewis's analysis, there is an obstacle. The theoretical virtue of simplicity is language-dependent. For example, suppose there is a predicate F that applies to all and only the things in the actual spacetime. Then consider the following system:

- System 4: $\{\forall x F(x)\}$

This is informationally equivalent to System 1 and more informative than System 3, and yet it is simpler than System 3. If we allow competing systems to use predicate F, there will be a system (namely System 4) that is overall better than System 3. Given the best-system analysis, the actual laws of the mosaic would not be Maxwell's equations but "$\forall x F(x)$." To rule out such systems, Lewis places a restriction on language. Suitable systems that enter into the competition can invoke predicates that refer to only natural properties. For example, the predicate "having negative charge" refers to a natural property, while the disjunctive predicate "having negative charge or being the Eiffel Tower" refers to a less natural property. Some properties are perfectly natural, such as those invoked in fundamental physics about mass, charge, spacetime location and so on. It is those perfectly natural properties that the axioms in the best system must refer to. The predicate F applies to all and only things in the actual world, which makes up an "unnatural" set of entities. F is not perfectly natural. Hence, System 4 is not suitable. The requirement that the axioms of the best system refers only to perfectly natural properties is an important element of Lewis's Humeanism.

Over the years, Lewis and his followers have, in various ways, extended and modified the best-system analysis of laws on Humean Reductionism. Let us summarize some of the developments and call the updated view *Reformed Humeanism about Laws*:

Reformed Humeanism about Laws The fundamental laws are the axioms of the best system that summarizes the mosaic and optimally balances simplicity, informativeness, fit, and degree of naturalness of the properties referred to. The mosaic contains only local matters of particular facts, and the mosaic is the complete collection of fundamental facts.

Reformed Humeanism can accommodate various kinds of laws of nature. Without going into too much detail, we note the following features:

1. *Chance*. Although chance is not an element of the Humean mosaic, it can appear in the best system. Humeans can introduce probability distributions as axioms of the best system (Lewis 1980). This works nicely for stochastic theories such as the Ghirardi-Rimini-Weber (1986) theory of spontaneous localization (GRW). Humeans can evaluate the contribution of the probability distributions by using a new theoretical virtue called *fit*. A system is more fit than another just in case it assigns a higher (comparative) probability than the other does the history of the universe. For certain mosaics, the inclusion of probability in the best system can greatly improve the informational content without sacrificing too much simplicity. Hence, fit can be seen as the probabilistic extension of

informativeness. Humeans can also allow what is called "deterministic chance" (Loewer 2001). Take a deterministic Newtonian theory of particle motion and add to it the Past Hypothesis and the Statistical Postulate (Albert 2000), which can be represented as a uniform probabilistic distribution, conditionalized on a low-entropy macrostate of the universe at t_0. The Humean account of chance (both stochastic and deterministic) is arguably one of the simplest and clearest to date.

2. *Particular facts.* Lewis (1983) maintains that "only the regularities of the system are to count as laws" (p. 367). However, there is no reason to limit the Humean account to laws about general facts. Physicists have entertained candidate physical laws about particular facts. For example, the Past Hypothesis is a candidate physical law about one temporal boundary of the universe ("t_0"). Such laws are uncommon, but conceptually we do not see any obstacle. If a particular place or a particular time in the universe is sufficiently significant, then it may be appropriate to have a physical law about the particular place or time. Other examples of such laws include Tooley's case of Smith's garden (1977) and the Aristotelian law about the center of the universe. Callender (2004) suggests that a Humean analysis can do away with Lewis's restriction to laws of general facts. In fact, this flexibility seems a significant advantage Humean Reductionism has over some other accounts of laws.
3. *Flexibility with respect to perfect naturalness.* For Lewis, perfect naturalness is a property of properties. Perfectly natural properties pick out the same set of things as Armstrong's theory of sparse universals (more on that in Sect. 2.2). However, the chief motivation of Lewis's use of perfect naturalness is to rule out systems that use "gruesome" predicates. If that is the issue, then perhaps, as Hicks and Schaffer (2017) suggest, we can simply require that "degree of naturalness" of the predicates involved be a factor in the overall ranking of competing systems, and the best system should also optimally balance degree of naturalness of the predicates together with the rest of the theoretical virtues, such as simplicity, informativeness, and fit. The flexibility with respect to perfect naturalness also allows the best system to refer to non-fundamental properties such as entropy, as may be necessary if the Past Hypothesis is a fundamental law.
4. *Theoretical virtues.* Humeans do not provide a full account of the theoretical virtues. There are certain theoretical virtues scientists do and should consider significant. With that in mind, perhaps Humeans can leave them open-ended. As such, there is also some vagueness in how systems are compared and in some cases there may be vagueness about which system is best.[8]

Reformed Humeanism is perhaps the most flexible view on the market for its accommodation with multiple kinds of candidate laws of physics. There is no

[8] Another issue concerning theoretical virtues is how we should use them to compare different systems. As noted earlier, simplicity is language relative. Cohen and Callender (2009) suggest that the comparisons should be relativized to languages. Their relativized account (called the Better Best System Account) perhaps can be used to support Fodor (1974)'s vision of the autonomy of the special sciences (e.g. biology, psychology, economics) from fundamental physics.

problem with giving lawhood status to non-dynamical facts such as the principle of least action, the Einstein equation, or even a version of the Past Hypothesis that refers to a particular time (t_0) and a non-fundamental property (entropy). Because of its accommodation of the Past Hypothesis and deterministic chance, Reformed Humeanism also accommodates reductionism about the direction of time. The ingredients for such a reduction can all be interpreted as axioms of the best system summarizing the mosaic. Hence, Humeans can do away with a fundamental direction of time (Loewer 2012).

2.2 *Platonic Reductionism*

With Humean Reductionism, nothing ultimately explains the patterns in the Humean mosaic. For illustration, suppose $F = ma$ is a fundamental law of our world. Humeans maintain that "$F = ma$" expresses a fundamental law in virtue of its being an axiom in the best system of the Humean mosaic. It merely summarizes what actually happens: the trajectories of all massive particles are solutions to $F = ma$. Those with a governing conception of laws may seek to find a deeper explanation. In virtue of what is every massive particle in the world behaving according to the same formula? What, if anything, enforces the pattern and makes sure nothing deviates from it? In other words, what provides the *necessity* or *oomph* that is usually associated with laws?

Dretske (1977), Tooley (1977), and Armstrong (1983) propose an intriguing answer based on a metaphysics of universals. The universals that they accept are in addition to things in the Humean mosaic. They are "over and above" the Humean mosaic. In traditional metaphysics, universals are repeatable entities that explain the genuine similarity of objects. Let us start with some mundane examples. Two cups are genuinely similar in virtue of their sharing a universal *being a cup*. The universal is something they both instantiate and something that explains their genuine similarity. A cup is different from a horse because the latter instantiates a different universal *being a horse*. Now, those universals are not fundamental, and they may be built from more fundamental universals about physical properties. Dretske, Tooley, and Armstrong use universals to provide explanations in science. For them, the paradigm examples are universals that correspond to fundamental physical properties, such as mass and charge. On their view, laws of nature hold because of a certain relation obtaining among such universals. This theory of laws has connection to Plato's theory of forms.[9] We thus call it *Platonic Reductionism*.[10]

---

[9] For an overview of Plato's theory of forms, see Kraut (2017).

[10] In the literature it is sometimes called the DTA account of laws or the Universalist account of laws. Calling it Platonic *reductionism* may be controversial. But see the discussion in (Carroll 1994, appendix A1).

Consider again the world where $F = ma$ holds for every massive particle. In such a world, any particle with mass m instantiates the universal *having mass* m, any particle under total force F instantiates the universal *being under total force* F, and any particle with acceleration F/m instantiates the universal *having acceleration* F/m. The universals are multiply instantiated and repeated, as there are many particles that share the same universals. Those universals give unity to the particles that instantiate them. The theory also postulates, as a fundamental fact, that the universal *having mass* m and the universal *being under total force* F necessitate the universal *having acceleration* F/m. Hence, if any particle instantiates *having mass* m and *being under total force* F, then it has to instantiate *having acceleration* F/m. It follows that every particle has to obey $F = ma$.[11] This adds the necessity and the oomph that are missing in Humean Reductionism.

With Platonic Reductionism, the regularity is explained by the metaphysical postulate of universals and the necessitation relation N that hold among universals. Following Hildebrand (2013), we may summarize it as follows:

Necessitation For all universals F and G, $N(F, G)$ necessitates the regularity that all Fs are Gs.

A few clarificatory remarks:

1. *Universals.* (i) The appeal to universals is indispensable in this theory of laws. The theory is committed to a fundamental ontology of objects (particulars) and a fundamental ontology of universals. Hence, Platonic Reductionism is incompatible with nominalism about universals. (ii) Defenders such as Armstrong appeal to a sparse theory of universals, where the fundamental universals correspond to the fundamental properties we find in fundamental physics. The sparse universals correspond to the perfectly natural properties that Lewis invokes in his account. Consider Lewis's example of the predicate F that corresponds to the property of all and only things in the actual world. For Armstrong,"$\forall x F(x)$" does not express a fundamental law because objects with property F are not genuinely similar, and F is a property that does not correspond to one of the fundamental, sparse universals.
2. *Necessity.* (i) The necessity relation among universals is put into the theory by hand. It is a postulate that such a relation holds among universals and does necessitate regularities. (It is also postulated that the relation among universals is itself a universal.) To some commentators, it is unclear why the postulate is justified.[12] In response, a defender of Platonic Reductionism may

[11] We note that this example about $F = ma$ does not exactly fit in Armstrong's schema of "All F's are G." See (Armstrong 1983, ch.7) for a proposal for accommodating "functional laws."

[12] In a famous passage, Lewis (1983) raises this objection: "Whatever N may be, I cannot see how it could be absolutely impossible to have N(F,G) and Fa without Ga...The mystery is somewhat hidden by Armstrong's terminology. He uses 'necessitates' as a name for the lawmaking universal N; and who would be surprised to hear that if F 'necessitates' G and a has F, then a must have G? But I say that N deserves the name of 'necessitation' only if, somehow, it really can enter into the

take the necessity relation simply as a primitive and stipulate its connections to regularities (Schaffer 2016).

(ii) The N relation, though called a *necessity* relation, holds contingently among universals. Thus, if $N(F, G)$ holds in the actual world, then in some possible world F is not connected to G via N. N is only nomologically necessary but metaphysically contingent. This has the consequence that laws of nature on Platonic Reductionism, while nomologically necessary, are metaphysically contingent. This respects a widespread judgment about the metaphysical contingency of laws. (In Sect. 2.3 we see that Aristotelian Reductionism violates it.)

(iii) Armstrong (1983) makes room for probabilistic laws as follows:

> Irreducibly probabilistic laws are also relations between universals. These relations give (are constituted by) a certain objective probability that individual instantiations of the antecedent universal will *necessitate* instantiation of the consequent universal. They give a probability of a necessitation in the particular case...Deterministic laws are limiting cases of probabilistic laws (probability 1). (p. 172)

What is "a probability of a necessitation?" Conceptually, whether F necessitates G seems like a matter that does not admit of degree. What does this probability mean, and how does it relate to actual frequencies and why should it constrain our credences? Even if one accepts the intelligibility of the necessitation relation, one may be unwilling to accept the intelligibility of objective probability of a necessitation and one may be puzzled by how the probability of a necessitation can explain the regularities. This may be an instance of the general phenomenon that it is difficult to give a unified and intelligible non-Humean account of probabilistic laws and non-probabilistic laws. It is much easier (if one sets aside the worry about the lack of governing) to do so on Humean Reductionism: just put them all in the best system.

3. *Explanation*. For those who are antecedently sympathetic to a theory of universals, Platonic Reductionism may offer an attractive metaphysical explanation of the patterns in nature. Its enlarged ontology provides extra explanatory resources. If two particles both have mass m, then there literally is something they have in common—the universal *having mass m*. That the two particles move in the same way can be partly explained by their genuine similarity to each other—their shared universals. The relation that obtains among such universals, the necessitation relation N, exists over and above the mosaic (the trajectories of particles in spacetime). Since the state of affairs that N obtains among universals of mass, force, and acceleration does not supervene on the objects, it can be said to *govern* the objects. In contrast, on Humean Reductionism, at the fundamental level there is nothing that exists except the Humean mosaic. However, the explanation on Platonic Reductionism may not be transparent to those who are not sympathetic to a theory of universals.

Because Platonic Reductionism analyzes laws in terms of universals and relations among them, it places certain restrictions on the forms of physical

requisite necessary connections. It can't enter into them just by bearing a name, any more than one can have mighty biceps just by being called 'Armstrong' " (p. 366).

laws. If universals are repeatable entities with multiple locations in space or time, Platonic Reductionism does not seem compatible with laws that are about particular places or times. In our view, that is a problem as it limits physical laws to general facts. For example, the account seems incompatible with taking the Past Hypothesis to be a fundamental law even though we have good arguments for doing so. We return to this point in Sect. 4.2.

On Platonic Reductionism, it is unclear how we should think about the direction of time. Even though there is a strong connection between the necessitation relation N and causation, it does not seem that the main defenders build the direction of time into N. However, Tooley (1997) seems to think that the direction of time is reducible to the direction of causation, and causal facts are fundamental in his metaphysics. If that is the case, then causal fundamentalism is true and the direction of time is close to being fundamental. Perhaps that is an optional add-on to his theory of laws. Nevertheless, if Platonic Reductionism does not have room for treating the Past Hypothesis as a fundamental law, it may need to invoke a fundamental direction of time for worlds like ours. Perhaps Platonic Reductionism is best paired with a primitivism about the direction of time.

2.3 *Aristotelian Reductionism*

The view about laws to which we now turn is most commonly associated with contemporary defenders of dispositional essentialism. On this view, laws, even if they exist, do not govern the world in any metaphysically robust sense. Laws do not push or pull things around. Instead, the patterns we see are explained by the fundamental properties that objects instantiate. Those properties are the seats of metaphysical powers, necessity, and oomph. Those properties make objects, in a certain sense, "active" (Ellis 2001, p. 1). Such properties are often called "dispositions," and also sometimes called "powers," "capacities," "potentialities," and "potencies."[13] Importantly, they are different from the universals in Platonic Reductionism or the natural properties in Humean Reductionism, which may be viewed as "passive." If there are any laws (and there is an internal debate about this question among defenders of this fundamental dispositional ontology), they derive from or originate in the fundamental dispositions of material objects.

Roughly speaking, objects with dispositions have characteristic behaviors (also called manifestation) in response to certain stimuli (Bird 2007, p. 3). For example, a glass has a disposition to shatter when struck; an ice cube has a disposition to melt when heated; salt has a disposition to dissolve when put into water. On this view, fundamental properties are similarly dispositional: negatively charged particles have a disposition to attract positively charged particles; massive particles have a disposition to accelerate in a way that is proportional to the total forces on them and inversely proportional to their masses. Moreover, a dispositional essentialist

[13] For an overview of the metaphysics of dispositions, see Choi and Fara (2021).

holds that some properties have dispositional essences, i.e. their essences can be characterized in dispositional terms.[14]

In contrast to Humean Reductionism and Platonic Reductionism, on this view the fundamental ontology is no longer "passive" but is "active and reactive" (Ellis 2001, pp. 1–2). We confess that we do not fully understand such locutions. Perhaps the idea is that material objects move in virtue of the dispositions they possess and not in virtue of something outside (such as a law) that governs them. Among those who endorse a dispositionalist fundamental ontology, not everyone accepts that fundamental laws, which are usually taken to be universally valid and always true, arise from dispositions. For example, Cartwright (1983, 1994b) and Mumford (2004) deny the need for laws. Nevertheless, the dispositional essentialists need not abandon laws. They can maintain that laws supervene on or reduce to dispositions. Because of its Aristotelian roots (Ellis 2014), we call such a view *Aristotelian Reductionism* about laws.[15] Bird (2007) characterizes it as follows:

> According to this view laws are not thrust upon properties, irrespective, as it were, of what those properties are. Rather the laws spring from within the properties themselves. The essential nature of a property is given by its relations with other properties. It wouldn't be that property unless it engaged in those relations. Consequently those relations cannot fail to hold (except by the absence of the properties altogether, if that is possible). The laws of nature are thus metaphysically necessary. (p. 2)

Aristotelian Reductionists maintain that (i) the metaphysical powers, necessity, and oomph reside in the fundamental dispositions; (ii) laws are metaphysically derivative of the dispositions; (iii) laws are metaphysically necessary.

How are laws derived from dispositions? Bird proposes that we can derive laws from certain counterfactual conditionals associated with dispositional essences. A more recent approach is that of Demarest (2017, 2021) and Kimpton-Nye (2017) that seek to combine a dispositional fundamental ontology with a best-system-analysis of lawhood. Here we focus on the approach of Demarest. She proposes that dispositions (she follows Bird and calls them potencies) do metaphysical work. They produce their characteristic behaviors, resulting in patterns in nature. Their characteristic behaviors, in different possible worlds, can be summarized in simple and informative axiomatic systems, and the best one contains the true laws of nature. That is like Humean Reductionism except that (i) Demarest's fundamental ontology includes potencies and (ii) the summary is not of just the actual distribution of potencies but also all merely possible ones. In this way, her proposal may be an elaboration of Bird's suggestion that we can derive laws from potencies, though she does not rely on counterfactuals. In her most recent work (2021), she proposes the following account:

Dynamic-Potency-BSA (DPBSA): The basic laws of nature at w are the axioms of the simplest, most informative, true systematization of all w-potency-

[14] Some, such as Bird (2007), go further and claim that all perfectly natural properties in Lewis (1986)'s sense or all sparse universals in Armstrong (1983)'s sense have dispositional essences.

[15] Many defenders of this view suggest that even though it has roots in Aristotle, it is not committed to many aspects of Aristotelianism.

distributions, where a w-potency-distribution is a possible distribution of potencies *that is generated by a possible initial distribution of only potencies appearing in w.* (p. 9, emphasis original)

In contrast to Humean Reductionism, here the patterns are ultimately explained by the potencies. How do potencies explain? Demarest provides this answer:

> I think the most promising solution is to appeal to production–dynamic, metaphysical dependence. According to my view, the fundamental ground includes spacetime and an initial arrangement of particles and potencies. And the subsequent behavior of the particles (further potency instantiations as well as trajectories through spacetime) is dynamically, metaphysically dependent upon that base. Since the potency-BSA systematizes those trajectories, the laws of nature are not fundamental, and do not govern, but rather depend upon the behavior of the particles and potencies. To summarize what (metaphysically) explains what: on my view, the initial distribution of particles and their potencies dynamically ground the subsequent behaviors of particles and subsequent property instantiations. And, all of the possible initial distributions and evolutions determine the (metaphysically inert) laws. (Demarest 2017, pp. 51–52)

The potencies at an earlier time explain how things move at a later time by dynamically producing, determining, or generating the patterns. We note that Demarest's view seems committed to a fundamental direction of time. The account of dynamic explanation presupposes a fundamental distinction between past and future, i.e. between the initial and the subsequent states of the world. The initial arrangement of particles and potencies metaphysically ground subsequent behaviors of particles. The commitment of a fundamental direction of time does not seem optional on her view.

Moreover, the metaphysical framework of fundamental dispositions already seems committed to a fundamental direction of time, independently of the issue of laws. For example, it is natural to interpret the discussions by Ellis, Bird, Mumford as suggesting that the manifestation of a disposition cannot be temporally prior to its stimulus, which presupposes a fundamental direction of time.[16] Therefore, although Aristotelian Reductionism does away with the governing conception of laws, the view seems committed to a fundamental direction of time twice over.

2.4 *Maudlinian Primitivism*

In his book *The Metaphysics Within Physics* (2007), Maudlin develops and defends a primitivist view about laws.[17] As a primitivist, he suggests that we should not analyze or reduce laws into anything else. Laws are metaphysically fundamental;

---

[16] In contrast, Vetter (2015) is open to a temporally symmetric metaphysics but assumes temporal asymmetry in her account of dispositions (which she calls potentialities).

[17] See Laudisa (2015) for a nice development of Maudlinian Primitivism. Carroll (1994) is often called a primitivist about laws, though recently Carroll (2018) distances his view from that of Maudlin and suggests a non-Humean reductive analysis of laws in terms of causation/explanation.

they are primitive entities that do not supervene on other entities. To have a sufficiently explanatory metaphysical theory, our fundamental ontology needs to include not only spatiotemporal objects but also laws that govern them. Maudlin rejects any reduction or deeper analysis of laws. He characterizes his primitivism as follows:

> My analysis of laws is no analysis at all. Rather I suggest we accept laws as fundamental entities in our ontology. Or, speaking at the conceptual level, the notion of a law cannot be reduced to other more primitive notions. (p. 18)

As a motivation for adopting primitivism over reductionism (especially Humean Reductionism), he writes:

> [Nothing] in scientific practice suggests that one ought to try to reduce fundamental laws to anything else. Physicists simply postulate fundamental laws, then try to figure out how to test their theories; they nowhere even attempt to analyze those laws in terms of patterns of instantiation of physical quantities. The practice of science, I suggest, takes fundamental laws of nature as further unanalyzable primitives. As philosophers, I think we can do no better than to follow this lead. (p. 105)

Maudlin is also committed to primitivism about the direction of time: that the distinction between past and future is metaphysically fundamental and not reducible to anything else. There is in effect a fundamental arrow or orientation at every spacetime point that points to the future. Maudlin combines the two commitments into a metaphysical package:

> Let's call the idea that both the laws of physics (as laws of temporal evolution) and the direction of time are ontological primitives *Maudlin's Non-Humean Package*. According to this package, the total state of the universe is, in a certain sense, derivative: it is the product of the operation of the laws on the initial state. (emphasis original, p. 182)

There are several reasons that Maudlin is committed to both. They become clear as we consider how laws explain on his account. For Maudlin, laws produce or generate later states of the world from earlier ones. In this way, via the productive power of the laws, subsequent states of the world (and its parts) are explained by earlier ones and ultimately by the initial state of the universe. It is this productive explanation that is central to his account. Production is closely related to causation, and just like (paradigm cases of) causation it is time asymmetric. Future states are produced from earlier states but not vice versa. This, for example, allows Maudlin's account to vindicate a widespread intuition about Bromberger's flagpole. The shadow is produced by the circumstances and the length of the pole (together with the laws). Although we can deduce from the laws the pole length based on the circumstances and the shadow length, the pole length is not produced by them. Hence, given the laws, the pole length and the circumstances explain, but are not explained by, the shadow length. Similar productive explanations can be given in more complicated cases.

The operation of the primitive laws depends on the primitive direction of time. Primitive laws act on past states to produce future states. Maudlin thinks that his package yields an attractive picture by being closer to our initial conception of the world:

> The universe started out in some particular initial state. The laws of temporal evolution operate, whether deterministically or stochastically, from that initial state to generate or produce later states. (p. 174)
>
> This sort of explanation takes the term initial quite seriously: the initial state temporally precedes the explananda, which can be seen to arise from it (by means of the operation of the law). (p. 176)
>
> The non-Humean package [described above] is, I think, much closer to the intuitive picture of the world that we begin our investigations with. Certainly, the fundamental asymmetry in the passage of time is inherent in our basic initial conception of the world, and the fundamental status of the laws of physics is, I think, implicit in physical practice. Both of the strands of our initial picture of the world weave together in the notion of a productive explanation, or account, of the physical universe itself. The universe, as well as all the smaller parts of it, is made: it is an ongoing enterprise, generated from a beginning and guided towards its future by physical law. (p. 182)

This intuitive picture of the world require certain restrictions on the form of fundamental laws. They have to be, what Maudlin calls, *fundamental laws of temporal evolution* (FLOTEs).[18] Examples include Newton's $F = ma$, Schrödinger's equation, and Dirac's equation on our first list in Sect. 1 but exclude most examples on our second list.

Let us summarize Maudlin's metaphysical package as follows:

Maudlinian Primitivism Fundamental laws are certain ontological primitives in the world. Only dynamical laws (in particular, laws of temporal evolution) can be fundamental laws. They operate on the universe by producing later states of the universe from earlier ones, in accord with the fundamental direction of time.

Maudlin allows there to be primitive stochastic dynamical laws—those laws that involve objective probability such as the GRW collapse laws. Hence, dynamic production need not be deterministic. An initial state can be compatible with multiple later states, determining only an objective probability distribution over those states. Perhaps the objective probability can be understood as *propensity*, with stochastic production implying variable propensities of producing various states, in proportion to their objective probabilities and in accord with the direction of time. However, even if deterministic production is an intelligible notion, it is not clear that stochastic production or propensity is as intelligible. (Recall the earlier point about "probabilistic necessitation" in Platonic Reductionism.) This may be another instance of the general phenomenon that objective probability (or chance) is conceptually murkier on non-Humean metaphysics than on Humean metaphysics.

---

[18] This reading of Maudlin is supported by the earlier passages as well as this one: "It was perhaps already clear when I wrote 'A Modest Proposal...' that the issue of time and the issue of natural laws were deeply intertwined: I noted in that essay that the fundamental laws of nature appear to be laws of *temporal evolution*: they specify how the state of the universe will, or might, evolve from a given initial state" (emphasis original, p. 172).

At first glance, Maudlin's view is intuitive. It is attractive to those who accept a fundamental direction of time. According to Maudlinian Primitivism, there is a fundamental distinction between past and future that is not reducible to entropic arrow of time, the distribution of matter in the universe, or special boundary conditions. This distinction picks out an initial state of the universe, in the literal sense of "initial," that is earlier than any other states. It is therefore surprising that Maudlin is committed to a "block universe" picture of time on which all times (past, present, and future) are equally real. Maudlin rejects presentism, the moving spotlight view, the growing block view, and the shrinking block view. So it is not the same as those pictures where spacetime is "dynamic" or the present moment is metaphysically privileged.

However, the view may not be as intuitive as it first seems. First, in relativistic spacetimes, there is no absolute simultaneity or a physically privileged notion of "now." The fundamental distinction between past and future needs to be understood without referring to a preferred foliation and should not involve an objective present. For any spacetime event, it requires an objective fact about which light cone points to the past and which one points to the future. Second, we may wonder how dynamic production extends to spacetimes with no "first" moment of time, such as those with an "initial" singularity or without temporal boundaries. If there is no initial state, perhaps the oomph of dynamic production, though having no beginning, always comes from earlier states. Third, the very notion of dynamic production is a bit unclear, especially in a block universe. We return to this issue in Sect. 3.2.

If one believes in Maudlinian Primitivism and its associated principle of (dynamic) productive explanation, then one needs to place restrictions on the form of laws. They can only take the form of FLOTEs. We return to this issue in Sect. 4.4.

3 Minimal Primitivism (MinP)

Having surveyed four existing approaches to laws, we propose our own view, which we call *Minimal Primitivism* (MinP).

3.1 The View

According to MinP, fundamental laws are ontological primitives that are metaphysically fundamental.[19] They do not require anything else to exist. They are

[19] Bhogal (2017) proposes a "minimal anti-Humeanism" on which laws are ungrounded (true) universal generalizations. It is compatible with primitivism, but it is less minimalist than MinP. For example, on Bhogal's view, laws cannot be singular facts about particular times or places.

not analyzable into (relations among) universals, powers, or dispositions. They are not reducible to (or supervenient on) the Humean mosaic. Rather, if the Humean mosaic describes spacetime and its contents, then the mosaic is governed by the laws, in a metaphysically robust sense. For laws to govern, they are not required to dynamically produce or generate later states of the universe from earlier ones, nor are they required to presume a fundamental direction of time. On MinP, laws govern by constraining the physical possibilities (often called nomological possibilities in the metaphysics literature). This places no in-principle demands on the form of fundamental laws. To summarize, the first part of our view is a metaphysical thesis:

Minimal Primitivism Fundamental laws of nature are certain primitive facts about the world. There is no restriction on the form of the fundamental laws. They govern the behavior of material objects by constraining the physical possibilities.

Even though there is no metaphysical restriction on the form of fundamental laws, it is rational to expect them to have certain nice features, such as simplicity and informativeness. On Humean Reductionism, those features are metaphysically constitutive of laws, but on our view they are merely epistemic guides for discovering and evaluating the laws. At the end of the day, they are defeasible guides, and we can be wrong about the fundamental laws even if we are fully rational in scientific investigations. The second part of our view is an epistemic thesis:

Epistemic Guides Even though theoretical virtues such as simplicity, informativeness, fit, and degree of naturalness are not metaphysically constitutive of fundamental laws, they are good epistemic guides for discovering and evaluating them.

Let us offer some clarifications:

(i) *Primitive facts*. Fundamental laws of nature are certain primitive facts about the world, in the sense that they are not metaphysically dependent on, reducible to, or analyzable in terms of anything else. If the concrete physical reality corresponds to a Humean mosaic, then fundamental laws are facts that transcend the mosaic. Many physicists may even regard fundamental laws as more important than the mosaic itself. Depending on one's metaphysical attitude towards mathematics and logic, there might be mathematical and logical facts that are also primitive in that sense. For example, arithmetical facts such as $2+3=5$ and the logical law of excluded middle may also be primitive facts that transcend the concrete physical reality and constrain the physical possibilities, since every physical possibility must conform to them. However,

However, Bhogal (p. 447, fn.1) seems open to relax the requirement that laws have to be universal generalizations. It would be interesting to see how to extend Bhogal's view to do so. In an arXiv preprint posted shortly after our paper, Adlam (2021) independently proposes an account that is, in certain aspects, similar to MinP; she also suggests we take seriously laws that do not have a time-evolution form. However, her account is not committed to primitivism and seems more at home in a structural realist framework. Moreover, simplicity does not seem essential to her account of nomic explanations.

we do not think that fundamental laws of nature are purely mathematical or logical. Hence, we stipulate that fundamental laws of nature are not such kinds of primitive facts.

(ii) *The governing relation.* We suggest that laws govern by constraining the world (the entire spacetime and its contents). We may understand constraining as a primitive relation between fundamental laws and the actual world. We can better understand constraining by drawing connections to physical possibilities. Laws constrain the world by limiting the physical possibilities and constraining the actual world to be one of them. In other words, the actual world is constrained to be compatible with the laws. To use an earlier example, $F = ma$ governs by constraining the physical possibilities to exactly those that are compatible with $F = ma$. If $F = ma$ is a law that governs the actual world, then the actual world is a possibility compatible with $F = ma$.

Constraint does not require a fundamental distinction between past and future, or one between earlier states and later states. What the laws constrain is the entire spacetime and its contents. In some cases, the constraint imposed by a law can be expressed in terms of differential equations that may be interpreted as determining future states from past ones. But not all constraints need be like that. We discuss this point in Sect. 3.2.[20]

For a concrete example, consider the Hamilton's equations of motion for N point particles with Newtonian masses $(m_1, \ldots, m_N)$ moving

[20] For those metaphysically inclined, here are some formal details. Consider w, the complete history of a possible world describable in terms of matter in spacetime. Let Ω_w be the non-empty set of worlds that are physically possible (from the perspective of w). It is *a priori* that $w \in \Omega_w$. Consider fact L, which may be Newton's equation of motion with Newtonian gravitation. Let Ω^L be the set of models generated by L. Now, suppose L governs w. Then the following is true:

Equivalence $\Omega^L = \Omega_w$

Equivalence makes precise the idea that on MinP governing laws limit the physical possibilities. Since $w \in \Omega_w$, it follows that:

Constraint $w \in \Omega^L$

If we let $w = \alpha$, the actual world, then Constraint makes precise the idea that, on MinP, laws constrain the actual world. For MinP, we postulate that the above notions and derivations make sense. A natural idea is to reduce or analyze physical possibilities and necessities in terms of fundamental laws and a notion of mathematical consistency. This makes physical possibilities a derivative notion rather than a fundamental one. However, we do not insist on it here. A few epistemological remarks: the fact that $\Omega^L = \Omega_w$ is knowable *a posteriori*; consequently, the fact that $w \in \Omega^L$ is also knowable *a posteriori*. A careful reader might raise a consistency worry here: what if a single world (history) w is compatible with two different laws L and L' with non-empty overlap in their solution spaces, such that $w \in \Omega^L \cap \Omega^{L'}$? The worry is handled by the earlier postulates. Having $w \in \Omega^{L'}$ is not sufficient for L' to be the governing law or for $\Omega^{L'}$ to be the set of physical possibilities. MinP assumes that, from the perspective each world, there is a single set of physical possibilities, given by the governing law(s). Hence, for w, if Ω^L and $\Omega^{L'}$ are different sets, then at most one of them is equivalent to Ω_w. Moreover, since Ω_w is non-empty, the laws that govern w must be consistent with each other.

in a 3-dimensional Euclidean space, whose positions and momenta are $(\boldsymbol{q_1}, \ldots, \boldsymbol{q_N}; \boldsymbol{p_1}, \ldots, \boldsymbol{p_N})$:

$$\frac{d\boldsymbol{q_i}(t)}{dt} = \frac{\partial H}{\partial \boldsymbol{p_i}}, \frac{d\boldsymbol{p_i}(t)}{dt} = -\frac{\partial H}{\partial \boldsymbol{q_i}} \tag{1}$$

where $H = H(\boldsymbol{q_1}, \ldots, \boldsymbol{q_N}; \boldsymbol{p_1}, \ldots, \boldsymbol{p_N})$ is specified in accord with Newtonian gravitation:

$$H = \sum_i^N \frac{\boldsymbol{p_i}^2}{2m_i} - \sum_{1 \le j < k \le N} \frac{G m_j m_k}{|\boldsymbol{q_j} - \boldsymbol{q_k}|} \tag{2}$$

Suppose Eqs. (1) and (2) are the fundamental laws that govern our world α. Let Ω^H denote the set of solutions to (1) and (2). Saying that (1) and (2) govern our world implies that α should be compatible with them. In other words, Ω^H delineates the set of physical possibilities, and $\alpha \in \Omega^H$. In this example, the dynamical equations are time-reversible. For every solution in Ω^H, its time reversal under $t \to -t$ and $\boldsymbol{p} \to -\boldsymbol{p}$ is also a solution in Ω^H. Since the concept of governing in MinP does not presuppose a fundamental direction of time, two solutions that are time-reversal of each other *can* be identified as the same physical possibility.[21]

We should not think that, in every case, a law is equivalent to the set of possibilities it generates. The two can be different. For example, there are many principles and equations that can give rise to the same set of possibilities denoted by Ω^H. But we expect laws to be simple. One way to pick out the set Ω^H is by giving a complete (and infinitely) long list of possible histories contained in Ω^H. Another is by writing down simple equations, such as (1) and (2), which express simple laws. Hence, the equivalence of physical laws is not just the equivalence of their classes of models. For two laws to be equivalent, it will require something more.[22]

Humeans might object that our notion of governing is entirely mysterious (Beebee 2000). The notion of governing seems derived from the notion of government and the notion of being governed. But laws of nature are obviously not imposed by human (or divine) agents. So isn't it mysterious that laws can govern? To that we reply that a better analogy for governing laws is not to human government, but to laws of mathematics and logic. Arithmetical truths

[21] If one prefers the representation where the set of physical possibilities contains each possibility exactly once, one can derive a quotient set Ω_α^* from Ω_α with the equivalence relation given by the time-reversal map.

[22] It is an interesting question, on MinP, what more is required and how to understand the equivalence of physical laws. Perhaps their equivalence is related to simplicity and explanations. In any case, we do not provide such an account as it is orthogonal to our main concerns in the paper. For a survey of the related topic of theoretical equivalence, see Weatherall (2019a, 2019b).

such as $2 + 3 = 5$ and logical truths such as the law of excluded middle can also be said to constrain our world. That is, the actual world cannot be a world that violates those mathematical or logical truths. In fact, every possible world needs to respect those truths. In a similar way, laws of physics constrain our world. The actual world cannot be a world that violates the physical laws, and every physically possible world needs to respect those laws. Those modal claims reflect physical laws and mathematical laws. We can also make sense of the difference in scope between those laws. Mathematical laws are more general than physical laws, in the sense that the former are compatible with "more models" than the latter. In any case, mathematical laws and logical laws can also be said to govern the universe in the sense of imposing formal constraints. They generate a class of models and constrain the actual world to be one among them. There is also a difference in epistemic access. In some sense, we discover mathematical and logical laws *a priori*, without the need for experiments or observations, but we discover physical laws *a posteriori*, empirically.

We do not claim that the analogy with mathematical and logical laws completely eliminates the mystery of how physical laws govern. However, we think it dispels the objection as previously stated, in terms of how something can govern the world without being imposed by an agent. If there is more to the mystery objection, it needs to be stated differently. On MinP, laws govern by constraining, and constraining is what they do. This provides the oomph behind scientific explanations. (We return to this shortly.) However, in contrast to other non-Humean accounts, such an oomph is minimalist. It does not require dynamic production, and it does not require an extra process supplied by a mechanism or an agent.

(iii) *Epistemic Guides.* On MinP, even though the Humean criteria for the best system are not metaphysically constitutive for lawhood, they are nonetheless excellent epistemic guides for discovering and evaluating them. Lewis is right that in scientific practice, in the context of discovery, we do aim to balance simplicity and informativeness (among other things).

Regarding Epistemic Guides, one might ask in virtue of what those theoretical virtues are reliable guides for finding and evaluating laws. This is a subtle issue, and we are not prepared to give a complete answer here. Unlike Humeans, we cannot appeal to a reductive analysis of laws.[23] We can offer an empirical justification: the scientific methodology works. In so far as those theoretical virtues are central to scientific methodology, they are good guides for discovering and evaluating laws, and we expect them to continue to work. Can they fail to deliver us the true laws? That is a possibility. However, if the true fundamental laws are complicated and messy, scientists would not be inclined to call them laws. Therefore, there are two aspects of Epistemic

[23] Humeans face a similar issue, as their account raises the worry as to why the fundamental Humean mosaic is so nice that it can be summarized in a simple way after all.

Guides: discoverability and believability. Whereas simplicity as a guide for discovering a law might raise the question as to why laws should be simple (so that simplicity would be a reliable guide), simplicity as a requirement for believability seems clearer: it may be the case that the law is not simple, but if it is not simple it will not be believable as the law. We return to this point in Sects. 3.2 and 4.1.

(iv) We now address several other questions that arise concerning MinP.

- According to MinP, can laws change with time? In particular, can fundamental laws be time dependent in such a way that different cosmic epochs are governed by different laws? In principle, we are open to that possibility. If there is scientific motivation to develop theories in which laws take on different forms at different times (or in different epochs), then that is sufficient reason to consider a set of laws that govern different times, or a single law that varies in form with time. As a toy example, if we have empirical or theoretical reasons to think that the laws of motion are different on the two sides of the Big Crunch, say Newtonian mechanics and Bohmian mechanics, then different sides of the Crunch can be governed by different laws, or by a single law with a temporal variation.
- According to MinP, can fundamental laws refer to non-fundamental properties, such as entropy? Most of the fundamental laws we discover refer only to fundamental properties. But it is reasonable to consider candidate fundamental laws that refer to non-fundamental properties. Our principle of Epistemic Guides allows for this, as long as the non-fundamental properties are not too unnatural (all things considered). In the case of the Past Hypothesis, for example, we may sacrifice fundamentality of the property involved but gain a lot of informativeness and simplicity if we invoke the property of entropy. The version of the Past Hypothesis that refers to entropy can still govern by constraining the physical possibilities. (Another strategy is to revise our definition of fundamental property such that any property mentioned by a fundamental law is regarded as fundamental, although it may be analyzable in terms of other fundamental properties. However, this may present a problem for certain views of fundamentality.)
- According to MinP, how are fundamental laws distinguished from non-fundamental laws? We prefer a reductionist picture where non-fundamental laws, when properly understood, are reducible to fundamental laws. We can distinguish them in terms of derivability: non-fundamental laws can be (non-trivially) derived from fundamental laws (Chen 2022, sect. 2.3). For example, the ideal gas law is less fundamental than Newton's laws of motion, in the sense that the ideal gas law can be derived from them in suitable regime. However, derivability may not be sufficient for non-fundamental *lawhood*, as other factors, such as counterfactual and explanatory robustness, may also be relevant.
- How does MinP compare to the other views in Sect. 2? We discuss this in Sect. 4.

3.2 MinP and Explanation

On MinP, laws explain, but not by accounting for the dynamic production of successive states of the universe from earlier ones. They explain by expressing a hidden simplicity, given by compelling constraints that lie beneath complex phenomena. A fundamental direction of time is not required for our notion of explanation.[24]

In a world governed by Newtonian mechanics, particles travel along often complicated trajectories because that is implied by the simple fundamental law $F = ma$. Laws explain only when they can be expressed by simple principles or differential equations. It is often the case that the complicated patterns we see in spacetime can be derived from simple rules that we call laws.

Fundamental laws need not be time-directed or time-dependent. They may govern purely spatial distribution of matter. For example, Gauss's law

$$\nabla \cdot \mathbf{E} = \rho \tag{3}$$

in classical electrodynamics—one of Maxwell's equations—governs the distribution of electric charges and the electric field in space.

Often the explanation that laws provide involves deriving striking, novel, and unexpected patterns from simple laws. The relative contrast between the simplicity of the law and the complexity and richness of the patterns may indicate that the law is the correct explanation of the patterns.

For a toy example, consider the Mandelbrot set in the complex plane (Fig. 1), produced by the simple rule that a complex number c is in the set just in case the function

$$f_c(z) = z^2 + c \tag{4}$$

does not diverge when iterated starting from $z = 0$. (For example, $c = -1$ is in this set but $c = 1$ is not, since the sequence $(0, -1, 0, -1, 0, -1, \ldots)$ is bounded but $(0, 1, 2, 5, 26, 677, 458330, \ldots)$ is not. For a nice description and visualization, see (Penrose 1989, ch.3).) Here, a relatively simple rule yields a surprisingly intricate and rich pattern in the complex plane, a striking example of what is called the fractal structure. Now regard the Mandelbrot set as corresponding to the distribution of matter over (a two-dimensional) spacetime, the fundamental law for the world might be the rule just described. What is relevant here is that given just the pattern we may not expect it to be generated by any simple rule. It would be a profound discovery in that world to learn that its complicated structure is generated by the aforementioned

[24] This type of explanation, sometimes called "constraint explanation," has been explored in the causation literature by Ben-Menahem (2018) and non-causal explanation literature by Lange (2016). Their accounts, with suitable modifications, may apply here. See Hildebrand (2013) for a critical discussion of primitive laws and explanations.

Fig. 1 The Mandelbrot set with continuously colored environment. Picture created by Wolfgang Beyer with the program Ultra Fractal 3, CC BY-SA 3.0, https://creativecommons.org/licenses/by-sa/3.0, via Wikimedia Commons

rule based on the very simple function $f_c(z) = z^2 + c$. On our conception, it would be permissible to claim that the simple rule expresses the fundamental law, even though it is not a law for dynamic production.

The previous examples illustrate some features of explanation on MinP:

1. Laws explain by constraining the physical possibilities in an illuminating manner.
2. Nomic explanations (explanations given by laws) need not be dynamic explanations; indeed, they need not involve time at all.
3. Explanation by striking constraint can be especially illuminating when an intricate and rich pattern can be derived from a simple rule that expresses the constraint imposed by a law.

On our view, more generally, there are two ingredients of a successful scientific explanation: a metaphysical element and an epistemic one. It must refer to the objective structure in the world, but it also must relate to our mind, remove puzzlement, and provide an understanding of nature. We suggest that a successful scientific explanation that fundamental laws provides should contain two aspects: (i) metaphysical fundamentality and (ii) simplicity.

The first aspect concerns the metaphysical status of fundamental laws: they should not be mere summaries of, or supervenient on, what actually happens; moreover, what the laws are should not depend on our actual practice or beliefs. This aspect is the *precondition* for having a non-Humean account of scientific explanations. On MinP, the precondition is fulfilled by postulating fundamental laws as primitive (metaphysically fundamental) facts that constrain the world. The constraint provides the needed *oomph* behind scientific explanations. Here lies the main difference between MinP and Humean Reductionism. (We return to this point in Sect. 4.1.)

The second aspect concerns how fundamental laws relate to us. Constraints, in and of themselves, do not always provide satisfying explanations. Many constraints are complicated and thus insufficient for understanding nature. What we look for in the final theory of physics is not just any constraint but simple, compelling ones that ground observed complexities of an often bewildering variety. The explanation they provide corresponds to an insight or realization that leads us to say, "Aha! Now I understand." Often, simplicity is related to elegance or beauty. As Penrose reminds us:

> Elegance and simplicity are certainly things that go very much together. But nevertheless it cannot be quite the whole story. I think perhaps one should say it has to do with *unexpected* simplicity, where one imagines that things are going to be complicated but suddenly they turn out to be very much simpler than expected. It is not unnatural that this should be pleasing to the mind. (Penrose 1974, p. 268)

The sense of unexpected simplicity is illustrated in the toy example of the Mandelbrot set as well as the physical laws discovered by Newton, Schrödinger, and Einstein.

Moreover, the second aspect of scientific explanation illuminates our principle of Epistemic Guides. It is obvious that fundamental laws should be empirically adequate and consistent with all phenomena. But why should we expect them to be simple? On our view, it can partly be answered by thinking about the nature of scientific explanations. If successful scientific explanations require simple laws, then laws should be sufficiently simple to perform the explanatory role. One might press further and ask why laws should perform such roles and why scientific explanations can be successful. But they can be raised *on any account of laws*. We note that it is a difficult issue, one that may be related to Hume's problem of induction (Henderson 2020), which may be regarded as the problem of giving a non-circular justification for the uniformity of nature. Indeed, if we accept MinP and that the fundamental laws are simple, we have eliminated the problem of induction. If laws are simple, they may be completely uniform in space and time or else provide a simple rule that specifies how they change over space or time (Sect. 3.1). Thus, the problem of induction reduces to the problem of justifying our acceptance of simple laws. In this sense, the problem of induction is not an additional problem over and above the justification of Epistemic Guides.

3.3 Examples and Further Clarifications

To further clarify MinP, we discuss some examples of dynamical laws and non-dynamical constraint laws. On our view, there is no difficulty accommodating them, as they can be understood as laws that constrain physical possibilities. We also consider laws that involve intrinsic randomness, as they are less easy to accommodate on our view. We offer five interpretive options for further consideration.

3.3.1 Dynamical Laws

We take a dynamical law to be any law that determines how objects move or things change. (Here we focus on non-probabilistic laws and leave probabilistic ones to Sect. 3.3.3.) Thus, our notion of dynamical laws is wider than Maudlin's notion of FLOTEs.

Hamilton's Equations. Consider classical mechanics for N particles, described by Hamilton's equations of motion (1) with a Hamiltonian specified in (2), a paradigmatic example of a FLOTE. Hamilton's equations are differential equations of a particular type: they admit initial value formulations. An intuitive way of thinking about dynamical laws is to understand them as evolving the initial state of the world into later ones. However, this view is not entirely natural for such a system. The view requires momenta to be part of the intrinsic state of the world at a time; but it seems more natural to regard them as aspects of extended trajectories, spanning continuous intervals of time. Regarding governing as dynamic production leads to awkward questions about instantaneous states and whether they include velocities and momenta.

The situation becomes even more complicated with relativistic spacetimes having no preferred foliation of equal-time hypersurfaces. If there is no objective fact about which events are simultaneous, there is no unique prior Cauchy surface that is responsible for the production of any later state. This seems to detract from the intuitive idea of dynamic production as a relation with an objective input, making it less natural in a relativistic setting.[25]

Instead of demanding that laws govern by producing subsequent states from earlier ones, we can regard laws as constraining the physical possibilities of spacetime and its contents. There is no difficulty accommodating the above example or any other type of dynamical laws. A dynamical law specifies a set of histories of the system and need not be interpreted as presupposing a fundamental direction of time. The histories the laws allow can often be understood as direction-less histories, descriptions of which events are temporally between which other events.

A dynamical law such as (1) governs the actual world by constraining its history to be one allowed by (1). And MinP requires no privileged splitting of spacetime into space and time, as the physical possibilities can be stated in a completely coordinate-free way in terms of the contents of the 4-dimensional spacetime.

Principles of Least Action Besides dynamical laws of Hamiltonian form, other kinds of equations and principles are often employed even for Hamiltonian systems.

[25] Christopher Dorst raised a similar point in personal communication. See also Dorst (2021).

Consider, for example, Hamilton's principle of least action: this requires that for a system of N particles with Cartesian coordinates $q = (\boldsymbol{q}_1, \boldsymbol{q}_2, \ldots, \boldsymbol{q}_N)$:

$$\delta S = 0 \tag{5}$$

where $S = \int_{t_1}^{t_2} L(q(t), \dot{q}, t)dt$, with $\dot{q} = q(t)/dt$, δ the first-order variation of S corresponding to small variation in $q(t)$ with $q(t_1)$ and $q(t_2)$ fixed, and L, the Lagrangian, is the kinetic energy minus the potential energy of the system of N particles. While mathematically equivalent to Hamilton's equations, the principle of least action feels very different from a law expressing dynamic production. For those who take dynamic production to be constitutive of governing, the principle of least action cannot be the fundamental governing law. They would presumably need to insist that the universe is genuinely governed by some law of a form such as (1), with the principle of least action arising as a theorem. For us, we have no problem regarding the principle of least action as a candidate fundamental law, with no need for it to be derived from anything else. For a universe to obey the principle, its history must be one compatible with (5). That is the sense in which it would govern our universe.

Wheeler-Feynman Electrodynamics Physicists have also considered dynamical equations that cannot be reformulated in Hamiltonian form. On MinP, there is no prohibition against laws expressed by such equations. For example, Wheeler and Feynman (1945, 1949) considered equations of motion for charged particles that involve both retarded fields (F_{ret}) and advanced ones (F_{adv}). On their theory, the trajectory of a charged particle depends on charge distributions in the past (corresponding to F_{ret}) as well as those in the future (corresponding to F_{adv}). Since the total field acting on particle j is $F_{tot} = \sum_{k \neq j} \frac{1}{2}({}^{(k)}F_{ret} + {}^{(k)}F_{adv})$, the equation of motion for particle j of mass m_j, charge e_j, and spacetime location q_j is

$$m_j \ddot{q}_j^{\mu} = e_j \sum_{k \neq j} \frac{1}{2}({}^{(k)}F_{ret}^{\mu\nu} + {}^{(k)}F_{adv}^{\mu\nu})\dot{q}_{j,\nu} \tag{6}$$

with the dot the time derivative with respect to proper time, ${}^{(k)}F_{ret}$ the retarded field contributed by the past trajectory of particle k, and ${}^{(k)}F_{adv}$ the advanced one, involving the future trajectory of particle k. (For more details, see Deckert (2010) and Lazarovici (2018).) It is unclear how to understand the above equation in terms of dynamic production. In contrast, it is clear on MinP: the fundamental law corresponding to such equations can be regarded as imposing a constraint on all trajectories of charged particles in spacetime.

Retrocausal Quantum Mechanics There have been proposed reformulations of quantum mechanics that involve two independent wave functions of the universe: $\Psi_i(t)$ evolving from the past and $\Psi_f(t)$ evolving from the future. (It is not entirely clear what this is supposed to mean, but something retrocausal is definitely contemplated.) Some such proposals, motivated by a desire to evade no-go theorems or preserve time-symmetry, implement retrocausality or backward-in-time causal influences (Friederich and Evans 2019). Consider Sutherland (2008)'s causally symmetric Bohm model, which specifies an equation of motion governing N particles moving in a 3-dimensional space under the influence of both $\Psi_i(t)$ and $\Psi_f(t)$:

$$\frac{d\boldsymbol{Q}_j(t)}{dt} = \frac{Re(\frac{\hbar}{2im_j a}\Psi_f^* \nabla_j \Psi_i)}{Re(\frac{1}{a}\Psi_f^* \Psi_i)}(Q(t), t) \tag{7}$$

with $Q(t) = (\boldsymbol{Q}_1(t), \ldots, \boldsymbol{Q}_N(t)) \in \mathbb{R}^{3N}$ the configuration of the N particles at time t, m_j the mass of particle j, and $a = \int \Psi_f^*(q,t)\Psi_i(q,t)dq$. It is unclear whether Sutherland's theory is viable; it also has many strange consequences. Nevertheless, MinP is compatible with regarding the above equation as expressing a fundamental law that constrains particle trajectories in spacetime (even though we have other reasons to not endorse the theory).

Similarly, MinP is compatible with Goldstein and Tumulka (2003)'s model involving two opposite arrows of time. To reconcile relativity (Lorentz invariance) and non-locality, the model involves a past hypothesis yielding the usual macroscopic arrow of time as well as a microscopic dynamical equation acting on future data to yield past data, thus providing a microscopic arrow of time from future to past. Together, they constrain the particle trajectories in spacetime.

The Einstein Equation In general relativity, the fundamental equation is the Einstein equation:

$$R_{\mu\nu} - \frac{1}{2}Rg_{\mu\nu} = k_0 T_{\mu\nu} + \Lambda g_{\mu\nu} \tag{8}$$

where $R_{\mu\nu}$ is the Ricci tensor, R is the Ricci scalar, $g_{\mu\nu}$ is the metric tensor, $T_{\mu\nu}$ is the stress-energy tensor, Λ is the cosmological constant, $k_0 = 8\pi G/c^4$ with G Newton's gravitational constant and c the speed of light. Roughly speaking, the Einstein equation is a constraint on the relation between the geometry of spacetime and the distribution of matter (matter-energy) in spacetime. On MinP, we have no problem taking the equation itself as expressing a fundamental law of nature, one that constrains the actual spacetime and its contents. If Eq. (8) governs our world in the sense of MinP, then (8) expresses a fundamental fact that does not supervene on or reduce to the actual spacetime and its contents.

There are ways of converting Eq. (8) into FLOTEs that are suitable for a dynamic productive interpretation. (A famous example is the ADM formalism of Arnowitt et al. 1962.) However, they often discard certain solutions (such as spacetimes that are not globally hyperbolic). For non-Humeans who take dynamic production as constitutive for governing or explanations, those reformulations will be necessary. For them, the true laws of spacetime geometry should presumably be expressed by equations that describe the evolution of a 3-geometry in time. In contrast, on MinP there is no metaphysical problem for taking the original Einstein equation as a fundamental law. The Einstein equation is simple and elegant and is generally regarded as the fundamental law in general relativity. We prefer not to discard or modify it on metaphysical grounds.[26]

The Einstein equation allows some peculiar solutions. A particularly striking class of examples are spacetimes with closed timelike curves (CTCs). For MinP, there would seem to be no fundamental reason why such a possibility should be precluded. But the possibility of CTCs is precluded by Maudlinian Primitivism, since they may lead to an event that dynamically produces itself (Maudlin 2007, p. 175). And it is hard to see how Demarest's version of Aristotelian Reductionism can allow them. Humean Reductionism should be compatible with CTCs, just as MinP is. It is unclear whether they are compatible with Platonic Reductionism.

3.3.2 Non-dynamical Constraint Laws

The examples mentioned earlier are explicitly related to time. There are also important equations and principles that are not. For example, some purely spatial constraints on the universe may be thought of as physical laws. We call them non-dynamical constraint laws. The minimal notion of governing easily applies to them. In Sect. 3.1 we considered two examples of such laws—(3) and (4). Here we consider a few more.

The Past Hypothesis In the foundations of statistical mechanics and thermodynamics, followers of Boltzmann have proposed a candidate fundamental law of physics that Albert (2010) calls the Past Hypothesis (PH). It is a special boundary condition that is postulated to explain the emergent asymmetries of time in our universe, such as the Second Law of Thermodynamics. Here is one way to state it:

PH At one temporal boundary of the universe, the universe is in a low-entropy state.

This statement of PH is vague. We may be able to make it more precise by specifying the low-entropy state in terms of the thermoydnamic properties of the universe or in terms of some geometrical properties (Penrose 1979). Penrose's version in general relativity renders it as follows: the Weyl curvature C_{abcd} vanishes at any "initial" singularity. Let us use Ω_{PH} to denote the set of worlds compatible with PH. If it is plausible that PH is a candidate fundamental law (Chen 2020), then the metaphysical account of laws should make room for a boundary condition to be a fundamental law. On MinP, such an account is no problem. Together, PH and dynamical laws can govern the actual world by constraining it to be one among

[26] Making a similar point, Callender (2017, p. 139) writes: "[The] ten vacuum Einstein field equations separate into six "evolution" equations $G_{ij} = 0$ and four "constraint equations," $G_{00} = 0$ and $G_{0i} = 0$, with $i = 1, 2, 3$. The latter impose nomic conditions across a spacelike slice. To decree that four of the ten equations that constitute Einstein's field equations are not nomic without good reason is unacceptable."

the histories compatible with all of them. They deem that the actual world (history) is a member of the intersection $\Omega^{PH} \cap \Omega^{DL}$, where the latter denotes the set of histories compatible with the dynamical laws. However, PH is not a governing law in the sense of dynamic production. So its fundamental lawhood is incompatible with Maudlinian Primitivism. And it is not a natural fit for Aristotelian Reductionism (but see Demarest (2019) for a recent proposal for how they might fit).

It is possible that the vagueness in PH cannot be eliminated. If PH (or something like it) turns out to be a fundamental law, then there can be vagueness in the fundamental laws. See Chen (2022) for a discussion of this possibility, called "fundamental nomic vagueness." On MinP, fundamental nomic vagueness implies that there are vague fundamental facts. Whether fundamental nomic vagueness exists in our world is a subtle question. (It is related to the issue of whether the ontic quantum state of the universe is pure or impure (Chen 2022, sect.4).)

Conservation Laws and Symmetry Principles According to a traditional perspective, symmetries such as those of rotation, spatial translation, and time translation are properties of the specific equations of motion. By Noether's theorem, those symmetries yield various conservation laws as theorems rather than postulates that need to be put in by hand. On that perspective, symmetries and conservation laws can be regarded as ontologically derivative of the fundamental laws, and are compatible with all metaphysical views on laws.

According to a more recent perspective, symmetries are fundamental. See for example: Wigner (1985, 1964) and Weinberg (1992). Lange (2009) calls them *metalaws*. For example, Wigner describes symmetries as "laws which the laws of nature have to obey" (Wigner 1985, p. 700) and suggest that "there is a great similarity between the relation of the laws of nature to the events on one hand, and the relation of symmetry principles to the laws of nature on the other" (Wigner 1964, p. 957). Here we do not take a firm stance on this perspective. Nevertheless, we note that it is compatible with MinP. If there is a symmetry principle K that a fundamental law of nature L must obey, then both K and L are fundamental facts, where K constrains L in the sense that the physical possibilities generated by L are invariant under the symmetry principle K, and any other possible fundamental laws are also constrained by K. This introduces further "modal" relations in the fundamental facts beyond just the constraining of the spacetime and its contents by L.

Methodologically, one might prefer theories with dynamical laws, especially FLOTEs, to those without them. MinP allows this preference. Even though MinP does not restrict laws to FLOTEs, the principle of Epistemic Guides suggests that we look for simple and informative laws. FLOTEs, when they admit initial value formulations, may come with such theoretical virtues (Callender 2017, ch.7–8). The preference for FLOTEs and dynamical laws more generally may be explained by a preference for laws that strike a good balance between simplicity and informativeness.

3.3.3 Probabilistic Laws

Candidate fundamental physical theories can also employ probability measures and distributions. Such measures and distributions can be objective, and they may be called objective probabilities. The probabilistic postulates in physical theories may well be lawlike, even though the nature of those probabilities is a controversial matter. As mentioned in Sect. 2, Humean Reductionism can accommodate those probabilistic postulates as axioms in the best system achieving the optimal balance of a simple and informative summary. It is not so clear what objective probability means in the non-Humean accounts. On MinP, the extension from non-probabilistic laws to probabilistic ones is also not straightforward. We propose several strategies for consideration, but which one is most promising depends on further interpretive questions about probabilities.

There are two types of probabilistic postulates in physics: (i) stochastic dynamics and (ii) probabilistic boundary conditions. We start with (i) as it is more familiar. Consider the GRW theory in quantum mechanics, a theory in which observers and measurements do not have a central place and in which the quantum wave function spontaneously collapses according to precise probabilistic rules. On the GRW theory, the wave function of the universe $\Psi(t)$ evolves unitarily according to the Schrödinger equation but is interrupted by random collapses. The probabilities of where and when the collapses occur are fixed by the theory. (For details, see Ghirardi et al. (1986) and Ghirardi (2018).) Another example is Nelson's stochastic mechanics that describes particle motion in accord with a stochastic differential equation. (See Nelson (1966) and Bacciagaluppi (2005).)

For an example of (ii), consider Albert and Loewer's Mentaculus theory of statistical mechanics, where they postulate in addition to the dynamical equations (such as (1) and (2)) and PH, a probabilistic distribution of the initial microstate of the universe:

Statistical Postulate (SP) At the temporal boundary of the universe when PH applies, the probability distribution of the microstate of the universe is given by the uniform one (according to the natural measure) that is supported on the macrostate of the universe (compatible with PH).

At first glance, it is not obvious what SP is intended to convey. One may understand it in terms of typicality: that we regard the initial probability distribution to pick out a measure of almost all or the overwhelming majority—a measure of typicality (Goldstein 2001, 2012). On this way of thinking, SP says the following:

SP' At the temporal boundary of the universe when PH applies, the initial microstate of the universe is typical inside the macrostate of the universe (according to the natural measure of typicality).

On the basis of SP', one can then explore what the theory says about typical histories and apply it to our universe. A similar probabilistic boundary condition appears in

Bohmian mechanics, where one can interpret the initial probability distribution of particle configuration as representing a typicality measure:

$$\rho_{t_0}(q) = |\Psi(q, t_0)|^2 \tag{9}$$

where t_0 is when PH applies and $\Psi(q, t_0)$ is the wave function of the universe at t_0. Based on this measure, almost all worlds governed by Bohmian mechanics will exhibit the Born rule. (For more details, see Dürr et al. (1992) and Goldstein (2017).)

In fact, it is also possible to interpret stochastic dynamics as yielding a typicality measure: the GRW theory specifies a probability distribution over entire histories of the quantum states, and what matters is the behavior of "almost all" of those possible histories. Although we are sympathetic to the typicality interpretation of both stochastic dynamics and probabilistic boundary conditions, we do not insist on it here.

Probability measures and typicality measures are not straightforwardly understandable in terms of MinP: it is not clear how they should be understood in terms of constraints. The difficulty is greater for stochastic dynamics. On the typicality approach, one has the option to regard the measures picked out by the probabilistic boundary conditions as referring to something methodological instead of nomological—how in practice one decides whether a law is supported or refuted by evidence. However, the probabilities in the stochastic dynamics are clearly nomological and not just a methodological principle of theory choice. Here we offer five interpretive options for how to understand probabilistic laws in MinP.

Option 1: Humeanism Probabilistic laws are second-class citizens that supervene on the distribution of matter. On this interpretation, probability and typicality measures are given a Humean best-system analysis. This is compatible with taking the non-probabilistic laws as fundamental facts (Hoefer 2019). This option is a Humean and non-Humean hybrid: it is Humean about probabilistic laws but non-Humean about other laws. We find this option, while viable, unsatisfactory as probabilistic laws and non-probabilistic laws seem on a par with respect to explaining patterns. Leaving the former supervenient on the mosaic renders the statistical patterns ultimately unexplained. The hybrid strategy goes against the non-Humean conviction that motivates our adoption of MinP.

Option 2: Primitivism Probabilistic laws are primitive facts but they are not directly related to constraints. On this interpretation, they are fundamental irreducible facts in the world, but they do not simply constrain the world. They are somehow connected to frequencies and credences but not via constraint. But then how are they connected? This option needs to be developed further for evaluation.

Option 3: Gradable Constraint Probabilistic laws are primitive facts that constrain the world, but their constraining admits of degree. This interpretation extends the concept of constraining from allowed/forbidden to degrees of constraints (between 0 and 1, inclusive), with the categorical ones (allowed/forbidden) taking the extreme values. This notion of degrees of constraint is not entirely clear. However, for those who are fine with probabilification (Platonic Reductionism) or propensities, perhaps this is also acceptable. This notion of degree of constraining is essentially propensity without a fundamental direction of time.

Option 4: Typicality Constraint Probabilistic laws are primitive facts about typicality that entail categorical constraint (which does not come in degree). If the right approach to probabilistic laws is in terms of typicality, and if typicality relates to categorical constraint, then we can relate both to the notion of constraint in MinP. A typicality statement is about particular kinds of behaviors that satisfy certain properties. We can regard such properties as imposing a constraint on possible histories. This makes typical histories the only physically possible histories. There are two potential problems. First, such a property may be complicated to specify and the complexity makes the statement a bad candidate for a law. Second, it seems to place too strong a requirement. We usually think that atypical histories are still physically possible. They are just expected not to occur.

Option 5: Dual Modalities Probabilistic laws are primitive facts about typicality, which is another kind of modality distinct from possibility. On this interpretation, there is a dualism between modal notions of possibility and typicality. Non-probabilistic laws govern by constraining the space of possibilities. Probabilistic laws govern by constraining which possibilities are typical. Neither is reducible to the other. Some worlds are possible but atypical. However, every typical world is possible. Both typicality and possibility should influence our expectations, and both play roles in scientific explanations. In this way, we may think of probabilistic laws as imposing a narrower kind of constraints on the world. The actual world must be a member of the physical possibilities delineated by non-probabilistic laws. The actual world must also be a member of the possibilities delineated by probabilistic laws.

Those issues above have not been much explored in the literature, and there are many open problems here. We do not take a firm stance, merely noting the above five strategies, with the acknowledged qualifications and subtleties, may be available to a defender of MinP. The problem of probabilistic laws is difficult for all non-Humean accounts of laws. Solving it may turn on questions about the relation between probability and typicality, and their relation to physical possibility.

4 Comparisons

We have argued that MinP is a minimalist version of non-Humeanism about laws that is flexible enough to naturally accommodate the diverse kinds of laws entertained in physics. In this section, we highlight some differences between MinP and the alternatives.

4.1 Comparison with Humean Reductionism

Although MinP is a non-Humean view, in several respects it is similar to Humean Reductionism. First, neither requires a fundamental direction of time, and both permit a reductionist understanding of it. MinP shows that anti-reductionism about laws does not require anti-reductionism about the direction of time. Second, both views are flexible enough to accommodate the distinct kinds of laws entertained in physics, although Humean Reductionism might have an upper hand in understanding probabilistic laws. Third, both views highlight the importance of simplicity (and other super-empirical virtues) in laws and scientific explanations.

We turn now to the key differences between the two. As mentioned earlier, the main differences are whether laws are reducible to the mosaic and whether laws depend on our practice and beliefs. Humean Reductionism answers Yes while MinP answers No to both.

Ultimate Explanation On Humean Reductionism, the patterns in the Humean mosaic have no ultimate explanation; after all, the mosaic grounds what the laws are. Many see this as a problem for the view. (For example, see (Armstrong 1983, p. 40) and (Maudlin 2007, p. 172) for further characterization of this worry.) Loewer (2012) responds by distinguishing between scientific explanations and metaphysical ones. He argues that Humean laws can scientifically explain the patterns even though they do not metaphysically explain them. On Loewer's view, a scientific explanation just requires that the explanans be simple, unifying, and exemplifying other theoretical virtues. A mere summary of the mosaic, satisfying those theoretical virtues, can be such an explanation. In contrast, metaphysical explanations go deeper and require more.

On MinP, suitable explanations of the patterns must not be merely summaries of the mosaic. On our view, fundamental laws are metaphysically fundamental facts that exist in addition to the mosaic. They govern the mosaic and explain its patterns by constraining it in an illuminating manner. Loewer's Humean scientific explanation, on our view, lacks the metaphysical element that provides the needed oomph. (We say more about the *oomph* at the end of this section.)

Non-supervenience Another metaphysical difference concerns non-supervenience. This has been much discussed in the literature (see for example Carroll 1994 and Maudlin 2007). On MinP, since fundamental laws are primitive facts, there can be a physically possible world corresponding to an empty Minkowsi spacetime governed by the Einstein equation. However, on Humean Reductionism, that world is one where the simplest summary is just Special Relativity, and it is impossible to have such a world where the law is the Einstein equation (Maudlin 2007, pp. 67–68). To allow two worlds with the same mosaic (empty Minkowski spacetime) but different laws (special relativity and general relativity), which is accepted in scientific practice, is to endorse non-supervenience of the laws on the mosaic.

Therefore, Humean Reductionism seems to be in conflict with scientific practice while MinP is not.[27]

Objectivity and Mind-Independence We take it that a hallmark of metaphysical realism about something is to believe in its objectivity and deny its mind-dependence. As metaphysical realists, we think that the fundamental reality is as it is, independent of what we human beings take it to be (Khlentzos 2021). We are fallible and can be wrong about the fundamental reality, including the fundamental laws of nature. This brings out an epistemological difference between MinP and Humean Reductionism. On Humean Reductionism, assuming that we are relying on the right theoretical virtues and have appropriate access to the mosaic, the best summaries will be the true laws. There is a certain sense that, in principle, we are guaranteed to be right. On MinP, even if we rely on the correct theoretical virtues and the correct scientific methodology, we can still be mistaken about what the true laws are. Epistemic guides are defeasible and fallible indicators for truth: they do not guarantee that we find the true laws (although we may be rational to expect to find them). There are fundamental, objective, and mind-independent facts about which laws govern the world, and we can be wrong about them. This is not a bug but a feature of MinP, symptomatic of the robust kind of realism that we endorse. For realists, this is exactly where they should end up; fallibility about the fundamental reality is a badge of honor!

Moreover, if the best systematization is constitutive of lawhood, and if what counts as best dependent on us, then lawhood can become mind-dependent. In a passage about "ratbag idealism,"[28] Lewis (1994) discusses this worry and tries to offer a solution:

> The worst problem about the best-system analysis is that when we ask where the standards of simplicity and strength and balance come from, the answer may seem to be that they come from us. Now, some ratbag idealist might say that if we don't like the misfortunes that the laws of nature visit upon us, we can change the laws—in fact, we can make them always have been different—just by changing the way we think! (Talk about the power of positive thinking.) It would be very bad if my analysis endorsed such lunacy....
>
> The real answer lies elsewhere: if nature is kind to us, the problem needn't arise.... If nature is kind, the best system will be *robustly* best—so far ahead of its rivals that it will come out first under any standards of simplicity and strength and balance. We have no guarantee that nature is kind in this way, but no evidence that it isn't. It's a reasonable hope. Perhaps we presuppose it in our thinking about law. I can admit that *if* nature were unkind, and *if* disagreeing rival systems were running neck-and-neck, then lawhood might be a psychological matter, and that would be very peculiar. (p. 479)

For Lewis, the solution is conditionalized on the hope that nature is kind to us in this special way: the best summary of the world will be far better than its rivals. That may be a generous assumption. Without a precise theory of which standards of simplicity and informativeness are permissible, and which are not, it is difficult to

---

[27] However, see Roberts (2008) for a Humean account of laws based on a contextualist semantics that may alleviate this worry.

[28] See also Gordon Belot's paper on ratbag idealism in this volume.

ascertain the assumption and determine what can be confirming evidence for it and disconfirming evidence against it.

In contrast, on MinP, fundamental laws are what they are irrespective of our psychology and judgments of simplicity and informativeness. Even though the epistemic guides provide some guidance for discovering and evaluating them, they do not guarantee arrival at the true fundamental laws. Moreover, changing our psychology or judgments will not change which facts are fundamental laws. Hence, MinP respects our conviction about the objectivity and mind-independence of fundamental laws.

The Package Deal In this paper, we mainly focused on fundamental laws. To be sure, they are related to the fundamental material ontology (fundamental entities and their properties). On MinP, we regard both fundamental laws and fundamental material ontology as metaphysical primitives and evaluate them in a package. In this respect, MinP is similar to Loewer's Package Deal Account (PDA), a descendent of Humean Reductionism that regards both as co-equal elements of a package deal (Loewer, 2021a, 2021b), but they also have significant differences. On PDA, we look for the best systematization in terms of a package of laws and (material) ontology; the package is supervenient on the actual world. Thus, fundamental laws and fundamental ontology enter the discussion in the same way, at the same place, and on the same level. MinP shares this feature, although fundamental ontology and fundamental laws are merely discovered by us and not made by us or dependent on us. On PDA, given the actual world (of which we have very limited knowledge), we evaluate different packages of laws + ontology, and we evaluate them based on our actual scientific practice. Hence, there will be some degree of relativism. Relative to different scientific practice or a different set of scientists, the judgement as to the actual laws + ontology would have been different. Consequently on PDA, fundamental laws and fundamental ontology are dependent on us in a significant way. On MinP, we may use the best package-deal systematization as a guide to discover the laws and ontology; given the actual world (of which we have very limited knowledge), we evaluate different packages of laws + ontology, and we evaluate them based on our actual scientific practice. Hence, there will be some degree of uncertainty. Relative to different scientific practice or a different set of scientists, the judgement as to the actual laws + ontology would have been different. Still, what they are is metaphysically independent of our belief and practice.

We quote a passage from Loewer (2021b). Although we disagree with him on the metaphysics, we agree on how the enterprise of physics should be understood:

> The best way of understanding the enterprise of physics is that it begins, as Quine says, "in the middle" with the investigation of the motions of macroscopic material objects e.g., planets, projectiles, pendula, pointers, and so on. Physics advances by proposing theories that include laws that explain the motions of macroscopic objects and their parts. These theories may (and often do) introduce ontology, properties/relations, and laws beyond macroscopic ones with which it began and go onto to posit laws that explain their behaviors...... The ultimate goal of this process is the discovery of a theory of everything (TOE) that specifies a fundamental ontology and fundamental laws that cover not only the

motions of macroscopic objects with which physics began but also whatever additional ontology and quantities that have been introduced along the way. (pp. 30–31)

From our perspective, this is an excellent description of how fundamental laws and ontology are discovered—in a package. We leave its analysis for future work.

A Sharper Contrast As a minimalist version of non-Humeanism stripped of the association with dynamic production, MinP forces us to think hard about the real difference between non-Humeanism and Humeanism. Here we draw a sharper contrast between MinP and the Humean best-system account of laws (BSA). (The contrast between MinP and PDA will be similar.) There are differences in simplicity, believability, and explanation. Suppose (granting Lewis's assumption) that given the mosaic ξ there is a unique best system whose axioms express the fundamental law L:

$$L = BS(\xi) \tag{10}$$

with $BS(\cdot)$ the function that maps a mosaic to its best-system law. To clarify the differences between BSA and MinP, let us stipulate that for both BSA and MinP, physical reality is described by a pair (L, ξ). For both, we must have that $\xi \in \Omega^L$, with Ω^L the set of mosaics compatible with L. On BSA, we also have that $L = BS(\xi)$. So in a sense, all we need in BSA is ξ; L is not ontologically extra. In contrast, on MinP, it is. And while ξ, which is all we have at the metaphysically fundamental level of BSA, is complicated, for MinP, we also have L, which is simple.

MinP regards L as an additional element of fundamental physical reality, above and beyond the concrete mosaic ξ, on which it is not merely supervenient or ontologically dependent. But then, why should we believe, take seriously, or have confidence in that law L as an additional element of fundamental reality constraining ξ? For this, the simplicity of L is crucial. If we take L seriously in the manner demanded by MinP, then we obtain, as a part of fundamental reality—regarded by many if not most physicists as the most important part—a simple element providing thereby a surprisingly simple account of the complex observed features of ξ that we wish to explain. (This may be compared to physicists proposing, say, new kinds of particles to account for new observations.)

At the same time, the MinP account of a law L, while making it believable, also makes it uncertain: L, as understood in MinP, is not a mere consequence of ξ and certainly not of any observations arising from ξ. This is to be contrasted with the certainty provided by the BSA account of L. Given ξ—and assuming for the sake of argument that not only is $L = BS(\xi)$ but that we humans are able to compute $BS(\xi)$—we can be certain about L, namely that $L = BS(\xi)$. The uncertainty in the MinP account of L is eliminated.

Someone might regard such a certainty to be an advantage of BSA over MinP. However, we are not given ξ. The nature of ξ is part of a theory about the physical world. Thus, on BSA we may also be uncertain about what the laws are.

Nothing that fundamentally exists in the BSA account of L is simple. So where is the simplicity in this account that would render L or any proposed ξ believable? On

BSA, L is simple *by accident*. If the simplicity is purely by accident, what good does it do for explaining the accident? The basic ground of believability is eliminated in the BSA account of L. So if the BSA account of a proposed law L is the correct account of L, it is hard to understand why we should then accept L, grounded as it is in an unknown and complicated mosaic ξ.

4.2 Comparison with Platonic Reductionism

MinP and Platonic Reductionism agree that there are governing laws that do not supervene on the Humean mosaic, but disagree on whether governing laws should be analyzed in terms of or are reducible to relations among universals.

While Platonic Reductionism is ontologically committed to fundamental universals, MinP is not. We do not think that universals offer additional explanatory benefits. The motivating idea of Platonic Reductionism is that universals are properties that genuinely similar objects share, and it is partly in virtue of the universals shared by those objects that the objects behave in the same way everywhere and everywhen. However, it is really the necessitation relation N that does the explanatory work in Armstrong's theory. It is crucial that the state of affairs $N(F, G)$ is understood as a universal. The metaphysics of N is a complicated business, and in our opinion it seems to create more mystery than it dispels. In contrast, on MinP we maintain that there are fundamental laws that govern the world by constraining the physical possibilities. Explanation in terms of simple laws seems clear enough to vindicate the non-Humean intuition that there is something more than the mosaic that governs it. Moreover, MinP is compatible with various metaphysical views about properties such as realism and nominalism. We do not think that a realist attitude towards laws requires a realist attitudes towards properties.

Platonic Reductionism places certain restrictions on the form of fundamental laws. On Platonic Reductionism, all laws need to be recast in the form of relations among universals, and it is unclear how to do so for the majority of laws in modern physics. (See also Hildebrand and Metcalf 2021, sect 3.2.) Here we agree with Wilson (1987)'s criticism that Armstrong's discussion is removed from concrete scientific practice and focuses mainly on schema of the "All F's are G" type. Consider a differential equation that expresses a candidate fundamental law such as (1). What are the universals that they actually relate? Assuming that velocity and acceleration are derived quantities, what are the universals that correspond to the derivatives on either side of the equations? Armstrong argues that universals must be instantiated in some concrete particulars. As Wilson observes (p. 439), differential equations conflict with Armstrong's principle about the instantiation of universals, as the values of the derivatives are calculated from values possessed by non-actual states (those in the small neighborhood around the actual one) that are not instantiated. In contrast, MinP has no difficulty accommodating laws expressed by differential equations.

Moreover, some candidate fundamental laws involve properties that do not seem to correspond to universals. For example, PH is a temporally restricted law that applies to only one moment in time. As such, it is a spatiotemporally restricted law that seems in tension with the approach involving universals (universal, repeatable, and multiply instantiated). Tooley (1977) considers an example of Smith's Garden, and there he seems open to accept spatiotemporally restricted laws if they are significant enough. But the tension needs a lot of work to remove, and we do not know if that is compatible with the metaphysics of universals that they are committed to. In contrast, spatiotemporally restricted laws can function perfectly as constraints on the universe that are about specific places or times. There is no in-principle obstacle of letting them be fundamental laws on MinP.

4.3 Comparison with Aristotelian Reductionism

There are several differences between MinP and Aristotelian Reductionism. Aristotelian Reductionists do not think that laws govern in a metaphysically robust sense.[29] In contrast, MinP vindicates the conviction that laws do so.

Aristotelian Reductionism is committed to a fundamental ontology of dispositions. MinP is not. Most physicists today may be unfamiliar with the concept of fundamental dispositions. In contrast, physicists are familiar with the concept of fundamental laws and how they figure in various scientific explanations. Hence, MinP seems more science-friendly.

It is natural to read dispositional essentialists such as Bird, Mumford, and Ellis as having an implicit commitment to a fundamental direction of time.[30] Demarest's account is more explicit in linking the dispositional essentialist ontology and the account of nomic explanations to that of dynamic metaphysical dependence, or what we call dynamic production. As discussed in Sect. 3.3.1, we do not understand how dynamic production works even in simple cases such as Hamilton's equations and much less in relativsitic spacetimes. Requiring dynamic production presumably rules out theories that permit closed timelike curves, as well as purely spatial laws, or even worlds for which spacetime is emergent. In contrast, MinP is not committed to a fundamental direction of time, and MinP is entirely open to those possibilities (even though we may have other considerations, beyond the conception of laws, to not consider them).

Finally, there are problems specific to accounts (such as Bird's) that analyze laws in terms of dispositions. Bird (2007) lists four problems (p. 211): (i) fundamental constants, (ii) conservation laws and symmetry principles, (iii) principles of least

[29] However, Bird (2007) talks about laws supervening on dispositions and allows that laws can still govern in a weaker sense.

[30] In a recent book, Vetter (2015) is open to the idea that there can be past-directed dispositions but still suggests that there is a temporal asymmetry: past-directed dispositions are trivial.

action, and (iv) multiple laws relating distinct properties. Problem (i) arises because slight differences in the constants do not require the properties to be different; problem (ii) because conservation laws and symmetry principles do not seem to be manifestations of dispositions; problem (iii) because the principles seem to commit to the physical possibilities of alternate histories, something not allowed on dispositional essentialism; problem (iv) because a third law relating two properties will not be the outcome of the dispositional natures of those properties. These problems may be solvable on Aristotelian Reductionism, and Demarest's version may be especially well posed to do that. In any case, such problems do not arise on MinP.

4.4 Comparison with Maudlinian Primitivism

MinP agrees with Maudlinian Primitivism that fundamental laws are metaphysically fundamental and that they govern. However, we disagree about how they do it. For Maudlin, dynamic production is essential, and every fundamental law needs to have the form of a dynamical law (in the narrow sense of a FLOTE) that can be interpreted as evolving later states of the universe from earlier ones. For laws to produce, they operate according to the fundamental direction of time, providing an intuitive picture close to our pre-theoretic conception of the world: "the universe is generated from a beginning and guided towards its future by physical law" (p. 182).

MinP is not committed to a fundamental direction of time; nor is it committed to dynamic production as how laws govern or explain. On MinP, explanation by simple constraint is good enough. Many candidate fundamental laws such as the Einstein equation are not (in and of themselves) FLOTEs that produce later states of the universe from earlier ones. For the same reason, the Past Hypothesis cannot be a Maudlinian law. And neither can a purely spatial constraint such as Gauss's law[31] or the simple rule responsible for the Mandelbrot world. On MinP, they can all be understood as fundamental laws that express simple constraints.

Our difficulty with dynamic production is not just that it precludes certain candidate fundamental laws. We also have difficulty understanding the notion itself. What does dynamic production mean and what are its relata? Does it relate instantaneous states or sets of instantaneous states of the universe? If it relates instantaneous states, we are unclear how to understand dynamic production even in paradigm examples of FLOTEs such as the one expressed by Hamilton's equations. (The initial data is not confined to a single moment in time, if we understand momentum as partly reducible to variations in positions over some time interval.) The notion becomes even less natural in relativistic settings.

[31] In personal communication, Maudlin suggests that he now regards (3) as expressing a metaphysical analysis or a definition of ρ in terms of the divergence of $\mathbf{E}$.

Moreover, on a simple understanding of dynamic production, the beginning of the universe does metaphysical work; it is what gets the entire productive enterprise started. However, for spacetimes with no temporal boundaries, it is unclear where to start the productive explanation. In contrast, constraints operate on the entire spacetime, regardless of whether there is an "initial" moment. Thus, MinP does not require a first moment in time. (Perhaps a more sophisticated understanding of dynamic production does not either.)

On MinP, even if the universe lacks a fundamental direction of time, we can still recover a notion of productive explanation at a non-fundamental level. For example, we can use PH to define a (non-fundamental) direction of time in the usual way: *earlier* is defined as being closer to the time of PH, while *later* is defined as being further away from that. We may regard FLOTEs as evolving earlier states of the universe into later ones (with respect to PH). In such a universe, dynamic production may be metaphysically derivative. Still, we can contemplate a (non-fundamental) productive explanation of Bromberger's flagpole and vindicate the intuition that the pole length and the circumstances explain, but are not explained by, the shadow length. Therefore, the intuitive picture behind Maudlinian Primitivism can be preserved even if there is no fundamental notion of dynamic production or a fundamental direction of time.

5 Conclusion

We suggest that MinP is an intelligible and attractive proposal for understanding fundamental laws of nature. It vindicates the non-Humean conviction that laws govern while remaining flexible enough to accommodate the variety of kinds of laws entertained in physics. In particular, it does not require that laws presume a fundamental direction of time. MinP illuminates metaphysics but is not unduly constrained by it.

Acknowledgments For helpful discussions, we thank David Albert, Jeff Barrett, Yemima Ben-Menahem, Craig Callender, Claudio Calosi, Lorenzo Cocco, Maaneli Derakhshani, Christopher Dorst, Joshua Eisenthal, Nina Emery, Veronica Gomez, Mara Harrell, Tyler Hildebrand, Christopher Hitchcock, Mario Hubert, Jenann Ismael, Marc Lange, Federico Laudisa, Dustin Lazarovici, Baptiste Le Bihan, Barry Loewer, Tim Maudlin, Giovanni Merlo, Kerry McKenzie, Jill North, Elias Okon, Ezra Rubenstein, David Schroeren, Charles Sebens, Ted Sider, Shelly Yiran Shi, Isaac Wilhelm, Ken Wharton, Christian Wüthrich, Nino Zanghì, audiences at California Institute of Technology, University of Geneva, Rutgers University, Metro Area Philosophers of Science, 2021 Annual Meeting of the California Quantum Interpretation Network, and participants in the graduate seminar "Rethinking Laws of Nature" at the University of California San Diego in spring 2021. EKC received research assistance from Shelly Yiran Shi and was supported by an Academic Senate Grant from the University of California San Diego.

References

Adlam, E. (2021). Laws of nature as constraints. Manuscript: arXiv 2109.13836.

Albert, D. Z. (2000). *Time and chance*. Cambridge: Harvard University Press.

Armstrong, D. M. (1983). *What is a law of nature?* Cambridge: Cambridge University Press.

Arnowitt, R., Deser, S., & Misner, C. W. (1962). The dynamics of general relativity. In L. Witten (Ed.), *Gravitation: An introduction to current research* (pp. 227–264). Wiley.

Bacciagaluppi, G. (2005). A conceptual introduction to Nelson's mechanics. In A. E. Rosolino Buccheri & M. Saniga (Eds.), *Endophysics, Time, Quantum And The Subjective: Proceedings of the ZiF Interdisciplinary Research Workshop, Bielefeld, Germany, 17–22 January 2005* (pp. 367–388). World Scientific.

Beebee, H. (2000). The non-governing conception of laws of nature. *Philosophy and Phenomenological Research, 61*, 571–594.

Ben-Menahem, Y. (2018). *Causation in science*. Princeton University Press.

Bhogal, H. (2017). Minimal anti-Humeanism. *Australasian Journal of Philosophy, 95*(3), 447–460.

Bhogal, H. (2020). Humeanism about laws of nature. *Philosophy Compass, 15*(8), e12696.

Bird, A. (2007). *Nature's metaphysics: Laws and properties*. Oxford: Oxford University Press.

Blanchard, T. (2016). Physics and causation. *Philosophy Compass, 11*(5), 256–266.

Callender, C. (2004). Measures, explanations and the past: Should 'special' initial conditions be explained? *The British Journal for the Philosophy of Science, 55*(2), 195–217.

Callender, C. (2017). *What makes time special?* Oxford University Press.

Carroll, J. W. (1994). *Laws of nature*. Cambridge University Press.

Carroll, J. W. (2018). Becoming Humean. In W. Ott & L. Patton (Eds.), *Laws of nature* (pp. 122–138). Oxford University Press.

Carroll, J. W. (2020). Laws of nature. In E. N. Zalta (Ed.), *The Stanford encyclopedia of philosophy*. Metaphysics Research Lab, Stanford University, Winter 2020 edition.

Cartwright, N. (1983). *How the laws of physics lie*. Oxford University Press.

Cartwright, N. (1994a). Fundamentalism vs. the patchwork of laws. In *Proceedings of the Aristotelian Society* (vol. 94, pp. 279–292). JSTOR.

Cartwright, N. (1994b). *Nature's capacities and their measurement*. Oxford University Press.

Chen, E. K. (2020). The past hypothesis and the nature of physical laws. In B. Loewer, W. E. & B. Weslake (Eds.), *Time's arrows and the probability structure of the world*. Harvard University Press, forthcoming.

Chen, E. K. (2022). Fundamental nomic vagueness. *The Philosophical Review, 131*(1).

Choi, S., & Fara, M. (2021). Dispositions. In E. N. Zalta (Ed.), *The Stanford encyclopedia of philosophy*. Metaphysics Research Lab, Stanford University, Spring 2021 edition.

Cohen, J., & Callender, C. (2009). A better best system account of lawhood. *Philosophical Studies, 145*(1), 1–34.

Deckert, D.-A. (2010). *Electrodynamic absorber theory: A mathematical study*. PhD Thesis, Ludwig Maximilian University of Munich.

Demarest, H. (2017). Powerful properties, powerless laws. In J. D. Jacobs (Ed.), *Causal powers* (pp. 38–53). Oxford: Oxford University Press.

Demarest, H. (2019). Mentaculus laws and metaphysics. *Principia: An International Journal of Epistemology, 23*(3), 387–399.

Demarest, H. (2021). Powers, best systems, and explanation. Manuscript.

Dorst, C. (2021). Productive laws in relativistic spacetimes. Manuscript.

Dretske, F. (1977). Laws of nature. *Philosophy of Science, 44*, 248–68.

Dürr, D., Goldstein, S., & Zanghì, N. (1992). Quantum equilibrium and the origin of absolute uncertainty. *Journal of Statistical Physics, 67*(5–6), 843–907.

Ellis, B. (2001). *Scientific essentialism*. Cambridge University Press.

Ellis, B. (2014). *The philosophy of nature: A guide to the new essentialism*. Routledge.

Fodor, J. A. (1974). Special sciences (or: The disunity of science as a working hypothesis). *Synthese, 28*, 97–115.

Friederich, S., & Evans, P. W. (2019). Retrocausality in quantum mechanics. In E. N. Zalta (Ed.), *The Stanford Encyclopedia of Philosophy*. Metaphysics Research Lab, Stanford University, summer 2019 edition.

Ghirardi, G. (2018). Collapse theories. In E. N. Zalta (Ed.), *The Stanford Encyclopedia of Philosophy*. Metaphysics Research Lab, Stanford University, fall 2018 edition.

Ghirardi, G., Rimini, A., & Weber, T. (1986). Unified dynamics for microscopic and macroscopic systems. *Physical Review D, 34*(2), 470.

Goldstein, S. (2001). Boltzmann's approach to statistical mechanics. In J. Bricmont, D. Dürr, M. C. Galavotti, G. Ghirardi, F. Petruccione, & N. Zanghì (Eds.), *Chance in physics* (pp. 39–54). Berlin: Springer.

Goldstein, S. (2012). Typicality and notions of probability in physics. In *Probability in physics* (pp. 59–71). Springer.

Goldstein, S. (2017). Bohmian mechanics. In E. N. Zalta (Ed.), *The Stanford encyclopedia of philosophy*. Metaphysics Research Lab, Stanford University, summer 2017 edition.

Goldstein, S., & Tumulka, R. (2003). Opposite arrows of time can reconcile relativity and nonlocality. *Classical and Quantum Gravity, 20*(3), 557.

Henderson, L. (2020). The problem of induction. In E. N. Zalta (Ed.), *The Stanford encyclopedia of philosophy*. Metaphysics Research Lab, Stanford University, Spring 2020 edition.

Hicks, M. T., & Schaffer, J. (2017). Derivative properties in fundamental laws. *The British Journal for the Philosophy of Science, 68*(2), 411–450.

Hildebrand, T. (2013). Can primitive laws explain? Philosophers' Imprint.

Hildebrand, T. (2020). Non-Humean theories of natural necessity. *Philosophy Compass, 15*(5), e12662.

Hildebrand, T., & Metcalf, T. (2021). The nomological argument for the existence of God. *Noûs*.

Hoefer, C. (2019). *Chance in the world: A Humean guide to objective chance*. Oxford University Press.

Khlentzos, D. (2021). Challenges to metaphysical realism. In E. N. Zalta (Ed.), *The Stanford encyclopedia of philosophy*. Metaphysics Research Lab, Stanford University, Spring 2021 edition.

Kimpton-Nye, S. (2017). Humean laws in an unHumean world. *Journal of the American Philosophical Association,3*(2), 129–147.

Kraut, R. (2017). Plato. In E. N. Zalta (Ed.), *The Stanford encyclopedia of philosophy*. Metaphysics Research Lab, Stanford University, Fall 2017 edition.

Lange, M. (2009). *Laws and lawmakers: Science, metaphysics, and the laws of nature*. Oxford University Press.

Lange, M. (2016). *Because without Cause: Non-casual explanations in science and mathematics*. Oxford University Press.

Laudisa, F. (2015). Laws are not descriptions. *International Studies in the Philosophy of Science, 29*(3), 251–270.

Lazarovici, D. (2018). Against fields. *European Journal for Philosophy of Science, 8*(2), 145–170.

Lewis, D. (1980). A subjectivist's guide to objective chance. In R. C. Jeffrey (Ed.), *Studies in inductive logic and probability* (Vol. 2, pp. 263–93). Berkeley: University of California Press.

Lewis, D. (1983). New work for a theory of universals. *Australasian Journal of Philosophy, 61*, 343–377.

Lewis, D. (1986). *Philosophical papers II*. Oxford: Oxford University Press.

Lewis, D. (1994). Humean supervenience debugged. *Mind, 103*, 473–490.

Loewer, B. (2001). Determinism and chance. *Studies in History and Philosophy of Science Part B: Studies in History and Philosophy of Modern Physics, 32*(4), 609–620.

Loewer, B. (2012). Two accounts of laws and time. *Philosophical Studies, 160*(1), 115–137.

Loewer, B. (2021a). The package deal account of laws and properties (PDA). *Synthese, 199*(1), 1065–1089.

Loewer, B. (2021b). What breathes fire into the equations. Manuscript.

Maudlin, T. (2007). *The metaphysics within physics*. New York: Oxford University Press.

Mumford, S. (2004). *Laws in nature*. Routledge.

Nelson, E. (1966). Derivation of the Schrödinger equation from Newtonian mechanics. *Physical Review, 150*(4), 1079.
Penrose, R. (1974). The role of aesthetics in pure and applied mathematical research. *Bulletin of Mathematical Analysis and Applications, 10*, 266–271.
Penrose, R. (1979). Singularities and time-asymmetry. In S. Hawking, & W. Israel (Eds.), *General relativity* (pp. 581–638). Cambridge: Cambridge University Press.
Penrose, R. (1989). *The emperor's new mind: Concerning computers, minds, and the laws of physics*. Oxford: Oxford University Press.
Roberts, J. T. (2008). *The law-governed universe*. Oxford University Press.
Schaffer, J. (2016). It is the business of laws to govern. *Dialectica, 70*(4), 577–588.
Strawson, G. (2015). 'Humeanism'. *Journal of the American Philosophical Association, 1*(1), 96.
Sutherland, R. I. (2008). Causally symmetric Bohm model. *Studies in History and Philosophy of Science Part B: Studies in History and Philosophy of Modern Physics, 39*(4), 782–805.
Tooley, M. (1977). The nature of laws. *Canadian Journal of Philosophy, 7*(4), 667–698.
Tooley, M. (1997). *Time, tense and causation*. Oxford University Press, Clarendon Press.
Van Fraassen, B. C. (1989). *Laws and symmetry*. Oxford University Press.
Vetter, B. (2015). *Potentiality: From dispositions to modality*. Oxford University Press.
Weatherall, J. O. (2019a). Part 1: Theoretical equivalence in physics. *Philosophy Compass, 14*(5), e12592.
Weatherall, J. O. (2019b). Part 2: Theoretical equivalence in physics. *Philosophy Compass, 14*(5), e12591.
Weaver, C. G. (2019). *Fundamental causation: Physics, metaphysics, and the deep structure of the world*. Routledge.
Weinberg, S. (1992). *Dreams of a final theory: The search for the fundamental laws of nature*. New York: Pantheon.
Wheeler, J. A., & Feynman, R. P. (1945). Interaction with the absorber as the mechanism of radiation. *Reviews of Modern Physics, 17*(2–3), 157.
Wheeler, J. A., & Feynman, R. P. (1949). Classical electrodynamics in terms of direct interparticle action. *Reviews of Modern Physics, 21*(3), 425.
Wigner, E. (1985). Events, laws of nature, and invariance principles. In A. Zichichi (Ed.), *How Far Are We from the Gauge Forces–Proceedings of the 21st Course of the International School of Subnuclear Physics, Aug 3–14, 1983* (pp. 699–708). Plenum.
Wigner, E. P. (1964). Symmetry and conservation laws. *Proceedings of the National Academy of Sciences of the United States of America, 51*(5), 956–965.
Wilson, M. (1987). What is a law of nature? (Book review). *The Philosophical Review, 96*(3), 435–441.

How to Make the World Safe for Autonomy; or, How to Fodor-Kitcher an Albert-Loewer

Marc Lange

Abstract Albert and Loewer have proposed an account of the origin of the regularities discovered by the special sciences. Their account purports to undercut frequent claims by non-reductive physicalists (such as Fodor and Kitcher) that the special sciences are autonomous. This paper argues that if the main ideas behind the Albert/Loewer picture are correct, then this picture can be tweaked to underwrite rather than to undercut the autonomy of the special sciences. In fact, the equipment posited by Albert and Loewer can be used to dispel what they regard as the mystery of the origin of metaphysically independent special-science laws. Anyone who likes the Albert/Loewer picture and special-science autonomy can have them both.

Keywords David Albert · Chance · Counterfactuals · Philip Kitcher · Laws of nature · Barry Loewer · Mentaculus · Necessity · Past Hypothesis · Statistical mechanics · Thermodynamics

1 Introduction

Albert (2000, 2015) and Loewer (2008, 2009, 2012) have proposed an account of the origin of the regularities discovered by the special sciences. Their account purports to undercut frequent claims by non-reductive physicalists (they cite Fodor, 1974 and Kitcher, 2001) that the special sciences are autonomous. In particular, Albert and Loewer argue that alongside the microphysical laws, "there are no additional ontologically *independent* dynamical or causal special science laws" (Loewer, 2012: 13, his emphasis). In Sect. 2, I will elaborate what Albert and Loewer mean by this claim, their argument for it, and their picture of the origin of special-science laws.

The purpose of this paper is to argue that if the main ideas behind the Albert/Loewer picture are correct, then this picture can be tweaked to underwrite

M. Lange (✉)
University of North Carolina, Chapel Hill, NC, USA
e-mail: mlange@email.unc.edu

Y. Ben-Menahem (ed.), *Rethinking the Concept of Law of Nature*,
Jerusalem Studies in Philosophy and History of Science,
https://doi.org/10.1007/978-3-030-96775-8_3

rather than to undercut the autonomy of the special sciences. In other words, my purpose is not to defend the main components of the Albert/Loewer picture, but rather to argue that they do not supply a strong argument against "additional ontologically *independent*" special-science laws. In fact, the equipment posited by Albert and Loewer can easily be used to dispel what they regard as the mystery of the origin of metaphysically independent special-science laws.

In particular, as I will elaborate in Sect. 3, the laws "PH" and "PROB" posited by Albert and Loewer could enable special-science laws to "transcend" the microphysical laws. Furthermore, as I will argue in Sect. 4, physicists have frequently regarded the laws of thermodynamics as having the same status as they have often ascribed to the conservation laws, kinematics, and the special theory of relativity – namely, as transcending the microphysical laws.[1] Since thermodynamics plausibly has this status, and since Albert and Loewer see PH and PROB as responsible for the second law of thermodynamics, the Albert/Loewer picture can plausibly be tweaked to make PH and PROB (and any special-science laws they ground) transcend the microphysical laws. In Sect. 5, I will conclude by arguing that the tweak helps the Albert/Loewer picture to better fit the example (discussed by Kitcher, Albert, and Loewer) of R.A. Fisher's explanation of Arbuthnot's regularity regarding the sex ratio of London births in 82 successive years.

My moral, then, will be that the Albert/Loewer picture does not supply a strong argument against the autonomy of the special sciences. On the contrary, anyone who likes both the Albert/Loewer picture and special-science autonomy can easily have them both.

2 The Albert/Loewer Picture and Its Account of Special-Science Laws

Loewer (2008: 149–150; 2009: 218) introduces non-reductive physicalists as committed to the following three claims:

(1) All items (whether entities or properties) belonging to the ontology of any special science are constituted by, realized by, or otherwise made up out of microphysical items – that is, items falling within the scope of fundamental physics;

[1] My arguments do not require that I claim that the conservation laws, kinematics, and the special theory of relativity actually possess this special status. Perhaps they will turn out not even to be true and so perforce not to be laws transcending the microphysical laws. I mention the conservation laws, kinematics, and special relativity only to argue that scientists have often ascribed this special status to various alleged laws, so our ascribing it to PH and PROB would not be unprecedented in science. By having these other examples in mind, we can better understand what this status is, what evidence would support ascribing it, and what role such ascriptions have sometimes played in science.

(2) The fundamental dynamical laws of microphysics completely specify the evolution of the microphysical state of the universe in that further laws add nothing to what they specify regarding that evolution; and.
(3) There are metaphysically independent special-science laws that underwrite special-science explanations that are irreducible to the explanations supplied by physics. (Loewer (2012: 14) cites Fodor (1974) giving Gresham's law as an example of such an irreducible special-science law.)

As we will see shortly, Loewer's (and Albert's) point is that (1) and (2) make (3) highly implausible. My aim is to argue that on the contrary, there is a way for (3) to be happily reconciled with (1) and (2).[2]

It is important to understand the sense in which (3) deems special-science laws to be "metaphysically independent". According to (3), the lawhood of some special-science laws does not derive (even partly) from the lawhood of the microphysical laws. Rather, "the lawfulness[[3]] of special science regularities is a fact about the world as basic as and independent of the lawfulness of the laws of fundamental physics" (Loewer, 2008: 150; cf. Loewer, 2009: 226–7).[4] (The metaphysical independence of the special-science laws would make the special sciences "autonomous".) Loewer continues (identifying the combination of (1), (2), and (3) as "Fodor's view"):

> Fodor's view can be illustrated with the help of a souped up version of Laplace's demon. The demon knows all the physical facts obtaining at all times and all the fundamental dynamical laws of physics, has perfect computational powers and also a "translation" manual connecting special science and physical vocabularies. The demon is thus able to tell which microphysical situations correspond to, for example, a philosophy conference and is able to determine which generalizations about philosophy conferences are true and which are false. It can do the same for all the special sciences. It will also be able to tell which special science regularities hold under counterfactual initial conditions and so which hold in all physically possible worlds (i.e. all the worlds at which the fundamental laws of physics obtain). But on Fodor's view the demon *will not* be able to discern which regularities are laws. Because of this "blindness" the demon will be missing those counterfactuals and explanations that are underwritten by special science laws and so will not have an understanding of special science phenomena. (Loewer, 2008: 150)

Despite knowing the truth of the special-science laws, the demon does not know their lawhood because their lawhood is not determined by the lawhood of the microphysical laws (which the demon does know).

Notice that the demon knows what would have obtained under a counterfactual antecedent *p* where *p* is logically consistent with all of the microphysical laws.

---

[2] Of course, (1) and (2) are open to question; for details of some of the issues involved, see Loewer, 2008, 2009. I will accept (1) and (2) for the sake of argument, since my aim is to show that contrary to Loewer, one who accepts them need not find (3) implausible.

[3] Loewer consistently uses "lawfulness" for what I call "lawhood" (a.k.a. "nomic status").

[4] On Loewer's own view, the lawhood of the fundamental microphysical laws is not metaphysically basic. Rather, their lawhood is grounded in the Humean mosaic. (See note 8.) But Loewer expresses (3) so that it does not presuppose a Humean account of laws.

(Loewer says above that *p* therefore qualifies as "physically possible".) So the demon knows which special-science regularities would have held under all physically possible circumstances. But then (Loewer argues) the demon should also know which special-science regularities are special-science laws, since those laws are exactly the special-science regularities that are physically necessary, i.e., would still have held under all physically possible counterfactual antecedents. In this way, Loewer argues, (1) and (2) make (3) extremely implausible.

Let's review this argument more carefully. The demon is just a device to put in epistemic terms what is really a metaphysical argument. So let's dispense with the demon and put Loewer's argument in purely metaphysical terms. The microphysical facts at all times together with the microphysical laws and the "translation" manual entail all of the facts about (as Loewer says above) "which special science regularities hold under counterfactual initial conditions and so which hold in all physically possible worlds". (This entailment holds because of (1) and (2).[5]) These facts about which counterfactual conditionals hold are precisely the facts that are supported by the special-science laws. So if (as (3) says) the special-science laws are not determined by the microphysical facts at all times together with the microphysical laws and the "translation" manual, then the counterfactual conditionals supported by the special-science laws are *independently* supported also by the microphysical laws. That is, the special-science laws are not *needed* to support any counterfactual conditionals; all of their counterfactual-supporting work is already being done by the microphysical laws. The special-science laws are therefore superfluous for supporting counterfactuals. This overdetermination of the special-science counterfactuals by two independent sets of laws seems weird. Loewer puts his argument as follows:

> The gist of my criticism [of the Fodorian combination of (1), (2), and (3)] is that if (1) and (2) are both true then, contra Fodor, special science counterfactuals *are* necessitated by fundamental physical laws and facts. So if there are metaphysically independent special science laws then they can only overdetermine counterfactuals. Such overdetermination is very puzzling. Why would there be a redundant system for some parts of nature? (Loewer, 2008: 153; cf. Loewer, 2009: 229–230)

Loewer's argument against (3) given (1) and (2) is not primarily that since the *truth* of special-science laws is entailed by the microphysical facts at all times together with the microphysical laws and the "translation" manual, then the *lawhood* of the special-science laws should also be so entailed. Loewer does himself a disservice by occasionally suggesting that his argument takes that form:

> It follows from (1) and (2) that special science regularities (including probabilistic regularities) are made true by physical facts and laws. It is hard to resist the conclusion that those special science regularities that are lawful derive their status as laws from the fundamental laws of microphysics. (Loewer, 2008: 151)

[5] To entail these counterfactual conditionals, the microphysical laws and the "translation" manual do not suffice. They must be supplemented by the microphysical facts at all times. For the same reason, "Had the match been struck, it would have lit" requires not just the microphysical laws, but also the match's chemical composition, the presence of oxygen, etc.

Nor is Loewer's argument primarily that if the lawhood of special-science laws does not come from the lawhood of microphysical laws (together with the microphysical facts at all times and the "translation" manual), then the ground of the special-science laws' lawhood is mysterious. (Yet Loewer (2008: 151) asks rhetorically, "Where else could it come from?") Rather, Loewer's main argument is that lawhood over and above truth is associated (as Goodman (1983) famously emphasized) with a distinctive role in supporting counterfactuals – that is, a distinctive degree of invariance under counterfactual antecedents. (The laws, unlike the accidents, would still have been true – i.e., are invariant – under any physically possible counterfactual antecedent.) But given (1) and (2), there are no counterfactuals for special-science laws to support that are not already being supported by microphysical laws. So "irreducible special science laws as Fodor proposed them would be otiose" (Loewer, 2009: 233). They would be metaphysically idle – which is certainly not what physicalist anti-reductionists intend.

It seems to me that Loewer's argument can even be strengthened somewhat so as not to rely on the premise that overdetermination of counterfactuals by two independent sets of laws is implausible. Since special-science laws support no counterfactuals that are not already supported by microphysical laws, special-science laws can inherit all of their invariance under counterfactual antecedents from the microphysical laws. Therefore, they can also inherit their lawhood from the microphysical laws, since lawhood is constituted by (as Lange, 2009 argues) or, at least, is associated with a distinctive degree of invariance under counterfactual antecedents. So "*irreducible* special science laws" (Loewer, 2008:154) as posited by (3) are made "immensely implausible" (Loewer, 2009: 217, 229) by (1) and (2). In securing that a given special-science regularity would still have been true under all physically possible counterfactual antecedents, the microphysical laws (together with the microphysical facts at all times and the "translation" manual) secure the special-science regularity's truth in all physically possible worlds. That is, they secure the special-science regularity's physical necessity, i.e., its lawhood. So the special-science laws are not metaphysically independent.[6]

Albert and Loewer offer not only this argument against non-reductive physicalism, but also a positive account of how special-science laws derive from microphysical facts and laws. According to Albert and Loewer, special-science laws are typically associated with temporally asymmetric processes – in particular, processes that lead to increased entropy and so manifest the second law of thermodynamics. To account for these processes despite the temporal (near-) symmetry of the fundamental microphysical laws already known or conjectured, Albert and Loewer posit two additional fundamental microphysical laws:

---

[6] The precise relation to counterfactuals that distinguishes laws from accidents is controversial in its details. (See Lange, 2009.) Loewer's argument (as I have just unpacked it) presupposes that it consists roughly in the laws' invariance under all physically possible counterfactual antecedents. I will simply adopt Loewer's presupposition since my aim is ultimately to show how easily Loewer's picture can be made to underwrite (rather than to undercut) the metaphysical independence of special-science laws.

> PH (the "Past Hypothesis"): that the universe's macrostate shortly after the Big Bang exhibited low entropy, and
>
> PROB: that the universe's objective chances of possessing various microphysical properties while occupying the macrostate posited by PH are given by a probability density distribution that is uniform (by the standard Lebesgue measure) over the various possible microrealizations of that macrostate.

It follows (according to Albert and Loewer) that the universe (or some isolated subsystem) likely evolves toward increased entropy. Albert and Loewer maintain that having been fortified by PH and PROB, the fundamental microphysical laws (and the "translation" manual) entail the second law of thermodynamics and the special-science laws (all perhaps amended to ascribe very high likelihood rather than certainty to various macroregularities). Again, I will simply assume that this entailment goes through, since my aim is ultimately to show how easily this picture can be made to underwrite (rather than to undercut) the metaphysical independence of special-science laws.[7] The lawhood of the special-science laws is thereby grounded in the lawhood of the microphysical laws (including PH and PROB).

As an example, Albert (2015: 15–21) and Loewer (2008: 161) discuss the regularity (discovered by Arbuthnot, 1710) that more boys than girls were born in London in each of the 82 years from 1629 to 1710. (Albert and Loewer are responding to Kitcher's (2001: 71) argument that this regularity – that each of the 82 years was a "male" year – cannot be explained microphysically, but only by evolutionary theory, and so illustrates special-science autonomy.) According to Albert, this special-science law can be derived (and thereby explained) by taking PROB's distribution over the microstates that can realize the initial low-entropy macrostate posited by PH. The microdynamical laws applied to each of these microstates entail the evolution of this probability distribution to January 1, 1629. Furthermore,

> ... conditionalize that evolved distribution on the existence of our galaxy, and of our solar system, and of the earth, and life, and of the human species, and of cities, and of whatever else is implicitly being taken for granted in any scientific discussion of the relative birth rates of boys and girls in London in the years following [1629]. (Albert, 2015: 20)

(After all, the why question is not "Why is each year from 1629 to 1710 a 'male' year rather than a year without earth or humanity or London in existence?" The question is "Why is each year from 1629 to 1710 a 'male' year rather than some of them being years with London inhabited by humanity but with no more boys than girls born.) Also conditionalize on whatever "ceteris paribus" proviso is implicit in the generalization (Loewer, 2012: 18). Having so conditionalized, we arrive at

[7] Callender (2011: 103) has doubts about the entailment. For that matter, Earman (2006) argues that PH is not applicable to the actual world, but only to a world governed by mechanics like the classical mechanics of point particles and conservative forces. Additional criticism of Albert and Loewer as unable to account for temporal asymmetry can be found in Frisch (2007) and Winsberg (2004). I set all of this aside since my concern is to show how easily the Albert/Loewer picture can be tweaked to yield the metaphysical independence of special-science laws.

a probability distribution; applied to this distribution, the microdynamical laws entail the chances of various subsequent microstates for London 1629–1710 and hence (from the "translation" manual) the chance that each of these years will be "male". That this chance is very high grounds the special-science law discovered by Arbuthnot. (That we could not in practice perform this calculation is irrelevant to whether this account gives the microphysical ground of the special-science law. Loewer's Laplacean "superdemon" could do it.)

3 How a Tweak to the Albert/Loewer Picture Yields Special-Science Autonomy

As we have just seen, Albert and Loewer posit PH and PROB to be laws on the grounds that their lawhood would "save the phenomena": the lawhood of the second law of thermodynamics (interpreted roughly as deeming entropy rise to be highly likely rather than certain) and special-science laws. The special-science laws and second law of thermodynamics can acquire lawhood by being entailed by PH and PROB (together with microphysical laws and the "translation" manual) only if PH and PROB are themselves laws. Since a given fact's lawhood is associated with its invariance under various counterfactual antecedents, Albert's and Loewer's motivation for positing PH's and PROB's lawhood in order to "save" the lawhood of special-science laws can be put in terms of "saving" various counterfactual conditionals' holding. Suppose that the laws of nature would still have obtained under any physically possible counterfactual antecedent. In other words, suppose that all laws are "held fixed" under any counterfactual antecedent under which they can without contradiction all be held fixed. Then if PH and PROB are laws, they are fixed under (e.g.) the counterfactual antecedent positing that a given perfume bottle was opened in a room. In this way, PH and PROB are available under this counterfactual antecedent to help entail that had the perfume bottle been opened, then perfume would (very likely) have spread quickly throughout the room. Thus, according to Albert and Loewer, if the laws have this "fixed" role under counterfactual antecedents positing physical possibilities, then the truth of the counterfactual conditionals given by special sciences is accounted for by the lawhood of the microphysical laws augmented by PH and PROB. No *ad hoc* time-asymmetry needs to be introduced into the logic of counterfactuals in order to capture temporal asymmetry under counterfactual antecedents (as in the spreading of the perfume). Rather, temporal asymmetry derives automatically from the laws' persistence under counterfactual antecedents together with the temporal asymmetry of the laws themselves (that is, PH's and PROB's lawhood).[8]

---

[8] Albert and Loewer have one additional motivation for positing PH's and PROB's lawhood: they maintain that PH and PROB qualify as laws on Lewis's "best-system account", once that account has been slightly modified (e.g., to allow macroproperties to figure in the laws despite not being

In this way, Albert and Loewer ground the lawhood of special-science laws in the lawhood of microphysical laws, PH, and PROB. However, I will now argue that if this strategy succeeds in grounding the lawhood of the special-science laws (because the truth of the special-science laws is indeed entailed by PH, PROB, the microphysical laws, and the "translation" manual), then a very similar strategy could instead "save" the *autonomy* of the special sciences. The same laws posited by Albert and Loewer (namely, PH and PROB) could ground the metaphysical *independence* of the special-science laws – as long as PH and PROB themselves have the appropriate sort of metaphysical independence from microphysical laws.

What sort of independence would this be? Let's find it in another example of the way in which certain laws of nature have often been plausibly believed to "transcend" others. Plausibly (in classical physics, at least), the reason why the conservation laws (of energy, linear momentum, angular momentum, etc.) hold is not because each of the various fundamental kinds of interactions separately conserves these quantities. Rather, the order of explanatory priority is widely believed to be the reverse; electromagnetic interactions (for example) conserve linear momentum (for instance) because its conservation is required by the general linear momentum conservation law covering all kinds of interaction. In other words, it is no coincidence that the various kinds of fundamental interaction all conserve these quantities. That is because each does so for the same reason as the others: because of general conservation laws to which all kinds of interaction must conform. As Feynman (1965: 49) says, the conservation laws are "wide principles which sweep across the different laws", constraining the fundamental forces there could have been. Even if there had been different (or simply additional) kinds of fundamental particles and interactions, the conservation laws would still have held.

---

perfectly natural properties on Lewis's account). I am not addressing this motivation on Albert's and Loewer's part; I intend my argument to be largely neutral between Humean and non-Humean analyses of natural lawhood. By contrast, Frisch (2014) uses the best-system account (again, as slightly modified) to argue that special-science laws earn their way into the best system as axioms, without their admission into the best system depending on PH and PROB being in the system and helping to entail them. Frisch thus aims to use a different approach from mine to accommodate special-science autonomy. Callender and Cohen (2009, 2010) also pursue a different approach to the same goal: they argue that if bestness is relativized to some or another basic vocabulary, then different sciences will have different best systems, and a special science's laws will belong to the best system for that science. Again, my argument differs in not presupposing any form of the best-system account.

Loewer (2008: 162) describes his and Albert's view as "reductionist" in that it takes the lawhood of the special-science laws as arising entirely from the lawhood of the laws of physics, including PH and PROB. As we have seen, Loewer contrasts his view with a "non-reductive" view according to which the special sciences are autonomous: the lawhood of special-science laws does not derive entirely from the lawhood of the laws of fundamental physics (including PH and PROB). Nevertheless, there remains a sense in which the Albert-Loewer picture is non-reductive regarding special-science laws: their lawhood arises partly from the lawhood of PH and PROB, which are fundamental laws. Moreover, there is also another, distinct sense in which the Albert-Loewer picture is reductive regarding every law's lawhood: as I just mentioned, they adopt a Humean analysis of natural lawhood, and on any such approach, a regularity's lawhood is reducible to a fact about the global Humean mosaic of sub-nomic facts.

The above counterfactual (indeed, counterlegal) conditional is not associated with any of the laws specifying or describing the kinds of fundamental particles and interactions there are (e.g., the gravitational-force law, the law giving the quantity of the electron's charge). That the conservation laws would still have held even if the microphysical laws had been different reflects the metaphysical independence of the conservation laws from the microphysical laws – the way in which the conservation laws "transcend" the microphysical laws. A counterfactual conditional such as "Had the figure skater stretched out her arms, angular momentum would still have been conserved" is overdetermined in exactly the sense that figured in the Albert/Loewer argument given above in Sect. 2. That is, the figure-skater counterfactual conditional's truth is secured by laws in two independent, individually sufficient ways. Firstly, it is secured by the microphysical laws specifying the kinds of fundamental interactions there are and the way that forces affect motion. These laws suffice to entail that every kind of fundamental interaction conserves angular momentum. Secondly, the figure-skater counterfactual is secured by the law of angular momentum conservation. The counterfactual is thereby "overdetermined" because the conservation law is metaphysically independent from the microphysical laws; although its lawhood follows from the lawhood of the microphysical laws, it also possesses a variety of natural necessity that the microphysical laws lack and that the conservation law therefore cannot inherit by virtue of being entailed by the microphysical laws. Its distinctive necessity is reflected in the fact that it would still have held even if there had been additional kinds of fundamental particles or interactions. This counterfactual conditional's truth (unlike the figure-skater conditional's truth) is not secured by the microphysical laws (since its antecedent is counterlegal). This conditional (unlike the figure-skater conditional) is not overdetermined. According to Feynman (1965), if we say that the reason why momentum is conserved is because its conservation can be derived from the various fundamental force laws (together, presumably, with the "closure law" that those are all of the fundamental forces there are) and the law relating force and motion, then "we take the derivation too seriously, and feel that [the conservation law] is only valid because [each of the force laws] is valid, [but] then we cannot understand the interconnections of the different branches of physics" (Feynman, 1965: 49). That is, we then fail to understand why every one of these kinds of interaction conserves momentum; we mistakenly regard this similarity as coincidental.[9]

[9] Admittedly, Feynman (1965: 46–55) emphasizes the value to physics in sometimes taking a given law as fundamental but on other occasions treating it as derivative. In this way, new laws can be discovered and a given law can sometimes be retained even after physics has discovered that another "law", from which the first law had sometimes been derived, is actually false. Feynman says that it is therefore not very "efficient" (47) to insist on certain laws as definitely more basic than others. However, Feynman emphasizes (49–50) that this "Babylonian" view (as he calls it) is exclusively an epistemic matter; physics should adopt it only as long as not all of the laws have been discovered. He emphasizes that the conservation laws' status as "wide principles which sweep across the different laws" (49) is *not* an artifact of the incompleteness of our knowledge or merely one of several equally accurate ways in which the laws of physics might be axiomatized. For instance, if we regard angular momentum conservation as holding merely because it can be

Even if the conservation laws transcend the microphysical laws, their lawhood (though not their transcendence) follows from the microphysical laws' lawhood; certain counterfactuals are overdetermined by being supported independently by conservation laws and microphysical laws. Nevertheless, the conservation laws are not superfluous. In transcending the microphysical laws, the conservation laws possess a variety of natural necessity that they cannot inherit from the microphysical laws (because those laws lack it) and there are counterfactuals that the conservation laws support but the microphysical laws cannot support (because those counterfactuals specify that the conservation laws would still have held, had there been different fundamental interactions).

The conservation laws (on the view that I have just sketched) illustrate a way for certain laws to "transcend" and thereby be metaphysically independent of other laws.[10] That a proper subset of the laws is metaphysically independent of the rest of the laws would consist roughly in the fact that the laws in the proper subset would still have been true, even if the other laws had not held. Here is one way in which this idea can be cashed out. (I have cashed it out more precisely in Lange, 2004, 2007, 2009.) Let's begin with the familiar thought that I mentioned in Sect. 2 as playing a role in Loewer's main argument: that lawhood over and above truth is associated with a distinctive role in supporting counterfactuals – that is, a distinctive degree of invariance under counterfactual antecedents. In particular, the laws, unlike the accidents, would still have been true (i.e., are invariant) under any physically possible counterfactual antecedent. That is, each truth *p* where it is a law that *p* is invariant under every counterfactual antecedent that is logically consistent with all of the truths *m* where it is a law that *m*. In other words, the deductive closure of the laws forms a "stable" set of truths: each of the set's members is invariant under every counterfactual antecedent that is logically consistent with all of the set's members taken together.[11]

---

derived from the various force laws, then "we take the derivation too seriously . . . " (49); we get the scientific explanations wrong.

[10] For my purposes here, it does not matter whether the conservation laws do in fact transcend the microdynamical laws. It suffices for my purposes that this conception of the conservation laws has been taken seriously in physics and so sets a precedent for the view of PH and PROB that I am describing. It is worth noting, though, that this picture of the conservation laws as transcending the microdynamical laws fits well with the modern view of the conservation laws as explained by symmetry meta-laws within a Lagrangian framework, rather than by the fundamental force laws that these meta-laws constrain to exhibit various symmetries. That the spacetime symmetry principles and Euler-Lagrange equations would still have held, had there been an additional kind of fundamental interaction, explains why the conservation laws would then still have held. For more on these topics and on the sort of evidence that scientists have offered for the stronger modal status of some laws relative to others, see Lange, 2009, 2016.

[11] A set's "stability" is not trivial just because the relevant range of counterfactual antecedents is restricted to those that are logically consistent with the set's members taken together. That a truth *p* is logically consistent with a given counterfactual antecedent *q* fails to guarantee that *p* would still have held, had *q* obtained. For instance, let the given set be the deductive closure of the fact *p* that none of the matches in this book is lit. Although *p* is logically consistent with my having struck a

With this notion of a "stable set" of truths, we can cash out more precisely what it would be for a deductively closed, proper subset of the laws to "transcend" the set of all laws. It would be for that proper subset likewise to be "stable" in that each of the set's members *p* would still have held under every counterfactual antecedent that is logically consistent with the set's members taken all together. Thus, the stability of this proper subset depends on exactly the counterlegals that I have been talking about, namely, those expressing the fact that the proper subset's members would still have held under counterfactual antecedents that are counterlegals in that they are logically inconsistent with some law – as long as those antecedents are logically consistent with every member of the given proper subset. Thus, the conservation laws (taken together with a few additional laws) transcend the remaining laws (such as the force laws) exactly when the conservation laws (and their colleagues) would all still have held under a specific broad range of counterfactual antecedents: each of the antecedents that is logically consistent with the conservation laws (and their colleagues). These antecedents include "Had the force laws been different" – but not "Had the force laws violated energy conservation", since this counterfactual antecedent is logically inconsistent with the conservation laws and so the subset's stability does not require its invariance under this counterfactual antecedent. If the conservation laws (and their colleagues) transcend the other laws (including the force laws), then had the force laws been different, they would still have upheld energy conservation.

The details of this account (as given, for instance, in Lange, 2009) do not matter for the purposes of this paper. My main argument depends only on there being some connection along roughly these lines between lawhood and invariance under a range of counterfactual antecedents, so that a given stratum of laws "transcends" the rest of the laws when it is invariant under the sort of counterlegals I have mentioned.[12]

Suppose we take the Albert/Loewer proposal that PH and PROB are laws and tweak it by specifying that PH and PROB transcend the microphysical laws in the manner that is often ascribed to the conservation laws. For it to be non-trivially the case that PROB would still have held even if the microphysical laws had been

match from the book (*q*), it may well be that had *q* obtained, *p* would not still have obtained (i.e., a match from the book would have been lit). The given set is "unstable". See Lange, 2009.

[12] Here is another fact on which my argument does not depend. Perhaps the metaphysical independence of the conservation laws is a brute fact because the conservation laws are fundamental. Alternatively, perhaps the conservation laws are explained (via Noether's theorem) by spacetime symmetry principles and the Lagrangian dynamical framework, which themselves transcend the microphysical laws and so make the conservation laws likewise transcendent. (In Lange, 2016, I have discussed in more detail this kind of scientific explanation.) Or perhaps the conservation laws have some other sort of explanation. For my purposes in this paper, it makes no difference. That is because my aim is only to say how the special sciences could be rendered autonomous by a tweak to the Albert/Loewer picture, and this tweak requires only that the relevant counterlegals hold, e.g., that energy conservation would still have held had the force laws been different. It does not matter to my argument whether such a counterlegal holds because certain symmetry principles would still have held, had the force laws been different, or whether that counterlegal holds for some other reason.

different, PROB must concern more than just the particular low-entropy microstates allowed by the actual microphysical laws. Instead, PROB must be a recipe for taking even some merely hypothetical microphysical laws and generating a probability density distribution over the microstates allowed by them. In other words, PROB must be extended from positing a particular uniform probability density distribution to positing that there is a uniform probability density distribution over whatever microstates the microphysical laws make available to realize the low-entropy initial state posited by PH.

Suppose that Albert and Loewer are largely correct about how special-science laws arise from PH and PROB. In particular, suppose that special-science laws are entailed by PH, PROB, certain other gross features of the microphysical laws that transcend the particular microphysical laws there happen to be, and the "translation" manual. Then the special-science laws are not "redundant", as we saw Loewer (2008: 153) say is the gist of his argument against non-reductive physicalism. Although some special-science counterfactuals are overdetermined, the counterlegals among them are not. (For instance, the fact that Gresham's law would still have held even if there had been an additional kind of fundamental interaction – or even if the electromagnetic force had been twice as strong – is not overdetermined. Loewer's Laplacean superdemon does not know this counterfactual conditional's truth.) The lawhood of the special-science laws then follows from the microphysical laws together with the "translation" manual, PH and PROB (just as the lawhood of the conservation laws follows from the microphysical laws[13]). But there is a stronger variety of natural necessity that the special-science laws possess that is *not* possessed by the laws specifying the kinds of fundamental particles and interactions there are (just as the conservation laws are often thought to possess a stronger variety of natural necessity than the force laws possess). The special-science laws are thus metaphysically independent of the details of the microlaws.

In short, just as Albert and Loewer posit that PH and PROB are laws in order to capture the status of special-science laws, non-reductive physicalists could posit that PH and PROB are laws transcending the microphysical laws in order to capture the metaphysical independence of special-science laws. There is thus a plausible way for Loewer's theses (1) and (2) – that all special-science items are microphysical and that the microphysical laws are complete – to hold without undermining thesis (3): the metaphysical independence of special-science laws. Therefore, Albert and Loewer lack a strong argument that (1) and (2) undercut (3). On the contrary, special-science autonomy can be achieved by the very apparatus that Albert and

[13] There is perhaps this disanalogy: the conservation laws follow from the microphysical laws without requiring any novel additions to the latter (unless one so considers the "closure law" mentioned in sect. 3 and note 14), whereas the special-science laws follow from the microphysical laws only when the latter have been augmented by PH and PROB. However, my strategy in this paper is to *suppose* that (as Albert and Loewer propose) PH and PROB are laws and then to show how it follows, once the Albert-Loewer picture is given a tweak for which the conservation laws provide a precedent (namely, a tweak making PH and PROB transcend the microdynamical laws), that the special-science laws are not redundant.

Loewer introduce to ground the special-science laws – once that apparatus is given a tweak.

4 Objection and Reply: Counterlegals in Scientific Reasoning

Loewer's Laplacean superdemon knows all of the counterfactuals associated with special-science laws as long as those counterfactuals' antecedents *p* are logically consistent with the microphysical laws. Once again:

> It will also be able to tell which special science regularities hold under counterfactual initial conditions and so which hold in all physically possible worlds (i.e. all the worlds at which the fundamental laws of physics obtain). But on Fodor's view the demon *will not* be able to discern which regularities are laws. Because of this 'blindness' the demon will be missing those counterfactuals and explanations that are underwritten by special science laws and so will not have an understanding of special science phenomena. (Loewer, 2008: 150)

Loewer's objection to Fodor, as we have seen, is that with these counterfactuals already secured by microphysical laws, there is no work that metaphysically independent special-science laws are needed to do. I have replied that such overdetermination is familiar from the case of conservation laws and that special-science laws still have work to do in securing various special-science counterfactuals with antecedents that violate microphysical laws.[14] Loewer has anticipated a reply along these lines. Considering that the demon knows which counterfactual conditionals are true only among those conditionals with antecedents logically consistent with the laws of physics, Loewer writes:

> The restriction to antecedents that are compatible with the laws of physics is required. It may be that there are worlds where the fundamental laws of physics fail (say worlds at which the laws are Newtonian rather than quantum mechanical) but at which [a Fodorian metaphysically independent special-science law] N(F,G) still obtains. It is plausible that such worlds are more similar to the actual world than worlds at which N(F,G) fails. If so a counterfactual like "If the fundamental physical laws were Newtonian rather than quantum mechanical than [sic] Gresham's law would still obtain" may be true. I think that Fodor believes this counterfactual is true. And this is a difference that N(F,G) can make. (Loewer, 2009: 231; cf. 2009: 225)

In other words, Loewer acknowledges that because metaphysically independent special-science laws would be associated with these counterlegals, the task of securing those counterlegals would constitute work for such laws to do.

[14] To posit a violation of the microphysical laws, a counterfactual antecedent does not need to posit that like electric charges attract or anything as specific as that. Even the counterfactual antecedent "Had there been additional kinds of fundamental forces" posits a violation of the microphysical laws since one of the microphysical laws is the "closure law" that all fundamental forces are either electromagnetic or strong nuclear or weak nuclear or... (filling in whatever kinds of fundamental forces there actually are).

Regarding this "difference that [metaphysically independent special-science laws] can make", Loewer comments (continuing the above passage):

> But it is a very slight difference. And how could we tell whether this counterfactual is true.[sic] There is no experiment we can perform since we can't alter the physics of our world to find out whether Gresham's law would still hold. (Loewer, 2009: 231)

This response seems to me deficient for two reasons.

Firstly, we do not have to "alter the physics of our world" to find out what would have happened if the physics had been different. Counterlegals are not ascertained that way any more than the truth or falsehood of other counterfactuals is ascertained by realizing their antecedents. Consider this counterlegal: "Had Coulomb's law been violated sometime after the universe was one billion years old, then Coulomb's law would still have been true until that time." I am fairly confident that this conditional is false; if Coulomb's law had been violated after the universe was one billion years old, then it wouldn't ever have been a law at all and so there would presumably have been nothing to prevent its also having been violated earlier. But I did not need to "alter the physics of our world" to discover that.

Secondly, the truth of these counterlegals would not be "a very slight difference." Their truth would capture the fact that the special-science laws are largely insensitive to the microphysical laws. The truth of these counterlegals would reflect something important: the reason why the special-science laws hold. The same applies to our earlier example of the counterlegals that reflect the conservation laws' independence from the microphysical laws. If momentum conservation would still have held even if there had been another kind of fundamental interaction, then the laws concerning the particular kinds of fundamental interaction there are (though helping to entail the truth and even the lawhood of momentum conservation) are not *responsible* (even partly) for momentum conservation; they do not *explain why* momentum is conserved, since momentum would still have been conserved even if the fundamental interaction laws had been different. This was Feynman's point in writing that if we regard the conservation law as explained by its derivation from the various force laws and the law relating force and motion, then we make a mistake: "we take the derivation too seriously, and feel that [the conservation law] is only valid because [each of the force laws] is valid, [but] then we cannot understand the interconnections of the different branches of physics" (Feynman, 1965: 49). Likewise, if Gresham's Law would still have held even if there had been another kind of fundamental interaction, then the laws concerning the particular kinds of fundamental interaction there are (though helping to entail the truth and perhaps even the lawhood of Gresham's Law) are not responsible (even partly) for Gresham's Law. A difference in why special-science laws hold is not "a very slight difference."

The conservation laws are not the only laws that physicists have often characterized as "transcending" the fundamental microphysical laws in the sense I have sketched – a sense associated with the counterlegals that Loewer characterizes as having slight importance. For instance, physicists commonly characterize the principle of relativity as a "super law" (Wigner, 1985: 700); Penrose (1987: 24) characterizes Einstein's insight as "that one should take relativity as a *principle*, rather than as a seemingly accidental consequence of other laws". It is thus no coincidence that all natural processes conform to it. Philosophers commonly follow suit; Earman (1989: 55), for example, says that the special theory of relativity "is not a theory in the usual sense but is better regarded as a second-level theory, or a theory of theories that constrains first-level theories." Its status as a constraint on the microlaws is expressed in counterlegals. For instance, Lévy-Leblond (1976: 271) asserts that if the force laws had been different so that photons, gravitons, and other kinds of particles that actually possess zero mass instead possessed non-zero mass, then the Lorentz transformations would still have held (though these particles would not have moved with the speed c figuring in those transformations). This is exactly the kind of counterlegal that, on my view, is associated with the transformation laws possessing a variety of natural necessity that the fundamental microlaws lack.

That relativity transcends the microlaws is at least part of Einstein's point in identifying special relativity as belonging to kinematics rather than dynamics. As Gross (2004) remarks, "Traditional physics distinguishes between kinematics (the framework for physics and its interpretation) and dynamics (the specific laws of nature and the forms of matter)." Kinematics has often been thought to transcend dynamics, imposing the framework within which the dynamical laws must operate. Likewise, that special relativity transcends the microlaws is at least part of Einstein's (1919) point in characterizing special relativity as a "theory of principle" rather than a "constructive theory." Principle theories take as their basis "principles that give rise to mathematically formulated criteria which the separate processes or the theoretical representation of them have to satisfy" (Einstein, 1919: 228). That the microlaws "have to satisfy" these constraints reflects the fact that whatever microlaws there might have been, they would still have satisfied these constraints. Einstein took the principle of relativity to be such a constraint:

> The principle of relativity is a principle that narrows the possibilities; it is not a model, just as the second law of thermodynamics is not a model (Einstein, 1911: 357)

In other words, the principle of relativity is not part of a model of microstructure. Rather, it "narrows the possibilities" regarding the microstructures; it limits the range of microphysical laws there could have been. Here we again have the counterlegals that Loewer deprecates.[15]

Of course, PH and PROB are not part of special relativity. But Albert and Loewer intend them to underwrite thermodynamics, and thermodynamics is Einstein's exemplar of a "principle theory". That thermodynamics would still have held, even if the microphysical laws had been different, is not an idle metaphysical excrescence on the body of science. Rather, that thermodynamics "transcends" microphysics often helps to explain derivative laws of thermodynamics.

[15] For more on the distinction between principle theories and constructive theories, see Lange, 2014 and Lange, 2016: 114–123.

Consider, for example, the standard textbook explanation (originating with Gibbs) of the entropy of a mixture of two non-interacting ideal gases. The explanation uses energy conservation to account for the expression for ΔS: the difference between the mixture's entropy and the entropy of the gases when separated. Suppose N_A molecules of gas A occupy volume V_A (the left side of a container) and N_B molecules of gas B occupy volume V_B (the right side); the container is isolated and the two gases have the same pressure P and temperature T. Suppose gas A is confined behind a freely moveable membrane permeable to B but not to A, and gas B is similarly confined behind a membrane permeable to A but not to B. Initially, the two membranes divide the container along the same plane, so the gases are entirely separated. Then the membranes are allowed to move slowly, each gas expanding quasi-statically, so that ultimately the two membranes reach opposite ends of the container and both gases fill the entire container (volume $V_A + V_B$). Each gas's expansion is a reversible isothermal process. Let W be the total work done on the system:

$$\begin{aligned} W &= -\int_{V_A}^{V_A+V_B} P\,dV - \int_{V_B}^{V_A+V_B} P\,dV \\ &= -\int_{V_A}^{V_A+V_B} N_A kT \frac{dV}{V} - \int_{V_B}^{V_A+V_B} N_B kT \frac{dV}{V} \\ &= N_A kT \ln\frac{V_A}{V_A+V_B} + N_B kT \ln\frac{V_B}{V_A+V_B}. \end{aligned}$$

By energy conservation (i.e., the first law of thermodynamics), the change ΔU in internal energy and the heat Q absorbed are related by

$$\Delta U = Q + W.$$

Since the gases expand isothermically, $\Delta U = 0$, so $Q = -W$. Thus

$$Q = N_A kT \ln\frac{V_A+V_B}{V_A} + N_B kT \ln\frac{V_A+V_B}{V_B}.$$

Then since $\Delta S = Q/T$,

$$\Delta S = N_A k \ln\frac{V_A+V_B}{V_A} + N_B k \ln\frac{V_A+V_B}{V_B},$$

which is the explanandum: the formula for the entropy of a mixture of two non-interacting ideal gases.

Crucially, this explanation does not presuppose that the microphysical laws make it possible for there to exist a pair of membranes, one permeable to A but not to B, the other permeable to B but not to A. Whether there are any materials that could constitute such membranes depends on the particular gases involved. Generally, such membranes are impossible. For instance, if molecules of A are small and uncharged whereas molecules of B are large and charged, then typically there is nothing that could form a membrane permeable to B but not to A according to the microphysical laws (which specify the molecular constitutions of A and B, as well as the behavior of physically possible membrane materials).

But remarkably, the thermodynamic explanation is not thereby undermined. That is because thermodynamics transcends the microphysical laws, so the explanation of this derivative law of thermodynamics must proceed entirely from other principles that transcend the microphysical laws. As far as those constraints are concerned, such membranes are possible for any molecular species. The relevant species of physical possibility here is not the one Loewer invokes – consistency with *all* of the natural laws – since the relevant species of physical possibility does not require consistency with the laws of microphysics. As Planck said in 1891 in commenting on Gibbs' derivation of this equation:

> The enormous generalization that Gibbs has given to this tenet and which must, in and of itself, appear irresponsibly daring, rests clearly on the self-evident thought that the validity of so fundamental a tenet as that of the entropy of a mixed ideal gas, cannot depend on the arbitrary circumstance of whether we really have available in each individual case a suitable semi-permeable membrane. (translated in Seth, 2010: 108)

Whether any such membranes are possible for a particular pair of gases depends on the microphysical laws. But the law regarding a mixture's entropy "cannot depend" on (that is, cannot be explained by) microphysical laws. So its explanation can afford to posit membranes that are microphysically impossible. The laws of thermodynamics would still have held, whether or not microphysical laws permit a suitable pair of membranes for a particular pair of gases – an "arbitrary circumstance", as Planck says. Thermodynamics's status as transcending microphysics thus plays an important role in thermodynamical reasoning.

5 Arbuthnot Revisited

I conclude by returning to the example of Arbuthnot's regularity (introduced in Sect. 2). Presumably, microphysical facts at the time that the first London child born in 1629 was conceived (together with microphysical laws and the "translation" manual) entail (or, at least, highly probabilify) the child's sex, and similarly for each London child subsequently born that year. Cumulatively they entail that more male than female children were born in London in 1629 – and similarly for all 82 years covered by Arbuthnot's regularity. However, Kitcher (2001: 71) writes, this account "would still not advance our understanding" – that is, would not explain why Arbuthnot's regularity holds. "By contrast," Kitcher writes,

> we can already give a satisfying explanation by appealing to an insight of R.A. Fisher. In considering sex ratios from an evolutionary point of view, Fisher recognized that, in a population in which sex ratios depart from 1:1 at sexual maturity, there will be a selective

> advantage to a tendency to produce the underrepresented sex.[[16]] It is easy to show from this that there should be a stable evolutionary equilibrium at which the sex ratio at sexual maturity is 1:1. In any species in which one sex is more vulnerable to early mortality than the other, this equilibrium will correspond to a state in which the sex ratio at birth is skewed in favor of the more vulnerable sex. Applying the analysis to our own species, in which boys are more likely than girls to die before reaching puberty, we find that the birth sex ratio ought to be 1.04:1 in favor of males – which is what Arbuthnot and his successors observed. We now understand why, for a large population, all years are overwhelmingly likely to be male. (Kitcher, 2001: 71).

Why does Fisher's account qualify as an explanation whereas the microphysical account does not?

Kitcher fails to say explicitly. He says that the microphysical account "would not show that Arbuthnot's regularity was anything more than a gigantic coincidence" (Kitcher, 2001: 71). By this, I think Kitcher means that it incorrectly depicts Arbuthnot's regularity as dependent upon microphysical states at each of the times that the children were conceived and microphysical laws governing those states. By contrast, immediately before this discussion, Kitcher says that the "fundamental Mendelian regularity [that] genes sufficiently far apart on the same chromosome or on different chromosomes assort independently" is explained not microphysically, but rather by the fact that "the transmission of genes to gametes is a process of pairing and separation" (2001: 70). This process, rather than the genes' microphysical composition, explains because "The regularity about genes would hold so long as they could sustain processes of this form, and, if that condition were met, it wouldn't matter if genes were segments of nucleic acids, proteins, or chunks of Swiss cheese" (2001: 70–1).[17] That is a counterfactual. Kitcher's point is that to explain the Mendelian regularity, we need facts that are as invariant under scientifically relevant counterfactual antecedents as the Mendelian regularity is.[18]

The same applies to Arbuthnot's regularity. It would still very likely have held (as long as boys remained more likely than girls to die before puberty, and human beings reproduced sexually, and there were no selection pressures besides the one figuring in Fisher's explanation) no matter what the microstates had been at the times that children were conceived. Indeed, it would likely still have held under various counterlegals (e.g., even if children had been made of continuous rigid material rather than atoms, or even if the microphysical laws had been different) as long as under those counterlegals, boys would have remained more likely than girls

[16] That is because members of the underrepresented sex will have more offspring on average than members of the overrepresented sex (since each offspring has a parent of each sex), so individuals with a hereditary tendency to have offspring of the underrepresented sex will (in the absence of other factors) tend to have offspring who themselves have more offspring on average.

[17] Weslake (2014: 253) also emphasizes this Kitcher passage – though not the counterlegals that I am about to emphasize.

[18] The same applies to explaining why PH holds. The fact C that the universe initially occupied a given microstate (together with some laws) entails PH since the microstate's details fix its low entropy. However, I presume that Loewer would say that this entailment does not explain why PH holds since PH is a matter of (fundamental) law whereas C is accidental.

to die before puberty, human beings would have continued to reproduce sexually, and no additional selection pressures besides the one figuring in Fisher's explanation would have been at work. The microphysical, copulation-by-copulation account fails to explain Arbuthnot's regularity because it incorrectly depicts that regularity as "coincidental" – that is, as dependent on accidents on which it does not depend.

As I described in Sect. 2, Albert and Loewer maintain that Arbuthnot's regularity is explained by the deduction of its high likelihood from PH, PROB, the microphysical laws, the "translation" manual, and whatever else is implicitly being taken for granted by the why question (e.g., London's existence). To the charge that this deduction, like Kitcher's copulation-by-copulation account, fails to explain by virtue of incorrectly depicting Arbuthnot's regularity as coincidental, Loewer replies that the deduction shows Arbuthnot's regularity to be highly likely and therefore not coincidental:

> Kitcher takes [Fisher's explanation] to show that there is a lacuna in the physical laws that is filled by special science laws. But if PROB is correct then it must also fill these lacunae. The reason is that if a regularity is lawful then it must also be likely and PROB is the arbiter of what is likely. If the super Laplace's demon knows PROB then it will be able to compute the probability of Arbuthnot's regularity given the macrostate of the world (or London) in [1623]. That probability must be close to 1. If it weren't then it would be a coincidence that it turned out to be true. (Loewer, 2008: 161)
>
> Adding dynamical special science laws clearly cannot be of any use in removing the coincidentalness of a regularity if that regularity already follows from the laws of physics. And if it doesn't follow from the laws of physics then either it conflicts with the laws of physics or the dynamical laws of physics are not, contrary to assumption, dynamically complete. (Loewer, 2009: 231)

I do not agree that in showing Arbuthnot's regularity to be highly likely, the Albert/Loewer derivation automatically shows the regularity to be non-coincidental in the sense I have just elaborated. Admittedly, the derivation shows Arbuthnot's regularity to be independent of the particular microstate preceding each copulation, independent of London's microstate at the start of the regularity's 82-year span, and independent even of the universe's initial microstate. Loewer's "demon" knows that the regularity would likely still have held (given the existence of humanity, London, etc.) even if London's initial 1623 microstate had been different – since that microstate would still have been governed by the probability distribution imposed by PH and PROB (since PH and PROB are laws and so would still have held under this antecedent).[19] However, as I have emphasized, that PH and PROB are laws on a par

[19] As I understand him, Frisch (2014: 236–39) argues (against Albert and Loewer) that their derivation fails to show that Arbuthnot's regularity is a law: what their derivation shows to be high is not the chance of Arbuthnot's regularity, but rather the chance of Arbuthnot's regularity *conditional* on various accidental facts (e.g., the existence of humanity, London, etc.). To his objection to Albert and Loewer, Frisch considers the reply that the relevant law is not that (probably) all 82 years are "male" (which I agree is accidental, since it requires that humanity, London, etc., exist, which is not a matter of law), but rather that if humanity, London, etc., exist, then (probably) all 82 years are "male". To this objection, Frisch replies (Frisch, 2014: 236–7) that if the Albert/Loewer derivation yields merely this law, then this derivation is no better than

with the microphysical laws does not suffice to ensure that Arbuthnot's regularity would still have been likely had the microphysical laws been different. The only counterfactuals that the demon knows are those with antecedents that Loewer terms "physically possible" (that is, logically consistent with the laws of microphysics, including PH and PROB). But as Fisher's argument reveals, Arbuthnot's regularity would still have held even under various microphysically impossible counterlegals; a tendency to produce the underrepresented sex would still have been selectively advantageous even if (e.g.) gravity had been stronger. The Albert/Loewer picture therefore incorrectly depicts Arbuthnot's regularity as coincidental: as reflecting microphysical details. PH and PROB can underwrite Fisher's explanation only if PH and PROB transcend the laws of microphysics.[20]

Recall (from Sect. 3) Feynman's contention that although momentum conservation can be derived from the various force laws (and the law relating force to

---

the copulation-by-copulation account that Kitcher mentions; the latter is not explanatory because it incorrectly depicts the run of 82 consecutive "male" years as dependent on the accident of London's initial microstate at the start of the 82 years, and the Albert/Loewer derivation (under this interpretation) is not explanatory because it incorrectly depicts the run of 82 consecutive "male" years as dependent on the accident of the existence of humanity, London, etc. I disagree with Frisch that the Albert-Loewer derivation (under this interpretation) is no better than the copulation-by-copulation account mentioned by Kitcher – for two independent reasons, Firstly, while the existence of humanity, London, etc. is accidental rather than a matter of law (just like London's initial microstate), their existence (unlike London's initial microstate) is (as I discussed earlier) typically presupposed by the question "Why does Arbuthnot's regularity hold?" Secondly, the Albert/Loewer derivation reveals that Arbuthnot's regularity would still (likely) have held even if London's initial microstate had been different (as long as London, humanity, etc. would still have existed) – Loewer's "demon" knows this counterfactual – whereas the copulation-by-copulation account does not show that Arbuthnot's regularity would still (likely) have held even if London's initial microstate had been different. (Nevertheless, as I am about to emphasize, I believe that the Albert/Loewer derivation fails to explain Arbuthnot's regularity by virtue of failing to capture the relevant range of counterfactual antecedents under which Arbuthnot's regularity is invariant.)

[20] Weslake (2014) characterizes Fisher's explanation as making its distinctive explanatory contribution by abstracting from microdetails. Frisch (2014: 237–9) similarly suggests that it unifies all species (that reproduce sexually, etc.) at all times and places by "singling out certain higher level structural features" that they share and that the Albert/Loewer explanation fails to identify. I agree with the spirit behind these remarks, but I offer an account of how this unification and abstraction relates to the explanatory power of Fisher's argument: in showing how Arbuthnot's regularity arises in a certain evolutionary model (transcending microphysical details), Fisher's argument unifies a broad range of actual and even counterlegal cases and abstracts even from microphysical laws.

Familiar arguments for the autonomy of special sciences (e.g., from Fodor, 1974) emphasize how higher-level kinds are multiply realized in lower-level kinds because higher-level kinds are functional kinds (as in Fodor's example of money). As Fodor cashes this out, the disjunction of lower-level kinds that realize a given higher-level kind is not a natural kind and so cannot appear in a natural law or scientific explanation. Rather than appealing to naturalness, I suggest that the disjunction of lower-level kinds covers only the microphysically possible realizers of the higher-level kind and so fails to include the microphysically impossible realizers of the functional kind. Therefore, a "translation" of the higher-level explanation in terms of a disjunction of microphysically possible lower-level kinds fails to do justice to the higher-level explanation, since that explanation would still have held even if the functional kind had been realized in a microphysically impossible way.

motion), that derivation incorrectly depicts momentum conservation as a coincidence – as dependent on the particular kinds of forces there happen to be. Of course, that derivation reveals momentum conservation to be physically necessary. But doing so is insufficient to reveal momentum conservation to be non-coincidental. By the same token, Loewer is mistaken in thinking that we can show Arbuthnot's regularity to be non-coincidental merely by showing it to be highly likely (by deriving its likelihood from PH, PROB, etc.).

In the previous section, I addressed the objection that the counterlegals in virtue of which Fisher's explanation transcends microphysics are scientifically irrelevant. Fisher himself used counterlegals to illustrate how most of his great book synthesizing evolutionary theory and population genetics, *The Genetical Theory of Natural Selection* (1930), is not an account of various particular events in the history of life, but rather concerns theorems of population genetics that transcend microphysics. He said that most of his book would still have been true had God miraculously created the world a few thousand years ago:

> ...if I had had so large an aim as to write an important book on Evolution, I should have had to attempt an account of very much work about which I am not really qualified to give a useful opinion. As it is there is surprisingly little in the whole book that would not stand if the world had been created in 4004 B.C., and my primary job is to try to give an account of what Natural Selection *must* be doing, even if it had never done anything of much account until now. (Fisher, 1930: 222)

The explanations invoke principles that have a variety of necessity (a *must*) that transcends the necessity of the microphysical laws. PH and PROB cannot underwrite such explanations unless they, too, transcend the microphysical laws.

The tweaked version of the Albert/Loewer picture that I have elaborated makes the special sciences autonomous without sacrificing either of the two premises in Loewer's argument (in Sect. 2 above) against the autonomy of special sciences: that the fundamental microphysical laws are complete and that the entities covered by special sciences are constituted entirely by entities governed by the fundamental microphysical laws. Therefore, the tweaked version of the Albert/Loewer picture does no worse than the original Albert/Loewer picture in resolving what Loewer regards as the "mystery" of special-science regularities: "How do the particles that e.g. constitute an economy 'know' that their trajectories are required (*ceteris paribus*) to enforce Gresham's law?" (Loewer, 2009: 220). Those particles "know" that their trajectories are required to accord with microphysical laws, and those laws (fortified by PH and PROB) yield special-science regularities. Likewise, in a microphysically impossible world in which the special-science regularities still hold (on the tweaked version of the Albert/Loewer picture), the alien microphysical laws in force there (together with PH and PROB) enable the microphysical entities (whatever they are) to "know" that they must yield the special-science regularities. There is no "mystery" to the special-science laws in any of these worlds.[21]

[21] This might suggest an objection to my earlier argument that there is work for autonomous special-science laws to do in connection with counterlegals such as "Had there been additional

Nevertheless, the tweaked version of the Albert/Loewer picture does not reduce the special-science laws' lawhood to the lawhood of PH, PROB, and the fundamental microphysical laws – unlike the untweaked Albert/Loewer picture:

> It is true that the account of special sciences I have described is reductionist in that it explains the lawfulness of special science laws in terms of the lawfulness of laws of physics including PROB. (Loewer, 2008: 162)
>
> . . . the central physicalist idea that every positive contingent fact (including nomological facts) obtains *in virtue of* the distribution of fundamental physical entities and fundamental laws of physics. In particular, facts about what special science laws . . . there are (i.e., the fact that a particular special science generalization is lawful . . .) obtain in virtue of physical facts and laws. (Loewer, 2009: 229)

With a simple tweak, the Albert/Loewer picture can be made to underwrite (rather than to undermine) the autonomy of special sciences. I see no reason to regard this tweak as conflicting with any "central physicalist idea", considering that physicalism remains unscathed under the common view that conservation laws, thermodynamics, kinematics, and relativity transcend the microdynamical laws.

References

Albert, D. Z. (2000). *Time and chance*. Harvard University Press.

Albert, D. Z. (2015). *After physics*. Harvard University Press.

Arbuthnot, J. (1710). An argument for divine providence, taken from the constant regularity observ'd in the births of both sexes. *Philosophical Transactions of the Royal Society, 27*, 186–190.

Callender, C. (2011). The past histories of molecules. In C. Beisbart & S. Hartmann (Eds.), *Probabilities in physics* (pp. 83–114). Oxford University Press.

Callender, C., & Cohen, J. (2010). Special sciences, conspiracy and the better best system account of lawhood. *Erkenntnis, 73*, 427–447.

Cohen, J., & Callender, C. (2009). A better best system account of lawhood. *Philosophical Studies, 145*, 1–34.

kinds of fundamental interactions . . . ". The objection is that had there been additional kinds of fundamental interactions, then the fundamental microphysical laws (whatever they would have been, including the laws governing the additional kinds of fundamental interactions) would have sufficed to determine the behaviors of all things (since all things would have been constituted exclusively microphysically, as Loewer's premise (1) says regarding the actual world, and the microphysical laws would have been complete, as his premise (2) says regarding the actual world). So (the objection concludes) there would have been no work for special-science laws to do, had there been additional kinds of fundamental interactions. In response, notice that my claim is not that had there been additional kinds of fundamental interactions, then there would have been events governed by special-science laws but no microphysical laws. Rather, my claim is that the special-science laws are associated with the holding of various counterfactual conditionals that have antecedents positing a different suite of microphysical laws from the actual world's. Special-science laws are associated with the fact that Gresham's Law would still have held, had there been additional fundamental interactions, even though had there been additional fundamental interactions, the microphysical laws there would have been would have sufficed to cover every event.

Earman, J. (1989). *World enough and space-time*. MIT Press.
Earman, J. (2006). The 'past hypothesis': Not even false. *Studies in History and Philosophy of Modern Physics, 37*, 399–430.
Einstein, A. (1911). 'Discussion' following lecture version of 'The theory of relativity'. In A. Beck (trans.), *The collected papers of Albert Einstein: Volume 3*. Princeton University Press, 1993, pp. 351–358.
Einstein, A. (1919, November 28). What is the theory of relativity? *The Times (London)*, Repr. Einstein, *Ideas and Opinions*. Bonanza, 1954, pp. 227–232.
Feynman, R. (1965). *The character of physical law*. MIT Press.
Fisher, R. A. (1930). Letter to J.S. Huxley, 6 may 1930. In J. H. Bennett (Ed.), *Natural selection, heredity, and eugenics. Including selected correspondence of R.A. Fisher with Leonard Darwin and others* (p. 222). Clarendon Press., 1983.
Fodor, J. A. (1974). Special sciences (or: The disunity of science as a working hypothesis). *Synthese, 28*, 97–115.
Frisch, M. (2007). Causation, counterfactuals, and entropy. In H. Price & R. Corry (Eds.), *Causation, physics, and the constitution of reality: Russell's republic revisited* (pp. 351–396). Oxford University Press.
Frisch, M. (2014). Why physics can't explain everything. In A. Wilson (Ed.), *Asymmetries of chance and time* (pp. 221–240). Oxford University Press.
Goodman, N. (1983). *Fact, fiction and forecast* (4th ed.). Harvard University Press.
Gross, D. (2004, October 9). *The future of physics* (lecture). Kavli institute for theoretical physics conference "The future of physics". Accessed 7 Oct 2013 http://online.kitp.ucsb.edu/online/kitp25/gross/rm/qt.html
Kitcher, P. (2001). *Science, truth, and democracy*. Oxford University Press.
Lange, M. (2004). The autonomy of functional biology: A reply to Rosenberg. *Biology and Philosophy, 19*, 93–101.
Lange, M. (2007). Laws and meta-laws of nature: Conservation laws and symmetries. *Studies in History and Philosophy of Modern Physics, 38*, 457–481.
Lange, M. (2009). *Laws and Lawmakers*. Oxford University Press.
Lange, M. (2014). Did Einstein really believe that principle theories are explanatorily powerless? *Perspectives on Science, 22*, 449–463.
Lange, M. (2016). *Because without cause: Non-causal explanation in science and mathematics*. Oxford University Press.
Lévy-Leblond, J.-M. (1976). One more derivation of the Lorentz transformations. *American Journal of Physics, 44*, 271–277.
Loewer, B. (2008). Why there is anything except physics. In J. Hohwy & J. Kallestrup (Eds.), *Being reduced: New essays on reduction, explanation, and causation* (pp. 149–163). Oxford University Press.
Loewer, B. (2009). Why is there anything except physics? *Synthese, 170*, 217–233.
Loewer, B. (2012). The emergence of time's arrows and special science laws from physics. *Interface Focus, 2*, 13–19.
Penrose, R. (1987). Newton, quantum theory, and reality. In S. Hawking & W. Israel (Eds.), *Three hundred years of gravitation* (pp. 17–49). Cambridge University Press.
Seth, S. (2010). *Crafting the quantum*. MIT Press.
Weslake, B. (2014). Statistical mechanical imperialism. In A. Wilson (Ed.), *Asymmetries of chance and time* (pp. 241–257). Oxford University Press.
Wigner, E. (1985). Events, laws of nature, and invariance principles. In A. Zuchichi (Ed.), *How far are we from the gauge forces?* (pp. 699–708). Plenum.
Winsberg, E. (2004). Can conditionalizing on the 'past hypothesis' militate against the reversibility objections? *Philosophy of Science, 71*, 489–504.

Is the Mentaculus the Best System of Our World?

Meir Hemmo and Orly Shenker

Abstract Barry Loewer and David Albert put forward a theory that they call "*The Mentaculus*", which they claim to be "arguably a complete scientific theory of the universe". Albert and Loewer's Mentaculus is an expanded version of statistical mechanics. On their view – as recently updated by Loewer's "*Package Deal Approach*" – The Mentaculus is the "best system" of our world in the Lewis-style sense of this term: it contains a partial description of the fundamental reality ("The Humean base") and provides the optimal balance between informativeness and simplicity. The Mentaculus has other advantages, in particular, it is reductionist in the sense that it unifies all the sciences, and it is physicalist in the sense that the account of everything is ultimately based on physics. In this paper we examine the extent to which The Mentaculus is reductionist and physicalist. We compare it with "*Flat Physicalism*", which is our version of expanded statistical mechanics, that is a reductive type-type physicalist identity theory of everything that there is. The Mentaculus is less reductionist than Flat Physicalism: both theories assume the fundamental microdynamics and suitable contingent facts, but The Mentaculus assumes the Past Hypothesis and a Statistical Postulate, as independent laws, whereas Flat Physicalism derives them from the microdynamics and the contingent facts. Additionally, Flat Physicalism derives from the latter all the special sciences kinds and laws, while The Mentaculus does not contain any explanatory account of such reduction. Therefore Flat Physicalism is arguably "better" than The Mentaculus in the "best system" sense of the term.

Keywords Best system approach to laws · Humean base · Multiple realizability (realization) · Past Hypothesis of low entropy · Physicalism · Probability in

---

M. Hemmo (✉)
Philosophy Department, University of Haifa, Haifa, Israel
e-mail: meir@research.haifa.ac.il

O. Shenker
Philosophy Department, The Hebrew University of Jerusalem, Jerusalem, Israel
e-mail: orly.shenker@mail.huji.ac.il

Y. Ben-Menahem (ed.), *Rethinking the Concept of Law of Nature*,
Jerusalem Studies in Philosophy and History of Science,
https://doi.org/10.1007/978-3-030-96775-8_4

statistical mechanics · Reduction to fundamental physics · Second Law of thermodynamics · Special sciences · Type-identity of physical and special sciences' kinds

1 The Mentaculus

Barry Loewer and David Albert put forward an "outrageously ambitious" theory that they call "*The Mentaculus*", which they claim to be "arguably a complete scientific theory of the universe" (Loewer, 2020a).

> "The Mentaculus is a probability map of the universe in that it determines a probability density over the set of physically possible trajectories of microstates emanating from M(0), and thereby probabilities over all physically specifiable macro histories. Once the dynamical laws are specified. The Mentaculus contains an answer to every question of the form "What is the objective probability of B given A?" for all physically specifiable propositions A and B. For example, it specifies the probability that an ice cube will completely melt in the next five minutes given a description of the macrostate of its environment, and the probability that Trump will be impeached given the current macrostate of the universe." (Loewer, 2020a, p. 6)

A central motivation of Loewer and Albert for endorsing The Mentaculus is that they think it forms the "best system", in the Lewis-style sense of this term: it is a partial description of the fundamental reality ("The Humean base", discussed below) which provides the optimal balance between informativeness and simplicity. It seems that on their view this "best system" has also other advantages, in particular, it is *reductive* in the sense that it unifies all the sciences, and *physicalist* in the sense that the account[1] of everything is ultimately based on physics.

In this paper we show that The Mentaculus is not the "best system" since a "better" theory is possible and available: our "Flat Physicalism" theory, described in Sect. 2, which is more reductive and more physicalist than The Mentaculus, and therefore offers a better balance of simplicity and informativeness. The Mentaculus can be improved along its lines, in a way that is compatible with Loewer's Package Deal Approach (Loewer, 2020b), but in this case the distinction between the Humean framework and reductive physicalism becomes less clear. We compare the two theories in Sects. 3 and 4.

The Mentaculus is essentially a version of statistical mechanics, *expanded* and *augmented* so as to account for all the phenomena described by the special sciences (e.g., thermodynamics, biology, psychology). "Statistical mechanics" was originally the name of the (ongoing, unfinished) project of accounting for the facts described by thermodynamics, on the basis of the fundamental theories of physics together with some auxiliary hypotheses. Various theories share this name as well as this

[1] By 'account' or 'explanation' we mean here anything compatible with the best system account of laws (see below).

aim, but differ in the details of how to achieve it.[2] Here is Albert and Loewer's theory of statistical mechanics, which is at the heart of The Mentaculus:

> "The Newtonian statistical-mechanical contraption for making inferences about the world consists, in its entirety, of three laws and one contingent empirical fact. The empirical fact is the one about what the macrocondition of the world currently happens to be (or rather, the empirical fact is the one about what the directly surveyable condition of the world currently happens to be; where the directly surveyable condition of the world – insofar as Mr. X is concerned – includes its macrocondition plus whatever, perhaps microscopic, features of Mr. X's brain he may happen to have direct and unproblematic introspective access to),[3] and the laws are:

1. The Newtonian law of motion (which is that F = ma).
2. The Past-Hypothesis (which is that the world first came into being in whatever particular low-entropy highly condensed big-bang sort of macrocondition it is that the normal inferential procedures of cosmology will eventually present to us).
3. The Statistical Postulate (which is that the right probability-distribution to use for making inferences about the past and the future is the one that's uniform, on the standard measure, over those regions of phase space which are compatible with whatever other information – either in the form of laws or in the form of contingent empirical facts – we happen to have)." (Albert, 2000, p. 96).

(For brevity, we use the following acronyms: NM for Newtonian mechanics, CF for the contingent fact, PH for the past hypothesis, SP for the statistical postulate. We stress that CF, which is not a law, but a contingent fact (if it is true), is a – as Albert writes – what the *macroscopic* condition of the world currently happens to be, as is *surveyable* by a given observer, and includes this observer's brain state which is[4] this observer's perceptions and memories.)

We will argue that our Flat Physicalism is "better" than The Mentaculus in that it retains the same degree of informativeness, but is simpler. In this paper as a working hypothesis we adopt the Humean framework of the Lewis-Loewer-Albert best system approach to laws (as in Loewer, 2020b; Albert, 2014; see Sect. 3 for a detailed discussion of this framework).[5] Famously, there are various ideas about what simplicity is, which lead to different "measures of simplicity". Also, there are various ideas concerning the features of the sentences that may be admitted to the best system as laws, for example, whether laws must be generalizations, and consequently, whether sentences such as PH and CF, which are not generalizations,

[2] The standard ones are described in (Frigg, 2008).

[3] Since here "introspection" must be a measurement of a system by itself (not one part of the system measuring another part, but one and the same part measuring itself), we are not sure that this can be part of a physicalist account of the universe. This detail of The Mentaculus can be amended by opting for a type-type reductive account of the mental, as in Flat Physicalism, described in Sect. 2.

[4] Or on which supervenes; we don't address this difference here.

[5] We don't strictly follow Lewis's best system approach; we say explicitly when and where we depart from it.

may or may not appear as separate independent laws in the best system.[6] Here we shall not dwell on these issues, because we will show that Albert and Lower assume (implicitly or explicitly) all the laws and facts that we assume, namely NM + the contingent fact CF (for brevity we write henceforth NM + CF), but add as separate laws PH + SP.[7] It seems to us that Albert and Loewer can arrive at their additional laws either in our way, namely by deriving PH + SP from CF + NM, in which case their Mentaculus will be improved in the sense that it will be simpler; or they must commit themselves to an extra source of information concerning these additional laws, and this extra source will make their system much more complex.

The system we call Flat Physicalism holds that all there is to the world is its actual microscopic trajectory as given by the specific equation of motion of the physical theory (quantum or classical, which includes all the parameters and boundary or initial conditions). In a Humean framework we take the micro-trajectory to be the *full* Humean base, where the laws of physics are supposed to give a partial simple and informative description of this base. It seems to us that this is also Albert and Loewr's view[8] (we discuss their Humean view in Sect. 3). Everything that is true about the world should be derived (see below) from facts we know to be true about the micro-trajectory and nothing else. That is, Flat Physicalism assumes NM and CF only, and we will show that one can derive from them PH (*if* it is true; see Hemmo & Shenker, 2012 Ch. 10 and Hemmo & Shenker, 2019a) and SP (*if* it is true; see Hemmo & Shenker, 2019b and below). This version of statistical mechanics is part of a reductive type-identity physicalist theory (see Hemmo & Shenker, 2019a, b, c, 2021a, b, c; Shenker, 2017a, which builds on Hemmo & Shenker, 2012, 2016; Shenker, 2017b, c, 2020a, b).

In order to be "arguably a complete scientific theory of the universe", both systems, Flat Physicalism and The Mentaculus, have to account not only for thermodynamics but also for all the (other) so-called special sciences. According to Flat Physicalism, since all there is to the world is the full micro-trajectory, Laplace's demon, who knows the actual trajectory of the universe, knows *ipso facto* all the special sciences laws and facts. However, in contemporary literature, most authors with physicalist inclination endorse so-called "non-reductive physicalism", that is: a view that allows for "multiple realizability" of special sciences kinds by physical kinds. Only a minority of thinkers endorse "reductive physicalism" and do not accept multiple realizability. Flat Physicalism is a full-fledged reductive type-identity theory, and therefore it rejects multiple realizability (and multiple realization, of course). We are not sure about Albert's and Loewer's current views on this matter: if they accept "multiple realizability" (as Lewis did), their metaphysics

[6] In particular the question arises with respect to PH, which is considered to be a law in the best system by Albert and Loewer. But our arguments in this paper don't bear on these issues, and so we set them aside.

[7] We thank Barry Loewer and David Albert for discussions about this and other points in this paper.

[8] Personal communication.

is not physicalist, but dualist. We will argue for this claim as we proceed and present what we take to be their views on this matter in Sect. 4.

A dualist metaphysics might be the "best system", but still it is of interest to know that this is the view endorsed. If so, we shall argue that Flat Physicalism is "better" in this context too, since it is physicalist, and moreover it accounts for the special sciences on the basis of CF + NM alone, thus retaining the informativeness of The Mentaculus in a simpler way.

In the above description of The Mentaculus (as well as in Flat Physicalism as we discuss it here), classical mechanics plays the role of fundamental physics. Of course, according to contemporary physics classical mechanics is false. Still, it is useful to consider statistical mechanics in the classical context because arguably the classical theory preserves some of the salient explanatory and predictive aspects of the true fundamental physics, under the appropriate relevant conditions[9] (see Wallace, 2001, Sect. 1; Ladyman & Ross, 2007, Ch. 1, for a critical discussion of the use of outdated physics in philosophical discussions; see Shenker, 2020a, b, c on the foundations of quantum statistical mechanics).

Let us describe already here what we take to be the crucial points of disagreement between Flat Physicalism and The Mentaculus approaches.

Both Flat Physicalism and The Mentaculus agree (we think) that we have available to us, as empirical data, the observations CF as described above: the macroscopic condition of the world as it is given to an observer. Since classical mechanics describes the world in terms of differential equations, there is no way to avoid introducing CF in addition to NM. The two theories differ on the following.

Flat Physicalism takes CF as a starting point, although it is open to the possibility that its content is false.[10] Suppose first that CF is by and large true: what is it true *of*? The answer to this question should be given by a theory about the world, which provides an explanation for CF. One good candidate for an explanation of CF is that the world is as described by NM. (In Flat Physicalism as well as – we think – in the The Mentaculus approach, NM is a placeholder for the laws of our best contemporary theories of physics.) In particular, this means that the best explanation of CF is that everything that there is to the world is its micro-trajectory in the sense of NM, of which CF + NM provide a partial description (CF provides only some of the parameters and initial and boundary conditions that are needed to provide a well defined micro-trajectory, and therefore it is compatible with a *set* of such

[9] In this paper we do not address Hempel's Dilemma concerning what is the physics that ought to be taken as fundamental in a reductionist project; see (Firt et al.2020).

[10] We focus here on the case in which CF is true, because it is the simplest way to present our main point. But importantly we are not committed to CF being true about the world or even to its being highly likely true. Theories of physics (and of special sciences which reduce to physics) are constructed so as to explain CF, but explaining CF does not mean vindicating CF as true about the world. Our scientific theories may or may not entail a correspondence between the content of CF and reality. It may turn out that the best explanation for the data of CF is that its contents are partly false; contemporary science gives us good reasons to think that this is the case (see Hemmo & Shenker, 2021d; Shenker, 2020c). We do not think that this result is catastrophic – unlike Albert, as we point out below. On an even more skeptical view concerning the status of the CF data, see also (Shenker, 2020b).

trajectories, only one of which is the actual one). According to Flat Physicalism, given CF + NM we can derive PH and SP (if they are true! See Sect. 2 below), since both PH and SP are partial descriptions of the actual micro-trajectory, that are more partial (and therefore compatible with) the partial description provided by CF + NM. PH is understood here as some macroscopic description of the state of the world at some past moment, and is *in this sense* on a par with the contents of CF (but contains *less* information than CF + NM); and SP is understood as expressing relative frequencies along the actual micro trajectory of the universe (and *not* as pertaining to counterfactual histories). Finally, there is no guarantee that CF is true: it may be that CF is false (we think that talking about the likelihood of its being true is circular, if meaningful at all).[11] It may even be that the best theories that explain the experience of CF are those that tell us that it is false to a larger or smaller extent. Why do we often adhere in practice to the option that CF is true? It is perhaps a more convenient choice, but it is not compelling. (Contemporary brain science gives good reasons to think that a substantial part of CF is false.)

In The Mentaculus, by contrast, the truth of CF has a much stronger status: the laws of nature are selected so as to guarantee its likelihood, given some understanding (which Flat Physicalism rejects) of what counterfactuals are and what probability in physics is (see some details in Albert, 2000, 2014; Hemmo & Shenker, 2012, 2015a, b, c, 2016). The crucial difference between the two approaches is this. In Flat Physicalism the schema is CF + NM → PH + SP (where → is material implication). In The Mentaculus, the schema is NM + PH + SP → CF (with high probability), so that CF is likely to be true, and this avoids what Albert (2000) calls the "skeptical catastrophe" (Albert, 2000). In our view, although a world in which NM + PH + SP obtains is nice and comforting in that – if you accept the right notions of counterfactuals and probability – the content of your experience and memories are likely to be true, this argument provides a less simple system. This can be appreciated by considering the following questions: why are NM + PH + SP accepted as part of the "best system" in The Mentaculus; and what kind of *support* do we have for postulating them? One option is that the truth of NM + PH + SP is *tested* empirically via its prediction of CF: it is in this sense that this theory is *empirical*. In this option The Mentaculus relies on CF to the *same* degree that Flat Physicalism does, and then the degree of simplicity is the same as in Flat Physicalism. Another option is this: since The Mentaculus wants something *stronger*, something that will guarantee *high probability* for the truth of CF, it needs to give PH + SP a status that is *stronger* than or at least *independent* (logically or metaphysically) of CF. But what is the source of PH + SP? Are they independent of NM + CF?

The short answer is that PH and SP are not independent of NM + CF. It is here that we think the disagreement between the two approaches lies. Both The

[11] It is circular because it requires postulating a probability measure (or a typicality measure) over counterfactual states, where the *choice* of measure, according to which the truth of CF is highly probable, if it is empirical at all and not given to us by some external source of information (see below), is determined by the facts along the *actual* micro-trajectory, described by CF itself (see our 2012, Chap. 8; 2015b).

Mentaculus and Flat Physicalism agree that NM (or F = ma) is part of the best system; and both agree that in order to make empirically adequate inferences about the universe using NM, one has to add the observations CF describing the macroscopic condition of the world as it is given to an observer. But in Flat Physicalism the deductive schema is CF + NM → PH + SP. We assume (as a working hypothesis) that CF is true, but it may be false (in which case PH and SP are false). Now: think of the derivation of PH and SP in the above schema as a calculation carried out by Laplace's Demon on the basis of CF + NM alone. We will show how precisely this calculation can be done in Sects. 2.3 and 2.4. If this is right, the question of whether or not SP and PH are true is completely settled by NM + CF.[12] By contrast, in The Mentaculus, the schema is NM + PH + SP → CF (with high probability), so that according to Albert and Loewer the high likelihood of CF turns out to be a consequence of NM + PH + SP. But here PH + SP are added as separate laws over and above CF + NM, where the idea is that by adding PH + SP to NM one may be able to prove the high likelihood of CF. We think that this is impossible, and that the arguments to this effect given in the literature are viciously circular (see above and Sect. 2). Therefore, it seems to us that Flat Physicalism (CF + NM) is better (simpler and more informative and more reductive) than The Mentaculus (NM + PH + SP + CF) in describing our world.

The paper is structured as follows. In Sect. 2 we describe Flat Physicalism, and in the following sections we compare it with The Mentaculus. Readers who are familiar with Flat Physicalism, or who are interested in first seeing in what sense we think Flat Physicalism is "better" (in the sense of "best system") than The Mentaculus, may skip Sect. 2 and return to it later. Section 2 is partitioned into subsections for ease of reference or of returning. We describe the basic ontology of this theory (Sects. 2.1 and 2.2); how probability arises and how to derive SP even when the underlying micro-trajectory is deterministic (Sect. 2.3); the status of the PH and how to derive it (Sect. 2.4); what it takes to reduce the special sciences to fundamental physics in the face of the idea of multiple realization (Sect. 2.5); the implications of this idea with respect to psychology (Sect. 2.6). In Sect. 3 we address the metaphysical framework in which Albert and Loewer formulate The Mentaculus: David Lewis' metaphysics and the best system approach to laws and Loewer's improvement of it in his (Loewer, 2020b) Package Deal Approach (PDA). In Sect. 4 we consider the non-reductive nature of the special sciences in the Mentaculus.

2 Flat Physicalism

In this section we present Flat Physicalism. While Flat Physicalism *resembles* The Mentaculus in that it is an "outrageously ambitious" theory that is "arguably a

---

[12] Note that PH and SP are logically independent.

complete scientific theory of the universe", and that its way to achieve this aim is by a generalization of statistical mechanics, it *differs* from The Mentaculus in important points concerning its conceptual basis. The difference fundamentally concerns the metaphysical basis: The Mentaculus is seen by its authors as a "best system" that is supposed to provide an informative and simple description of the Humean base, which is the fundamental reality, whereas in Flat Physicalism the fundamental reality is the physical one, and there is nothing else and in particular nothing more fundamental than physics.

One important consequence of this difference is that in Flat Physicalism everything is supposed to be accounted for by NM and CF only (see Sects. 2.1 and 2.2), whereas in The Mentaculus additional fundamental elements are assumed, namely PH and SP. In this sense Flat Physicalism, if successful, is simpler (see Sects. 2.3 and 2.4). In addition, Flat Physicalism derives from NM and CF the special sciences' laws and kinds, including those of psychology (Sects. 2.5 and 2.6), as well as a notion of chance in a deterministic world (see Sect. 2.3).

Due to this difference, in Flat Physicalism – unlike The Mentaculus – the PH, SP, the special sciences and the notion of chance are all derived from CF + NM. This is the main reason why it is simpler, and hence "better", than The Mentaculus. Perhaps The Mentaculus can be changed so as to incorporate some ideas from Flat Physicalism, to become simpler, more reductionist, and more physicalist, but then it becomes so close to reductive physicalism that the difference may cease to be significant. We will discuss this consequence later.

As we said above, in this paper we present statistical mechanics as the project of accounting for the thermodynamic (and other) regularities by *classical* mechanics, as is usual (and as was done with respect to The Mentaculus in Sect. 1 above). The justification for doing so is the conjecture that although classical mechanics is strictly speaking false, it preserves some of the salient explanatory and predictive aspects of the true fundamental physics, under the appropriate conditions. (See Wallace, 2001 Sect. 1, and Ladyman & Ross, 2007. See Shenker, 2020a, b, c on quantum statistical mechanics.)

2.1 Microstates: Fundamental Physics

Statistical mechanics, as the project of accounting for thermodynamics by classical mechanics, consists of two (related) stages. First, express the thermodynamic quantities (such as volume and temperature) in terms of mechanical quantities (like mass, position and velocity). Second, prove that due to the laws of mechanics (e.g. F = ma) and some contingent fact (i.e. together with some auxiliary hypotheses[13]) these mechanical quantities evolve in a way that mirrors the thermodynamic regularities (such as the ideal gas law or the Second Law). In this paper we focus on

[13] These auxiliary hypotheses are described in (Shenker, 2017b, c).

the first stage (for the second stage see Hemmo & Shenker, 2012, 2016, Shenker, 2017b, c), and therefore we begin with the ontology of fundamental physics.

According to classical mechanics the universe consists of particles each having properties or parameters such as mass or charge, that are subject to certain constraints and limitations such as total volume available to the particles or the total energy available to all of them, and each of them has, at each point of time, precise and well defined position and velocity.[14] The positions and velocities of all the particles at a time is called the universe's "*microstate*" at that time.[15] The microstate, together with the parameters and constraints, is everything that there is at every moment, according to mechanics. For brevity, henceforth by the term "microstate" we will refer also to the parameters and constraints.

The following terminological remark is of utmost importance for understanding statistical mechanics. There are various notions of *'microscopic'* in the literature: the term sometimes means small, and sometimes part of a whole. But in statistical mechanics it is customary to use the term 'microstate' to denote the complete state of the system of interest, or of the world, according to the fundamental theory, which is here classical mechanics. For instance, in classical mechanics the precise positions and velocities of the particles of the entire universe is a microstate, despite the fact that it is neither small nor part of anything.

Albert (2000) Ch. 1) points out that the microstates are not independent of each other: since instantaneous velocity is part of the microstate, the microstate in a given moment imposes constraints on the positions in the microstates in its temporal vicinity. The requirement of independence may be important for Lewis's original characterization of the Humean base (see Sect. 3), but (we think) not for Loewer's (2020a, b) recent PDA version of it (that we address in Sect. 3). In any case, according to Flat Physicalism the fundamental reality is as described by physics.

Given the microstate at some moment, together with the parameters and constraints, the equations of motion of classical mechanics determine the microstates at all other moments, that is, an entire history or an entire *trajectory* of the universe.[16] While we can perhaps conceive of counterfactual microstates, and hence on counterfactual histories, *the actual micro-history is everything that there is*, according to classical mechanics.

2.2 Aspects: Macrovariables and Macrostates

Normally, we are not interested in the full details of the microstate of a system or its micro-history: we are interested in certain sub-systems of the universe, and in certain

[14] Zeno's arrow paradox needs to be solved here; see (Arntzenius, 2000).

[15] Microstates can be thought of as events, not states. We don't expand on this point.

[16] In quantum mechanics the micro-history of the universe is determined by the equations of motion of the quantum state. Some interpretations postulate "collapse" of the quantum state, either deterministically or non deterministically.

properties of those sub-systems. In particular, we are interested in properties of systems that exhibit regularities that we can follow. For example, in thermodynamic systems we are interested in the volume of a gas, or the degree of uniformity of its temperature. We just said that the microstate is everything that there is; and this entails that properties like volume or temperature cannot be facts over and above the microstate. Indeed, those properties are *aspects* of the microstates, sometimes called *macrovariables*.[17] For example, volume is an aspect since it is a function (certain coarse grained distribution) of the positions of the system's particles, and does not pertain to all the details of these positions nor to the velocities of the particles. (The description of an aspect (of a macrovariable) is a partial description of the microstate; but we stress that we focus on what there is, about *ontology*, and not about how it can be known or described.)

It has been discovered – and this is a highly non-trivial discovery by the creators of statistical mechanics – that certain aspects stand in lawful relations to certain other aspects, either at a time (synchronically) or over time (diachronically), and these aspects and their lawful relations have been *identified* with thermodynamic magnitudes and the laws governing them. Here is what we mean by "identified": Since the microstates and micro history are everything that there is in the universe, the thermodynamic magnitudes and regularities are *nothing over and above* aspects of the mechanical microstates and micro histories. That is to say: The thermodynamic magnitudes do not only "supervene" on the mechanical ones; they are *them*.[18]

A famous example of an aspect is the Maxwell-Boltzmann energy distribution. To know whether or not this aspect obtains in the particular sample of diluted gas in front of us, we carry out indirect measurements of *other* aspects, e.g., to see whether the gas is uniformly spread in the container and its temperature is distributed uniformly in it, or whether when we increase the pressure on it, its temperature increases in direct proportion, in accordance with the ideal gas law. And so, the main idea of statistical mechanics, which makes it so important and useful, is that in order to provide a successful and informative account of thermodynamic phenomena one needs only partial information about the microstate of a system, that is, one needs only to specify its *aspect*.

[17] The notion of aspect is in some sense perspectival. But, since this perspective is derived from the complete microstate by partial description there is no danger of inconsistency between the perspectives, nor of not knowing "perspective of what", especially where theoretical entities are concerned (see Arabatzis, 2018).

[18] Gibbs (1902) talked about "analogies" and only later – and not always – his quantities came to be understood in terms of identity statements. In Boltzmann's thought (see Uffink, 2017) this is less clear, and Boltzmann and Maxwell came to identify certain macrovariables with certain thermodynamic quantities due to the similarity in the regularities that they exhibit (see overviews in Frigg, 2008; Uffink, 2007; Sklar, 1993). The Gibbsian picture of statistical mechanics differs from the Boltzmannian one in important details (see Callender, 1999; Frigg, 2008; Goldstein et al., 2020; Uffink, 2007, 2017; Wallace, 2015), and therefore the predictions of these two approaches agree only approximately (see Jaynes, 1965; Werndl & Frigg, 2017). However in both traditions thermodynamic magnitudes are associated with macrovariables or functions of macrovariables (see more on the connection between these pictures in (Hemmo & Shenker, 2012, Chap. 11).

The notion of aspect has two characteristics, and their interplay is the core of statistical mechanics and the basis of expanding it to a general theory of reductive physicalism.

(i) The aspect (macrovariable) of the actual microstate is what we observe. When we measure the volume of a gas, for example, what we interact with is the actual particular distribution of the positions (or the actual sequences of particular positions) of the actual particular sample of gas in front of us. There isn't anything else in the world for us to observe. (We don't measure sets of counterfactual states; see (ii) below.)
(ii) The aspect (macrovariable) of the actual microstate describes an equivalence set of microstates, called "a macrostate". When we measure the volume of a gas, for example, and find that it is (say) V, this partial information is all that we have about this sample of gas, and as far as we are concerned it could have been in any number of microstates, provided that they share this aspect, this coarse- grained partial description of the molecules' positions. Thus, this ignorance concerning the other details of the gas' microstate (everything except V) gives rise to an equivalence set of microstates, namely, those that share the aspect V and differ in all other aspects. The set is an epistemological notion. Ontologically, one microstate in the set is actual at the time of measurement, and all the others are counterfactual; some will become actual in other times, most will not.

The set of counterfactual microstates that share an aspect is sometimes called *macrostate*, and this is how we shall use this term here. The *aspect* (*macrovariable*) of an actual individual microstate (or micro history) is "out there in the world", as the entire actual microstate is; but the *macrostate* – as a set of *counterfactual* microstates – is not "out there in the world"; it is only a theoretical construct. *Microstates and aspects (macrovariables) are in the world; macrostates are not.*

Of course, macrostates are extremely important theoretical constructs, and serve various epistemic purposes: they enable us to talk about *entropy* (i.e. the degree to which the energy of a system can be exploited by manipulating the system via external constraints) and *probability* (i.e. predicting the future values of the aspects (macrovariables) of the system's microstates). We discuss probability in Sect. 2.3. We discuss the notion of entropy only very briefly in Sect. 2.4 (see Hemmo & Shenker, 2012 Ch. 7). (In the literature the terms macrostates and macrovariables are sometimes used interchangeably,[19] or other terms are used. See various examples in Ehrenfest & Ehrenfest, 1912; Sklar, 1993; Lebowitz, 1993; Goldstein & Lebowitz, 2004; Frigg, 2008; Uffink, 2007).

[19] In (Hemmo & Shenker, 2016) the terminology is a bit different: *macrovariable* refers to any aspect of a microstate, and *macrostate* refers to those that appear in our experience.

2.3 Probability in Flat Physicalism vs. The Mentaculus

Since Flat Physicalism is committed to the idea that the microstate and micro-trajectory of the universe are everything that there is in the universe, if there are probabilistic facts in the universe then those have to be part of this fundamental ontology, namely of CF + NM, for there is nothing else that they can be. In this section we show how probabilities come about in a deterministic world in Flat Physicalism. This reductive approach is by contrast to the non-reductive Mentaculus proposed by Albert and Loewer.

Albert and Loewer's Statistical Postulate (SP) posits a probability distribution over the past low entropy macrostate of the universe: the postulate is that this probability distribution is uniform relative to the Lebesgue measure. According to Albert, as we saw above,

> "the right probability distribution to use for making inferences about the past and the future is the one that's uniform, on the standard (i.e., Lebesgue) measure, over those regions of phase space which are compatible with whatever other information – either in the form of laws or in the form of contingent empirical facts – we happen to have." (Albert, 2000, p. 96)

Since SP forms an independent part of The Mentaculus, it is taken to be a fact (or a law; we don't go into this question here) over and above CF + NM. Since SP is defined over *counterfactual* microstates (or micro-trajectories), it denotes a fact (or a law) in addition to the *actual* trajectory of the universe, which is fixed by CF + NM. The justification for adding SP to The Mentaculus is that together with it (but not without it) The Mentaculus becomes the "best system" (Loewer, 2001; Albert, 2014, pp. 23–24). Albert explains this move by an "audience with God" who upon our request for the "best system" of our world adds SP over and above NM:

> "I have something more to tell you as well ... Something about the initial condition of the world ... it was one of those which is typical with respect to a certain particular probability distribution – the Boltzmann–Gibbs distribution, for example. The best I can do by way of a simple and informative description of that initial condition is to tell you that it was precisely the sort of condition that you would expect, that ... you would have been rational to bet on, if the initial condition of the world had in fact been selected by means of a genuinely dynamically chancy procedure where the probability of this or that particular condition's being selected is precisely the one given in the probability distribution of Boltzmann and Gibbs." (Albert, 2014, pp. 23–24)

God in this parable just informs us, top-down, that *if* we assume this particular probability measure then the *actual* matters of fact, the actual observed regularities, will turn out to be *likely*, so that it would be rational to expect them (perhaps in the framework of Lewis's concept of probability, see Lewis, 1980b; Hoefer, 2019). Albert takes this to justify endorsing SP as part of The Mentaculus. This kind of justification for choosing a probability measure, and for explaining the actual matters of fact, is assumed by supporters of the so-called "typicality" approach (although for some of them the measure is not one of probability; see Dürr, 2001). The typicality approach is the subject of an ongoing debate (see Callender, 2007; Dürr, 2001; Dürr et al., 1992; Goldstein, 2012; Maudlin, 2007a). Some think that the probability measure can be derived from theorems of mechanics or by dynamical considerations: were this the case it would provide a way to derive the SP from CF + NM, but we have shown that those considerations are irrelevant for determining the measure in question. This means that the typicality approach is non-reductive and introduces the measure as an additional fact, not derivable from the mechanical principles and the contingent facts. (On why the typicality approach is viciously circular, see Hemmo & Shenker, 2015b.)

In the "best system" approach, within which Loewer and Albert work, adding a postulate such as SP to the laws of nature is justified if it makes the system of laws the "best" balance between informativeness and simplicity with respect to the Humean base, even if it comes about via divine revelation. (See Sect. 3 for more on the relation between The Mentaculus and the Humean base in a Lewis-style approach.)

Can the endorsement of SP be justified on the basis of CF + NM, without a decree from God? Arguably, such a possibility would lead to a simpler and "better" system. We now turn to showing how this can be done. More generally, we will show how the probability in classical statistical mechanics can be derived from NM + CF, so that there is no need for an independent postulate like SP. This means that probability in statistical mechanics is a feature of the actual (deterministic) trajectory of the universe, and in this sense it is an objective feature of the world despite the underlying determinism. We describe the steps of this derivation below. Our point here is metaphysical, *not* epistemic: one should think about this derivation as if it is carried out by Laplace's Demon in which case (as we show below) SP follows deductively from NM + CF.

Consider Fig. 1 which illustrates the state space of some system. The microstates in this state space are partitioned into sets (M_0, M_1, M_2) that are macrostates: in each set all the microstates share some aspect (macrovariable) of interest. Other details in the figure are explained later.

Suppose we prepare a system G with a thermodynamic property associated with the macrovariable M_0. This means that the initial *actual* microstate $x(t_0)$ of G – whose macrovariable M_0 we *actually* observe and prepare – belongs to

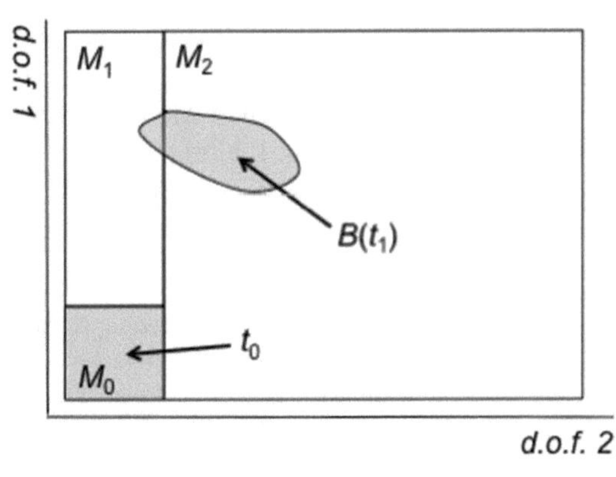

Fig. 1 Microstates, macrostates and probability in statistical mechanics

the macrostate-set M_0 in Fig. 1, in which all the microstates share the same macrovariable M_0; all of them except $x(t_0)$ are counterfactual and do not obtain in reality at t_0. We now follow the evolution of each of the continuous infinity of microstates in M_0, tracking its trajectory from its starting point within M_0 at t_0 until time t_1, and call the set of all the end points $B(t_1)$.[20] The Lebesgue measure of $B(t_1)$ is the same as that of M_0, satisfying Liouville's theorem.

Needless to say: for us the calculation of $B(t_1)$ in this way is possible only under *huge idealizations*. In practice, however, *there are shortcuts* for some special interesting cases (see discussion of these shortcuts in Hemmo & Shenker, 2012, Sects. 6.5 and 6.6). Despite these shortcuts, discussing the idealization is important in order to understand how probability emerges in the theory and thus helps to see what exactly is involved in those shortcuts. We shall see below additional shortcuts, and those need to be understood as such; without understanding the idealization it is hard to see that our actual practices are shortcuts and those may be given an exaggerated ontological status. This is why we emphasize them here and below.

In general, region $B(t_1)$ partly overlaps with more than one macrostate, and in our example it partly overlaps with M_1 and partly with M_2. In this sense, the following if-statement is an objective fact about the system G: *if* it is prepared in a microstate with the mechanical macrovariable M_0, *then* its final microstate will be one with either macrovariable M_1 or M_2. And the transition (or conditional) probability that a system prepared in M_0 at t_0 will end up in M_1 at t_1 is determined by the relative measure of the overlap of $B(t_1)$ with M_1; and similarly for M_2: That is, the transition probability (or conditional probability) is given by the following probability rule:

$$\text{Probability Rule}: P\left(M_1, t_1 \mid M_0, t_0\right) = m\left(B\left(t_1\right) \& M_1\right) \tag{1}$$

Here "$P(M_1, t_1 \mid M_0, t_0)$" is the probability that the microstate of the system at t_1 will turn out to be in M_1, conditional on its being in M_0 at t_0; "$B(t_1)$ & M_1" denotes the *intersection* of the phase space regions $B(t_1)$ and M_1; and $m(x)$ is the normalized measure of the region x in the phase space, given some appropriate measure m; we discuss below how to determine the measure.[21] Obviously, the empirical adequacy of this rule should be tested against the relative frequencies we observe.

How is m(x) fixed? According to Flat Physicalism, since the micro-history is *everything that there is*, the task is to derive m(x) from it. Briefly and ideally this is the way to do it.

Step 1: given some particular M_0 of interest, calculate the continuous sequence of regions like $B(t_1)$ in Fig. 1, along the entire infinite continuous trajectories. (Also here, for us, this calculation is feasible only under another huge idealization, for

[20] The region $B(t_1)$ is a sort of Poincare section: Normally Poincare sections are used to describe semi-periodic systems, but the principle is the same.

[21] If the states are confined to, for example, an energy hypersurface, on which all regions have measure zero relative to the entire phase space, then the above rule is modified such that the measures of sets are relative to this hypersurface.

which there are useful shortcuts, as well.) Step 1 is based on CF + NM (but not on SP or PH). We assume here that CF is true (see Sect. 1). Also we stress that we (unlike Albert and Loewer) don't seek to derive the truth of CF (with high probability) from anything, since we don't think that this is possible (see below).

Step 2: follow the *actual* micro-trajectory as it passes through these B(t) regions, and count (by calculation) the *actual* relative frequencies of the different Mi along the infinite continuous micro-trajectory. Again the resources here are only CF + NM. (For us, of course, this involves another huge idealization. The *shortcut* that we use *in practice* is to measure the M_i values in actual experiments on similar systems in the actual short history through which we live, and to generalize the observed relative frequencies by induction, hoping that it works; the usual problem of induction applies.) But for Laplace's Demon, there is no question of induction, and the entire calculation up to here is strictly deductive based on CF + NM alone.

Step 3: choose a measure over the B(t), and thus in practice over the entire accessible region, that will be compatible with these relative frequencies. Step 3 is a derivation from CF + NM only. In general there will be more than one such measure; any of them will do; choose any of them, either arbitrarily or the most convenient one. Since the measure is chosen such that it fits the observed relative frequencies, it cannot be taken to account for the high probability of those same relative frequencies; this is why, as we said, one cannot come up without vicious circularity with an empirical theory which gives high probability for the truth of CF. This point stands at the heart of the question concerning the origin and nature of the probability measure (or the measure of typicality) in statistical mechanics.

Notice that the probability measure obtained in this approach may turn out to be time-dependent. Indeed, even SP is a time dependent probability distribution, since distributing probability uniformly in a time-independent way yields empirically inadequate results, which is a known problem for Boltzmann's approach; see (Hemmo & Shenker, 2015c).

Step 4: apply the above calculations by tracing backwards the *history* of the universe until the putative "big bang" moment, and then apply the chosen measure. The result will be SP only if it is true. In this way SP is not a postulate at all, but a theorem that follows deductively from CF + NM.

By adding SP to The Mentaculus Albert and Loewer *assume* that Steps 1–3 will yield the particular probability distribution that it mentions. In Flat Physicalism, by contrast, there is no guarantee that the probability distribution described in SP is true. Flat Physicalism also brings out the fact that the measure is not a fact out there in the world, but a mere convenient tool for calculations. In The Mentaculus, SP and with it the measure it mentions are given the status of an *additional* law of nature, *irreducible* to the rest (Loewer, 2001, 2008); in Flat Physicalism, this law is derived from CF + NM. (See more details, including shortcuts to overcome the difficulties of the idealizations involved in the Flat Physicalist account of probability, in Hemmo & Shenker, 2012, 2019b and Shenker, 2017b, c.)

Notice that since our description of how the chances come about is reductive and is in fact a generalization of our experience of relative frequencies of certain macrovariables along the actual trajectory of the universe, the probability law (1) itself is not fundamental, but a derived law.

In the Flat Physicalist concept of probability described above the probability measure is fixed by the harmony between two elements: (I) The dynamics, (II) The partition of the state space into macrostates. Let us comment on these two.

(i) **The dynamics**. On the Flat Physicalist concept of probability, a different dynamics, or different initial conditions, will yield different probabilities. The probability rule is general, in this respect, and should it be discovered that the dynamics of the universe is different from what we thought, the probability rule will be able to accommodate this fact.
(ii) **The partition**. On the Flat Physicalist concept of probability, a different partition of the state space into macrostates yields different probabilities. It is useful to see the significance of this point by comparing Flat Physicalism with The Mentaculus. Albert and Loewer's SP pertains to a certain macrostate, the "big bang" one. But, the *complete ontology* of the big bang is given by a particular (unknown) actual microstate, which is supposed to be a member of the "big bang" macrostate set. The big bang *actual micro*state – as any microstate – has (infinitely) many aspects, (infinitely) many macrovariables, and each of them gives rise to a different partition of the state space, and thus to a different big bang *macro*state. Which of them is the one that features in PH and SP? In Albert and Loewer's approach a certain partition of the state space into macrostates has a preferred status, and it is not clear where exactly this preferred status comes from, given that all the partitions of the states space (all the macrovariables) into macrostates are on a par. There seems to be something arbitrary or a-priori that comes from outside of physics that fixes this preferred partition. Flat Physicalism, by contrast, accepts that according to NM all the partitions to macrostates are on a par, and the probability rule of Flat Physicalism gives rise to a host of probability statements, each pertaining to a different partition of the state space. The only way to explain why it is that *we*, human beings, experience certain macrovariables rather than others (e.g., the thermodynamic macrovariables), pertains to contingents facts about our physiology, the way that we interact with our environment and facts about the physics of the mind (see Hemmo & Shenker, 2016, on "*Ludwig's problem*"). However, those are not additional facts: they are all part of CF + NM. There is nothing else in the universe, according to Flat Physicalism. (This is why reductive physicalism about the mind is an essential element of statistical mechanics; see Shenker, 2017a; Hemmo & Shenker, 2021a, b, c)

What is the fate of the Second Law in Flat Physicalism? Suppose that the probability rule (1) yields high transition probabilities for entropy increase as required by the Second Law. This would mean that along the *actual* trajectory of the universe the relative frequencies of macrostates conform to the Second Law probabilities. But one cannot derive from this result anything about the behaviour of other counterfactual trajectories. In general, the probability rule is compatible with Second Law behavior along one trajectory as well as with Maxwellian Demons along other trajectories.[22]

As we said above concerning other elements of The Mentaculus, if one accepts the Lewis-style "best system" approach, then the only criterion for whether or not one needs to introduce an additional postulate to one's laws of nature, e.g. adding SP to CF + NM, and the way that one should understand the notion of probability in a deterministic world, is if it contributes to the system being "better" (see Hoefer, 2019 on Humean chance). We think that since Flat Physicalism derives the notion of probability from CF + NM, then it is "better" (the other option is that there is an external source of support outside or in addition to the micro-trajectory; see Sect. 1; and this will make both the theory and the Humean base much more complex).

2.4 *The Past Hypothesis in Flat Physicalism*

In Sect. 2.3 we saw how probability comes about in Flat Physicalism as a consequence of a certain harmony between the dynamics and the partition of the state space into macrostates, which is a consequence of CF + NM. Assuming CF + NM, focus now on a certain partition, e.g., some partition into thermodynamic macrostates. And suppose that we can prove that (either generally or for certain conditions) a mechanical counterpart of the Second Law of thermodynamics will continue to obtain. Then, as is well known, it is a theorem that a time-reversed "Second Law" obtains as well, and to avoid this consequence Feynman (1965) and Albert (2000) postulate the PH. On this view the PH is a postulate, it is not derived from CF + NM. However, Flat Physicalism seeks to derive everything from CF + NM. How is the PH derived?

Here we employ three steps.

First, the temporal directionality. It is a point of logic that asymmetric consequences are not derivable from symmetric assumptions. Hence to break the symmetry between past and future a symmetry breaking assumption needs to be introduced. There is no way out of this. The only question is whether there is a trick that will enable it without adding anything to CF + NM. We think there is; our solution is that the psychological arrow of time, namely what is felt in our mind to be a temporal asymmetry, is *reduced* to a non-temporal asymmetry of the structure of our brain, which may be some local *in-brain* entropy gradient, or some other *non-temporal* magnitude, which is part of CF (see Hemmo & Shenker 2019a). By contrast, Albert (2000, 2014) and Loewer (2001, 2012, 2020a) argue that the symmetry-breaking fact is the PH.

[22] See (Leff & Rex, 2003) for an overview and various ideas about Maxwell's Demon. For why Maxwell's Demon is compatible with classical statistical mechanics, see Albert (2000, Chap. 5); for a different argument in the classical case, see (Hemmo & Shenker, 2010, 2012, 2016; in quantum mechanics, see (Hemmo & Shenker, 2020).

The second step, once we have a temporal asymmetry, is to retrodict the past. Predictions are carried out using probability statements as illustrated in Fig. 1 in Sect. 2.3, and they contain ignorance concerning the macrostates (or more precisely macrovariables) that will actually obtain when the time comes. Retrodictions, by contrast, involve some ignorance as well as some knowledge concerning the macrostates (or more precisely macrovariables) that actually obtained in certain moments in the past. (Here we need a concept of records, that we don't discuss here; our concept is different from Albert's, 2000.) Conceptually, retrodictions are measurements: we retrodict the past possible macrovariable in a manner similar to the prediction of Fig. 1, and then "measure", by recollecting, which of the macrovariables actually obtained (for details see Hemmo & Shenker, 2012, Ch. 10). Measurements, in this framework, end by the measurer having a certain brain state in which she believes that a certain state of affairs obtained, and therefore this notion is perfectly reducible to CF + NM.

Notice that on this view Flat Physicalism explains retrodictions in a way that does not guarantee that the PH obtains! Postulating the PH may solve what Albert (2000) has called the "skeptical catastrophe", according to which The Mentaculus *minus* the PH implies that our memories of past events, including the past events that we take to support NM and CF in the first place, are not true (or unreliable). But – as desirable as this result may be – it is not guaranteed (not even probabilistically) by CF + NM. Flat Physicalism accommodates the fact that CF + NM are compatible with violations of the Second Law as well as with the possibility that the PH is false. The "skeptical catastrophe" has to be dealt with by NM and CF alone; We assume as a working hypothesis or a starting point that CF is true, but Flat Physicalism is open to the possibility that its contents is false (see Sect. 1).

Flat Physicalism also accommodates the fact that there are infinitely many different partitions of the state space into macrostates, all of which are on a par, while the thermodynamic partition is just one of them. The PH in The Mentaculus concerns the "low-entropy highly condensed big-bang sort of macrocondition it is that the normal inferential procedures of cosmology will eventually present to us." (Albert, 2000, p. 96). That is, it elevates the thermodynamic partition to the status of a law, despite the fact that given NM all partitions of the state space have equal status. As we said, the only way to explain the preference of the thermodynamic partition pertains to contingent facts about our physiology, the way that we interact with our environment and facts about the physics of the mind (see Hemmo & Shenker, 2016, on "*Ludwig's problem*"). Although in Flat Physicalism these facts are all part of CF + NM, they do not give the PH a preferred status, let alone the status of a law of nature.

Finally, the PH is given in terms of "entropy". What is entropy, in Flat Physicalism? Again, we will not go here into the full details, and just make the following remarks concerning what entropy is and how it is to be expressed in terms of CF + NM.

(1) While the meaning or reference of thermodynamic entropy is not always clear (see e.g. Uffink, 2001) one of the ways it is often understood is as quantifying

the degree to which energy can be exploited. In our view this idea should be translated, or rather transformed, into the notion of controlling the system's microstate. Intuitively, the smaller the macrostate-set corresponding to the macrovariables we can manipulate is, the closer we are to zooming in on the actual microstate and controlling it. This is the intuition behind associating the measure of macrostates with their entropy (in a Boltzmannina framework; for the connection to the Gibbsian approach see Hemmo & Shenker 2012 Ch. 11).

(2) But how should the measure be chosen? We saw in Sect. 2.3 that this is a subtle and problematic issue that invites fallacies or petitio principii. To avoid these dangers we note the following.

(3) In thermodynamics, entropy quantifies the degree of exploitablity of energy *only if* the Second Law is true (see Fermi, 1936, and Hemmo & Shenker 2012 Ch. 2).

(4) Entropy and probability are two distinct concepts: the measure of entropy should quantify exploitability, and the measure of probability should provide correct predictions of future macrostates. Therefore there is no reason at all – elegance notwithstanding – that the measure in the two cases should be the same. We propose (in Hemmo & Shenker 2012, Ch. 7) to choose the measure of entropy such that the phenomena of the Second Law of thermodynamics will be recovered; we do that in view of comment (3) above.

(5) Finally, notice that entropy is a number without units, a hint that it is not out there in the world: there isn't, and there cannot be, an entropy measuring device. Entropy is the measure of a set of counterfactual states of affairs. The only way to measure entropy is to measure some macrovariable, e.g., the degree of uniformity of temperature in a given sample of gas, and then to define the macrostate set of counterfactual microstates that share this macrovariable, and then introduce the measure as in remark (4) above, and *calculate* the entropy. Entropy is not a fact "out there in the world", but a very useful theoretical construct. The PH inherits this nature of entropy.

2.5 Special Sciences: A Reductive Account

Sometimes it seems that the special sciences, such as biology or geology, describe facts about the world that are over and above those described by physics. The commitment of Flat Physicalism is that this is not the case. If the microstates and micro histories are everything that there is, then the magnitudes and regularities described by the special sciences must be features of the microstates and micro histories; for there is nothing else they can be. *Reductive type-identity physicalism* is the view that the magnitudes and regularities of the special sciences are nothing but aspects (macrovariables) and the regularities that they may exhibit along the micro trajectory, while the complete microstate evolves according to the equation of motion. In order to come up with the best system of the laws of a certain special science (e.g. biology, economics), one needs to follow the same kind of derivation we presented above with respect to the laws and facts of thermodynamics (which is one example of a special science, after all) from CF + NM. This means that according to Flat Physicalism the best system of laws for the entire world which includes the laws of *all* the special sciences is NM.

Our example of a special sciences is thermodynamics. In our view the science of thermodynamics, despite the fact that it is studied by physicists, has the status of a "special science": the very project called "statistical mechanics" is the attempt to account for the thermodynamic magnitudes and regularities in term of those of fundamental physics (be it Flat Physicalism or The Mentaculus), and this is precisely the way that the (other) special sciences are treated in Flat Physicalism. In Flat Physicalism, then, the microstate and the micro history are *everything that there is* in the universe, and therefore the magnitudes of thermodynamics (such as volume or temperature) *are* (*identical to*) aspects of the microstates, and the regularities exhibited by the thermodynamic magnitudes *are* (*identical to*) the ways these aspects evolve along the micro-history of the universe and its subsystems. According to Flat Physicalism this is the way – *the only way* – to account for the other special sciences as well, for example biology or geology, because the microstate and the micro history are *everything that there is*. (We return later to the subtle question of the status of thermodynamics in The Mentaculus.)

The picture that Flat Physicalism proposes is flat: there are no levels of reality. So-called "high-level" kinds (such as temperature) are understood here as aspects, or macrovariables, given by partial descriptions of the microstructure of a single level of entities and their interactions, as described by the fundamental theories of physics. This is the essence of Flat Physicalism.

One challenge for Flat Physicalism in this context is the following. It is a prevalent view in contemporary philosophy of science that certain special sciences kinds are *multiply realized* (or at least *multiply realizable*) by physical kinds. Allowing for multiple-realizability is more or less a standard view in contemporary thinking, especially in functionalism (since Putnam, 1975; Fodor, 1974; Davidson, 1970; and of course Lewis, 1966, 1980a). When taken together with supervenience of the special sciences kinds and properties on physical ones, this view is called "*non-reductive physicalism*". Whether or not experience supports this claim is an open question (see e.g., Polger & Shapiro, 2016); We shall not argue for nor against the factual existence of multiple realizability. Our task, rather, is to show how Flat Physicalism can account for the appearance of multiple realizability, should it turn out to be an empirically verified fact. Here is an interesting example. The macrovariables associated with the thermodynamic magnitude called "temperature" in an ideal gas (in which there are no interactions between the molecules) is different from the one in a van der Waals gas (in which there are the simplest interactions between the molecules), and the two cases agree quantitatively only in certain limiting conditions. "Spin temperature" is yet another aspect of the (quantum!) microstate, and there is supposed to be quantitative agreement (in relevant scale) between it and the other macrovariables associated with the notion of "temperature". Frigg and Hoefer (2015) see this as a case of multiple realization. They write:

> "[M]any non-fundamental properties are multiply realisable, meaning that the same non-fundamental property can be realised by a number of distinct fundamental kinds. This is true not only in psychology (the classical example being pain); we find multiply realised properties even in close-to-fundamental physics: temperature in gases and temperature in spin systems have *completely different* micro-realisers." (Frigg & Hoefer, 2015; our italics.)

For Frigg and Hoefer there are high-level facts that are not in the microstate or micro history: for example, the fact that "temperature in gases" and "temperature in spin systems" are both cases of "temperature" cannot be explained by the micro structure of the world, because at the micro level they are "completely different" – they do not share any aspect (any macrovariable) that could explain this fact. By "*genuine multiple realization*" we mean a case in which the fact that the different "low-level" cases appear to be multiple realizations of the same "high-level" fact cannot be explained by a hidden or unnoticed "low-level" fact that they share. According to Frigg and Hoefer (if we understand them correctly) the case of temperature is one of genuine multiple realization.

Flat Physicalism, insisting that there is nothing over and above the microstate and micro history, has to offer a different account of the case of temperature. Two options are open for the Flat Physicalist in this case.

Option 1: First impression notwithstanding and even contemporary scientific knowledge notwithstanding, there is an *aspect* of the microstate that is shared between "temperature in gases" and "temperature in spin systems", and this shared aspect is (identical with) "temperature".

Option 2: When this is not the case, that is, when the microstates of the "temperature in gases" and "temperature in spin systems" are "completely different" or *heterogeneous*. In this case Flat Physicalism makes the following observation. When *we* say that "temperature in gases" and "temperature in spin systems" are both cases of "temperature" what happens is precisely this, namely, that upon interacting with either "temperature in gases" or "temperature in spin systems", *we* enter a brain state that says "this is a case of temperature". The fact that in both heterogeneous cases (of "temperature in gases" and "temperature in spin systems") our brain enters the same state is not a result of us measuring the quantity of "temperature" in those objects, since – by assumption – they do not share anything the measuring of which will yield the same outcome. We just happen to enter the brain state saying "this is temperature" for no reason at all – that is, for no reason about the "temperature in gases" and "temperature in spin systems". It is a case of pure coincidence. We give the same name, if you will, to two distinct things, in a way similar to giving the same name to two different people: giving them the same name does not indicate that they share any property.

However, even in this case there is a shared physical fact between the cases of "temperature in gases" and "temperature in spin systems", and it is *our* brain state that says "temperature". In other words, if we *extend* the microstate, to include – in addition to the gas or the spins system – also the observer, then the microstates share an aspect, namely, the state of the brain of the observer. Consider an alien observing a gas and a spin system: if the microstate and micro history are all that there is, then the alien will not see them as sharing anything, because they don't share anything.

This is why Option 2 is a case of *apparent but not genuine* multiple realization: Flat Physicalism opens our eyes to the fact that the *seeming* of multiple realization (if indeed it so seems) is a result of ignoring our role as physical observers that partake in the extended microstate.

Of course, this latter option is highly counter-intuitive, for it means that the observer does not *discover* which things belong to the special science kind, but rather *creates* that kind. The things out there do not share anything to be discovered (since the physics is all that there is, and the things out there do not share anything physical) and the only thing that makes them a kind is the shared brain state of the observer, which just happens to be the same, by pure coincidence and for no reason at all – again, in a similar way that different people happen to have the same name. But we must allow for this option, which may explain some contingencies of human history.

Is there another option? Supporters of *genuine* multiple realization claim that neither Option 1 nor Option 2 obtain, and nevertheless aliens will sense that "temperature in gases" and "temperature in spin systems" are of the same kind, "temperature". How can this case be explained? Flat Physicalism will then *have* to admit that the physical facts are not everything that there is; there are non-physical facts as well, and those are the facts that the alien can sense, and in virtue of which it will discover empirically that "temperature in gases" and "temperature in spin systems" are of the same kind, "temperature". This is *token dualism. Flat Physicalism will, in this case, fail and turn out to be false.* And so, the claim of our Flat Physicalism approach is that the name "non-reductive physicalism" is a misnomer: an approach that allows for genuine multiple realization is not physicalism at all, but is an approach of *token dualism.*[23]

Another way to see this is the following. If, as Flat Physicalism conjectures, the facts of fundamental physics are everything that there is, then by looking at the token microstate or micro history one is (ideally) able to know everything. In particular, one is (in principle) able to know, given the microstate, to which high-level kinds a system belongs. By contrast, if multiple realizability obtains, this is not possible, since the very fact that certain heterogeneous physical kinds (but not others) all under the same special-science kind, is not derivable from the fundamental physical facts. This partition needs to be given top-down, that is, from some source that is not part of the physics of the system or of the world. Importantly, *even if supervenience obtains*, unless one receives information "from above" concerning *which* physical

[23] By the way, we propose here that Putnam's famous Twin Earth thought experiment, in which two substances, XYZ and H_2O, bring about in the observer the same experience, should be analyzed along these lines, and similarly Kripke's (1980) "theoretical identifications" such as "water is H_2O" (see Hoefer & Marti, 2019). Importantly, water is not a mere "Lego"-like composition of particles (see Chang, 2012, 2015): the interactions responsible for the existence of each molecule and for their combination to form water are described by the theories of fundamental physics, and form part and parcel of the aspects of the fundamental physical state of affairs that we call "water"; there is no residue that calls for explanation in terms of other metaphysical relations like grounding, determination, realization, etc.

kinds are subsumed under *which* high-level kind, this information is not, and *cannot* be, available by looking at the system of interest. This is clear *token dualism.*

For these reasons, Flat Physicalism is the conjecture that the hypothesis of multiple realizability is false. The special sciences kinds and regularities are those of fundamental physics: this is a reductive type-identity physicalist view.

2.6 Mental Kinds

We said that Flat Physicalism can account for all the cases of *apparent, non-genuine* multiple realization in the special sciences such as biology or geology, by appealing to the observer and extending the microstate to include it (Option 2 above). But this line of explanation does not apply for the science of psychology. The reason is this. (And the reason has *nothing* to do with the so-called "hard problem of consciousness" and notions like subjectivity, qualia, etc. The problem is totally *different*, and is the following.)

It is often said in contemporary literature that psychological kinds (e.g. pain) are realized by different creatures, e.g.: by different human brains, and *ipso facto* by non-human brains, a famous example being Putnam's (1975) octopus pain. Let us assume here, for the sake of the argument and without arguing for or against it, that indeed octopuses and human beings share something about their mentality (not only about their behaviour!), and that the shared thing is that they are both "in pain". A point that we find important is this: admittedly there are different kinds of pain; we know this to be the case in humans, and *ipso facto* it is the case with respect to non-human animals; still, the assumption here is that all these different cases share the fact of being "in pain" and not something else. How can this (assumed) fact be explained?

Option 1 (of Sect. 2.5 above) is type-type identity reductionism, according to which all the cases of pain *are (identical with)* shared aspects (macrovariables) of the microstate of the creature that feels pain, be it a human being or an octopus or an alien. The shared aspect is *exactly* the same, it is reproduced *precisely* in all the cases in which there is pain (as Putnam, 1975 realized from the start). This is the Flat Physicalist account of the mental. Friends of multiple realization of the mental by the physical acknowledge that Option 1 is coherent, but think it is unreasonable (famous examples are Putnam, 1975 and Fodor, 1974).

Friends of multiple realizability, i.e. of non-reductive physicalism, reject Option 1, insisting that whatever human pain and octopus pain share is not an aspect of their microstate, that is, these cases are *genuinely physically heterogeneous*. They need to opt for either Option 2 or Option 3 (of Sect. 2.5 above).

According to **Option 2** it should be possible to account for this case by introducing an observer that, when any of the heterogeneous cases takes place, is in the brain state in which it "believes that this creature is in pain". But this option raises two problems.

(1) It starts an infinite regress, since the question re-arises concerning the belief of the observer.
(2) It can't work since human beings (and perhaps Octopi) *feel* – directly and intrinsically – that they are in pain, so that this pain must be about *them*, rather than about the *observer*.[24]

So Option 2 is not open to friends of multiple realizability of mental kinds by physical kinds.

We are left with **Option 3**: since the creatures that are said to share the feeling of pain do not share a physical fact, they share a *non*-physical fact, a non-physical feature that obtains in *each and every token* of "being in pain". This is *token dualism.* "Non-reductive physicalism" is a misnomer: an approach that allows for genuine multiple realization is not physicalism at all. This is token dualism in disguise.[25]

3 The Metaphysical Framework: The Mentaculus and the Humean Base

Having presented The Mentaculus (in Sect. 1) and Flat Physicalism (in Sect. 2) we now return to examine The Mentaculus in some more detail, comparing it with Flat Physicalism. Albert and Loewer work in the framework of a version of a Lewis-style metaphysics concerning laws and chance (see e.g., Albert, 2014 pp. 23–24, Loewer, 2020a, and especially Loewer's recent "package deal" version in his Loewer, 2020b). This framework guides the construction of The Mentaculus and explains some of its features. In this section we explore this metaphysical framework.

David Lewis is often seen as one of the most important reductionist philosophers (see for example Weatherson 2021). But it is not clear whether he endorsed metaphysical reductionism in a straightforward way. Hall (2020) writes:

> "What he in fact recommends is a holistic approach: we start with the total body of claims we are inclined to believe – whether on the basis of "common sense" (an oft-invoked category, for Lewis) or of science – and try our best to systematize it in accordance with standards of theoretical goodness that are themselves endorsed by common sense and/or

---

[24] Some thinkers deny that first person reports are good evidence concerning the mental realm. We set aside this issue; see Shenker, 2020b for references and discussion.

[25] By the way, we propose here that Putnam's famous Twin Earth thought experiment, in which two substances, XYZ and H_2O, bring about in the observer the same experience, should be analyzed along these lines, and similarly Kripke's (1980) "theoretical identifications" such as "water is H_2O" (see Hoefer & Marti, 2019). Importantly, water is not a mere "Lego"-like composition of particles (see Chang, 2012, 2015): the interactions responsible for the existence of each molecule and for their combination to form water are described by the theories of fundamental physics, and form part and parcel of the aspects of the fundamental physical state of affairs that we call "water"; there is no residue that calls for explanation in terms of other metaphysical relations like grounding, determination, realization, etc.

> science (and so are themselves, to some extent, also up for grabs). ... Here is an especially succinct description of this approach: "One comes to philosophy already endowed with a stock of opinions. It is not the business of philosophy either to undermine or to justify these preexisting opinions, to any great extent, but only to try to discover ways of expanding them into an orderly system." (Lewis 1973, p. 88)... Still, while Lewis's method of philosophical inquiry is certainly not "bottom-up", in my opinion it is best to present the results of that inquiry in a bottom-up fashion." (Hall, 2020).

And this is the line we shall pursue. To the extent that such a bottom-up description recovers Lewis's main results we would see it as (for all practical purposes) reductionist. It seems to us that Loewer's (2020b) PDA is in line with the holistic spirit that Hall ascribes to Lewis. We shall interpret Loewer's approach and examine its reductive nature along the same lines.

Generally speaking, Lewis's metaphysics can be described as consisting of four main elements, which we describe by bringing four quotations from Lewis (marked (i), (ii), (iii) in Sect. 3 and (iv) in Sect. 4).

> (i) "Humean supervenience is named in honor of the great denier of necessary connections. It is the doctrine that all there is to the world is a vast mosaic of local matters of particular fact, just one little thing and then another. (But it is no part of the thesis that these local matters are mental.) We have geometry: a system of external relations of spatiotemporal distance between points. Maybe points of spacetime itself, maybe point-sized bits of matter or aether or fields, maybe both. And at those points we have local qualities: perfectly natural intrinsic properties which need nothing bigger than a point at which to be instantiated. For short: we have an arrangement of qualities. And that is all." (Lewis, 1986, pp. ix–x)

Lewis's "vast mosaic", which is "all there is to the world", is often called the "*Humean base*". Loewer's (2020b) PDA introduces substantial amendments to Lewis's conception of the Humean base: both to what the Humean base *is*, and to how we can *know* it. In Lewis's approach the properties of the elements of the Humean base are categorical, and their distribution among the elements is contingent. This contingency is what makes it "Humean". Loewer writes:

> "Most contemporary Humeans ... agree with Hume that fundamental necessary connections between distinct entities are metaphysically and epistemologically mysterious and not required by an account of the metaphysics of laws found in science." (Loewer, 2020b)

In Lewis's Humean base nothing is necessary about the distribution of the qualities; they just happen to be the way that they are. Several questions arise: What determines the elements of the Humean base? What are they? And how can we know them? We examine Lewis's view on this and Loewer's (2020b) PDA.

First question. In Lewis's account, the qualities of the elements of the Humean base are categorical: brute, unexplained and unexplainable facts. In a sense, nothing "determines" them; they are what they are, and that's the end of it. In particular, they are not defined by any laws governing them: the fact that there are no such laws is, as we said, a key feature of Lewis's Humean base. It is precisely this nature of these qualities that provides Lewis with the possibility of making their distribution completely contingent. But categorical qualities ("quiddities") are considered by many, Loewer (2020b) included, as metaphysically problematic (see e.g. Maudlin, 2007b). To avoid them and provide some justification for the postulation of certain qualities rather than others, Loewer (2020b) opts for properties that are characterized under laws. He writes:

> "The PDA is a descendant and I believe improvement over Lewis' [best system approach]. Like Lewis' Humean [best system approach] it rejects governing laws and law determining powers. But unlike Lewis' [best system approach] its account follows more closely the aims of physics and the criteria physicists appeal to in order to achieve them than a priori metaphysics. By rejecting Lewis' reliance on metaphysically given perfectly natural properties while maintaining the basic idea of his [best system approach] the PDA results in an account of laws and fundamental properties that is friendlier to contemporary physics." (Loewer, 2020b)

Loewer acknowledges that this introduces into the Humean base an element of necessity – the very thing that Lewis tried to avoid. But Loewer thinks that this kind of necessity is compatible with the main aims, gist, and principles of a Lewis-style metaphysics. The reason might be that the properties and the laws are parts of the same "package deal", in which the parts impose constraints on one another, but as a whole the package is contingent.

Next questions: What are Lewis's categorical qualities? And how can we know them? For Loewer (2020b), contemporary physics provides a key for discovering the nature of the Humean base, in line with Lewis's following words.

> (ii) "It is a task of physics to provide an inventory of all the fundamental properties and relations that occur in the world...We have no a priori guarantee of it. but we may reasonably think that present-day physics goes a long way toward a complete and correct inventory... And we may reasonably hope that future physics can finish the job in the same distinctive style. We may think, for instance, that mass and charge are among the fundamental properties... we may provisionally accept that all fundamental properties and relations that actually occur are physical. This is the thesis of materialism." (Lewis, 1994, p. 412–13)

Notice that for Lewis "the thesis of *materialism*" is subtly different from what is often taken to be the thesis of *physicalism* (or of *materialism* standardly understood). According to physicalism, the world is fundamentally physical, and there is nothing over and above whatever physics describes. According to Lewis, by contrast, the world is fundamentally the Humean base, and physics – which is (part of) the "best system" (a notion discussed in a moment) – is a theory that provides some information about it, that enables useful predictions about the phenomena. We write that physics is "*part of*" the "best system" (and not the "best system" *simpliciter*) because on the approach of Lewis as well as of Loewer and Albert the best system may include additional elements, as we discuss below and in the next section.

Why is physics the key to discovering the properties of the Humean base elements, in Lewis's approach? The answer is that physics is (part of) the "best system" with respect to the Humean base; Lewis writes:

> (iii) "Laws of nature, whatever else they may be, are at least exceptionless regularities. Not all regularities are laws, of course. But, following the lead of... Ramsey, I suggest that the laws are the ones that buy into those systems of truths that achieve an unexcelled combination of simplicity and strength. That serves the Humean cause." (Lewis, 1986, p. xi)

Albert and Loewer endorse the general Lewisian framework for understanding laws of nature as "best systems". Albert explains this idea thus:

> "You get to have an audience with God. And God promises to tell you whatever you'd like to know. And you ask Him to tell you about the world. And He begins to recite the facts: ... you explain to God that you're actually a bit pressed for time, ... And you ask if maybe there's something ... that will serve you well, or reasonably well, or as well as possible, in making your way about in the world. And what it is to be a law, and all it is to be a law, on this picture of Hume's and Lewis's and Loewer's, is to be an element of the best possible response to precisely this request – to be a member (that is) of that set of true propositions about the world which, alone among all of the sets of true propositions about the world that can be put together, best combines simplicity and informativeness."[26] (Albert, 2014, pp. 23–24)

As Loewer (2020b) emphasizes, in order to quantify the degree of simplicity and informativeness in describing the distribution of the properties among the elements of the Humean base, we need to know *which* those properties *are*. For Lewis, those are "perfectly natural" categorical properties; but which are they?

The "best system" is a theory that bears a particular kind of relation to the Humean base: it *is* (or *contains*) the set of true propositions *about* the Humean base, and this means that if the best system is physics (i.e. assuming that physics contains true statements about the Humean base) then the Humean base has to be *amenable* to such a partial description by physics, and this provides some constraints, and thereby some information, about the nature of the Humean base. As we saw in quotations (i) and (ii) from Lewis tries to make his Humean base amenable to a description by classical physics. But why constrain the Humean base to fit classical physics which is not our best physical theory? Lewis would say that classical physics may give wrong predictions, but (a) some of its fundamental ideas (such as the idea of point-like particles) may be still true; and (b) there are good reasons to think that quantum mechanics in its present form is fundamentally flawed, so we can't take lessons from it about the base, either. Loewer disagrees: for all the reasons for accepting contemporary physics as the best available theory, he wants a Humean base that will be amenable to a description by our best contemporary theories. Therefore he proposes that the properties of the elements of the Humean base should be those that appear in the laws of contemporary physics (but perhaps not exclusively; see Sect. 4 below on the special sciences).

Damerest expresses the following worry concerning Loewer's approach:

> "For the Lewisian Humean, the actual world is made up of a spacetime with perfectly natural properties distributed throughout, and it is this distribution that we systematize with our logical law-statements. However, on Loewer's account, there are no perfectly natural properties—or if there are, they may well be irrelevant to scientific theorizing. So, there is the question of what the world is made up of, fundamentally. What is meant to stand in for the 'Humean mosaic' that other Humeans are happy to posit? What, exactly, do our best logical law-statements – formulated in terms of useful predicates – actually systematize?" (Demarest 2019 p.393)

---

[26] Here we take these latter properties to have objective meanings, so that there is a matter of fact concerning their degree in each "set of true propositions" (see Weatherson 2021).

Loewer's response to Demarest is this:

> "The PDA accomplishes this trick [by requiring] that the optimal package enables accounting for the macroscopic in terms of the microscopic. While macroscopic entities and properties are not fundamental, one of the main aims of physics is to account for macroscopic truths in terms of fundamental properties and laws. This requirement helps nail down the optimal." (Loewer, 2020b)

(We are not sure what is meant here: perhaps Loewer alludes to an "account" by multiple realization of the properties and laws of the special sciences (sciences that are "not fundamental", perhaps those are the "macroscopic") by the properties and laws of physics (which is "fundamental", perhaps this is the "microscopic"). We explained the notions of "microscopic" and of microstates and macrostates in physics in Sect. 2 above; in Sect. 4 we discuss the special sciences.)

Lewis's idea (in quotations (i) and (ii) above) was that in order for physics to be able to say something significant about the categorical qualities of the elements in the Humean base, those qualities must appear (at least partly) in physics, and hence from physics we can learn something about the Humean base. Those are the statements of physics that are the "set of true propositions about the world" (Albert, 2014, p. 24).

Which "physics" are we talking about? Suppose, with Loewer and Albert, that physics is The Mentaculus:

> "My proposal is that The Mentaculus (with the correct dynamical laws) is the best system for our world." (Loewer, 2020a).

(We may assume that here the NM and CF are *placeholders* for contemporary physics, for example: they might be some suitable version of quantum mechanics; and PH and SP are *placeholders* for suitable auxiliary hypotheses.) In order to understand the relations between The Mentaculus and the Humean base it is useful to consider to what extent they overlap.

Are there elements of The Mentaculus that are not in the Humean base?

Loewer (1996) says that there are:

> "[Humean Supervenience] doesn't entail physicalism since it is compatible with there being Humean properties that are not physical. Physicalism doesn't entail [Humean Supervenience] since there is no guarantee that the fundamental properties posited by physics are intrinsic properties of spatio-temporal locations. In fact, it seems pretty clear that contemporary physics does dream of non-Humean properties." (Loewer, 1996, p. 103–4)

Here is an example of an element of The Mentaculus that is not in the *Lewisian* (not PDA!) Humean base. In NM – which is part of The Mentaculus – there are velocities (and other time derivatives) in the microstate, and since these entail that the time slices are not independent of each other, as Albert (2000), Ch. 1) emphasizes, the instantaneous velocities, which play an important role in NM, are not elements of the Humean base.

Of course, the fact that there are elements of The Mentaculus that are not part of the Humean base, is by itself not a problem in a Lewisian framework, as long as they are conducive to the overall "best system" balance between simplicity and informativeness *about* the world, that is, *about* the Humean base.

On the other hand, this raises the following problem: *How* do we know that (or whether) velocities, for example, are not in the Humean base? We know that only if we are sure that the elements of the Humean base are independent. But how do we know *that*? *For Lewis* some criterion for this is provided by the motivation for coming up with the whole Humean framework, namely, eschewing the notion of necessity from the fundamental metaphysical picture. (We address Loewer's PDA in a moment.) But while this Lewisian motivation may help with respect to deciding whether or not velocities are in the Humean base, it is less helpful in other cases. Here is an example. Consider an "upgraded" version of The Mentaculus in which NM is a placeholder for some version of quantum mechanics in which the quantum state is interpreted realistically. This requires a non-trivial *alteration* of Lewis's description of the Humean base. Loewer writes:

> "The lesson for a defender of [Humean supervenience] to take from quantum mechanics is to count a property as Humean in a world iff it is an intrinsic quality of points in the fundamental space of that world. If Bohm's theory (or any other version of QM that construes the wave function realistically) is correct then that space is configuration space. Given this account of Humean properties quantum non-locality poses no threat to [Humean supervenience]." (Loewer, 1996, p. 104)

But this lesson is controversial. For example, Dürr et al. (1992) argue that the wavefunction in Bohmian Mechanics is not a physical state of anything, but rather a law defining the motion of particles, since it is defined over the high-dimensional 3 N-configuration space, which they take to be a state space (of all possible Bohmian positions), not the real physical space. Albert (2014) criticizes this view arguing that the wavefunction in the 3 N space is (in John Bell's words) "a physically real field, as real here as Maxwell's fields were for Maxwell." (Albert, 2014, p. 126, quoting Bell 1982). This idea, however, violates features of the space of elementary events proposed by Lewis.

So, there are elements in The Mentaculus that are not in the Humean base: in order to provide the "best" system that balances informativeness and simplicity, The Mentaculus includes a partial description of the Humean base as well as additional elements. But Does a "best system" *have* to be like that? Couldn't there be a "best system" that consists only of a partial description of the Humean base with no additional elements? As the Humean base becomes closer to physics (and vice versa) the whole metaphysical framework becomes closer to physicalism, as for example in Flat Physicalism. Will the in-principle difference between these metaphysical frameworks cease to be interesting at some point? We leave this question open.

Are there elements of the Humean base that are not in physics?

Trivially there are elements of the Humean base that are not in physics, since if physics is to be a "best system" that *balances* simplicity and informativeness, it will in general not provide the full information about the Humean base (as is made clear in Albert's parable above).

But more generally, are there elements of the Humean base that *cannot* (in principle) be accounted for by physics? To answer this question it is important to distinguish between two approaches to the term "physics":

(1) "Physics" is the *entire* Mentaculus, where CF + NM may be placeholders for contemporary physics, and PH and SP may be placeholders for suitable *auxiliary* hypotheses. PH and SP are included in The Mentaculus since it is assumed that they *cannot* be deduced from CF + NM. So here, the corresponding elements of the Humean base can be deduced from "physics" *qua* The Mentaculus, but not from "physics" *qua* CF + NM..
(2) "Physics" is *only* CF + NM, where again those are placeholders for contemporary physics, and PH and SP (or the statements for which they are placeholders, if they are true,) are derived from CF + NM. Flat Physicalism is a theory of this structure (as we saw in Sects. 2.3 and 2.4).

In each of these two cases the question of whether there are elements in the Humean base that are not in physics may be different.

In Lewis-style metaphysics, "physics" is (part of) the "best system", and therefore the question of which of option (1) or (2) is "physics" depends on the question which of them is at least "better", if not the "best". *If* there is a theory of kind (2) which provides the same degree of informativeness with respect to the Humean base as The Mentaculus does, *then* it is "*better*", at least in the sense of "best system", than The Mentaculus which is a theory of kind (1). It seems to us that both Albert and Loewer accept this idea for which theory would be "better". For example, Albert (2000), Ch. 7; Albert, 2014, Ch. 1) argued that his "GRW Mentaculus", in which NM is replaced with the collapse theory by Ghirardi et al. (1986; see also Bell, 1987), is better than the classical Mentaculus (or any other quantum theory), since given a certain hypothesis he introduced about the GRW dynamics, the Second Law probabilities are reduced to the Born probabilities, without assuming any probability distribution over quantum states. The GRW Mentaculus is better presumably because it gets closer to a kind (2) theory.

Flat Physicalism is a theory of kind (2): as we saw in Sect. 2 it assumes only CF + NM (seen as placeholders for our best fundamental theory of physics) and it accounts for PH, for SP, for the notion of objective chance in a deterministic universe, and for the special sciences including psychology, all on the basis of CF + NM. Relative to Flat Physicalism, The Mentaculus is not the "best system".

4 The Mentaculus and the Special Sciences

In the previous section we addressed the relation of "best system" between The Mentaculus and the Humean base. This is not a reductive relation, in any of the directions, and moreover we expressed the view that The Mentaculus is not the "best system" of the Humean base, and that there is a "better" one: Flat Physicalism. In this section we set this relation aside and turn to another question concerning The Mentaculus: suppose that we take The Mentaculus to be our "physics" (type (1) theory in the terms just discussed in Sect. 3). Given this assumption, according to the metaphysical framework endorsed by either Lewis or Albert and Loewer, do all the phenomena and facts in the universe, and in particular those described by the special sciences, reduce to – or are otherwise accounted by – The Mentaculus? *In this sense: Is everything "physical"?*

Albert writes on this matter:

> "... the special sciences must all, in some principled sense, be deducible from the fundamental laws of physics." (Albert, 2014, p. 10)

Let us clarify **(a)** what these "fundamental laws of physics" are, and then **(b)** what those "special sciences" are, and are they indeed "deducible" from the former in the account of Loewer and Albert.

(a) The fundamental laws are, for Albert and Loewer, The Mentaculus (i.e. all of its elements, not only CF + NM):

> "If anything along the lines of the picture we are trying to imagine here should turn out to be true, then any correct special-scientific explanation whatsoever can in principle be uncovered, can in principle be descried, in the fundamental physical theory of the world, by the following procedure. Start with The Mentaculus. Conditionalize The Mentaculus on whatever particular features of the world play a role in the special-scientific explanation in question – conditionalize The Mentaculus (that is) on whatever particular features of the world appear either explicitly or implicitly among the explanantia of the special-scientific explanation in question. And check to see whether or not the resultant probability distribution – the *conditionalized* probability distribution, makes the explanandum *likely.* If it does, then we have recovered the special-scientific explanation from the fundamental physical theory – and if it *doesn't,* then either the fundamental theory, or the special-scientific explanation, or both, are *wrong*." (Albert, 2014, p. 17)

(By the way, Albert endorses here the idea that showing that a certain fact is likely explains why this fact obtains. *If* (as we believe) the SP of The Mentaculus involves assuming a probability measure just because it makes the actual facts likely, *then* the SP cannot explain why those actual facts obtain. This point is crucial with respect to some of the typicality arguments, see Hemmo & Shenker, 2012, 2015b; we do not expand on this issue here.)

(b) What does The Mentaculus explain? What are the special sciences and how are they explained by The Mentaculus? It seems to us (we are not sure) that on this matter, in particular with respect to multiple realization, Loewer's and Albert's views may differ, and therefore we address their views separately.

4.1 Loewer's View

Loewer says that The Mentaculus is "arguably a complete scientific theory of the universe" (Loewer, 2020a p.6), but then he adds:

> "The Mentaculus provides an account not only of the laws of thermodynamics, but also an important *ingredient in* an account of *other* special science laws." (Loewer, 2020a p. 12; our italics)

Importantly, for Loewer The Mentaculus is sufficient to account for the laws of thermodynamics, but *not* for "*other* special sciences": there, it is only an *ingredient* in the account. Why is that so?

We suspect that the reason for this is that Loewer accepts that the special sciences are *not* deducible from The Mentaculus, because on his view special sciences properties are multiply realizable by physical properties. In earlier writings Loewer (2008) endorsed multiple realizability *explicitly*. He expressed the "*physicalism*" element in "non-reductive physicalism" thus:

> "Question: "Why is there anything except physics?" Answer: "Because there is physics!"" (Loewer, 2008, p. 162)

And expressed the "*non-reductive*" element in "non-reductive physicalism" thus:

> "It is true that the account of special sciences I have described ... isn't reductionist in some other ways. It doesn't entail that special science properties are identical to properties of fundamental physics and it allows for the multiple realizability, temporal asymmetry and so on of special science laws." (Loewer, 2008, p. 162)

In this Loewer followed the footsteps of Lewis:

> (iv) "My reductionism about the mind begins as part of an a priori reductionism about everything... the very same fundamental properties and relations, governed by the very same laws occur in the living and the dead parts of the world, and in the sentient and the insentient parts, and in the clever and the stupid parts. ... When we describe mental state M as the occupant of the M-role,... [i]t says nothing about what sort of state it is that occupies the role. It might be a non-physical or a physical state, and if it is physical it might be a state of neural activity in the brain, or a pattern of currents and charges on a silicon chip, or the jangling of an enormous assemblage of beer cans." (Lewis, 1994, p. 412–3, 418; see also Lewis, 1966, 1980a)

In Sects. 2.5 and 2.6 above, when we discussed the Flat Physicalist account of the special sciences, we showed that allowing for multiple realizability amounts to endorsing token dualism, that is, endorsing the view according to which *not* everything is physical, and in the present context, that not everything is deducible from The Mentaculus. *To the extent* that Loewer accepts the multiple realizability thesis, he endorses a dualist metaphysics, in which indeed The Mentaculus is at best an *ingredient* in accounting for the special sciences. While Lewis acknowledges that both supervenience and multiple-realizability are not a priori requirements, he takes them to hold as a matter of fact in our world, and in this sense *Lewis endorses a dualist metaphysics*, whatever his "physics" is and whatever his account of the Humean base is.

Notice that the "non-physical" facts, that must be assumed as part of the multiple realizability thesis, need not be supernatural facts; they are merely non-physical ones, that is – in our context – *not in The Mentaculus*. Whether the "best system" laws, that optimally balance informativeness and simplicity with respect to the Humean base, turn out to describe a world that is *unified* in being reducible to "physics", or turn out to be some *patchwork*, is a question of fact. Presumably, however, if a theory according to which everything is "physical", and a theory according to which some things are "physical" and some are not, provide comparable informativeness, we conjecture that the wholly physical one will be "better", in the sense of "best system". In this sense it seems to us that Flat Physicalism is "better" than The Mentaculus *á la* Loewer, due to its account of the special sciences.

Notice that for Loewer, in the above quotation, The Mentaculus does provide a full account of thermodynamics, unlike the "other" special sciences (he takes thermodynamics to be a special science, albeit with a special status). The reason might be that all the facts that are needed to account for thermodynamics, have been inserted by hand into The Mentaculus: this is why it includes PH and SP in addition to NM and CF. We are not sure why thermodynamics merits this special treatment: Why insert into The Mentaculus only the facts that explain the special science of thermodynamics, and not the facts that explain the other special sciences? In Flat Physicalism, at any rate, all the special sciences are on a par and are given the same analysis and the same explanation.

Why does Loewer require *supervenience* of the special sciences kinds on the physical kinds, in his "best system"? In the literature it is often said that supervenience expresses the primacy of physics, that many accept (e.g. Kim, 2012; Davidson 1993), but in the "best system" approach this criterion is secondary, or even irrelevant: *the sole criterion* for being a system of laws of nature is whether or not the system is "best" in providing the optimal balance between informativeness and simplicity. Whether requiring supervenience while allowing for multiple realization is "best" in this sense we are not sure. Our Flat Physicalism, at any rate, requires supervenience as part of its reductive physicalist principle, and because of the primacy it gives to physics, and its commitment to there being nothing over and above the physical, it rejects the multiple realizability thesis.

In sum, the situation for Loewer seems to be this. CF + NM cannot account for any of the special sciences. To account for thermodynamics, you need to add the extra facts (auxiliary hypotheses) of PH and SP. Those are declared to be part of The Mentaculus. But because of *multiple realization* to account for other special sciences one needs *additional* extra facts, and those are not made part of The Mentaculus. We are not sure why this distinction between thermodynamics and the other special sciences is made in Loewer's approach. In Flat Physicalism (see Sects. 2.3, 2.4, 2.5 and 2.6) all the special sciences are on a par, and all are derivable from CF + NM.

4.2 Albert's View

Albert, unlike Loewer, thinks that "something's funny" about the idea of multiple realizability. He brings two arguments by way of examples; let's analyze them briefly, by comparison with Flat Physicalism.

As we saw (in Sects. 2.5 and 2.6) Flat Physicalism identifies special sciences properties with aspects (macrovariables) of microstates, stressing that at each moment everything that exists, and therefore everything we observe, is the *actual particular* microstate (or the actual particular sequence of microstates), and with it of course its aspects. This is the ontology. Epistemologically, if everything that we know about the actual microstate is one of its aspects, and we remain ignorant about the rest of the aspects, then we cannot distinguish between the actual microstate and a continuous infinity of counterfactual microstates that have the same aspect, and the set of these indistinguishable microstates, all but one of which are counterfactual, is the macrostate. This set is a theoretical construct, and is not "out there in the world".

By contrast, Albert (2014) identifies features of the world that are described by the special sciences, with *sets* of microstates. He writes:

> "Those explanantia [i.e. the features of the world] are initially going to be given to us in the language of one or another of the special sciences. And so, in order to carry out the sort of conditionalization [i.e. the account in terms of The Mentaculus] we have in mind here, we are going to need to know which of those special-scientific explanantia correspond to which *regions* of the space of possible exact physical microconditions of the world. And those correspondences can be worked out – not perfectly (mind you), but to any degree of accuracy and reliability we like – by means of the super-duper computational techniques alluded to in Section (i)." (Albert, 2014 p. 17 footnote 4; our emphasis)

(The "super-duper computation" is akin to Laplace's Demon.) Note that Albert's identification of special science magnitudes with sets of microstates (that occupy state space regions) is unclear and somewhat ambiguous. Supporters of multiple realization agree that by following the equation of motion for each initial state the super-duper computer can work out which "special science explanantia" correspond to which phase space regions of possible microstates. But what they reject is that the super-duper computer will find a *single macrovariable* (not disjunctive!) that is shared by the microstates in all those regions corresponding to the same special science kind (that figures in the "explanantia"). Obviously, if there is such a type-type-mapping between macrovariables and special sciences kinds, the super-duper computer will find it, and then the multiple-realization thesis will break down. With this point in mind, let us now proceed to examine Albert's argument against the multiple realizability thesis.

To illustrate his objection to the thesis of multiple realizability Albert describes the following example concerning the "science of epidemiology":

> "there are ... physical systems in the world ... which are capable of distinguishing, in a more or less reliable way, under more or less normal circumstances, between those possible fundamental physical situations of the universe in which there is (say) a flu going around, and those in which there isn't. And so there must be a fully explicit and fully mechanical technique for coordinating epidemiological situations with their fundamental physical equivalents – or (at any rate) for doing so in a more or less reliable way, under more or less normal circumstances – because there are (after all) mechanical devices around, right now, that can actually, literally, get it done." (Albert, 2014, p. 10)

To understand this argument we need to recall that when discussing the Flat Physicalist account of the special sciences we distinguished between three options. Applied to the flu example, **Option 1** says that several cases are all of the kind "flu" because there is an *aspect* shared by the microstates of the observed systems "out there", and it is this aspect that is (identical with) flu. The epidemiologist interacts with this aspect in each and every one of the cases, and thereby comes to have the brain state in which she believes that it is a case of flu. In Option 1 there is no multiple realization; it is the option of reductive type-identity physicalism.

According to **Option 2**, by contrast, the microstates "out there" do not share any physical aspect; they are physically heterogeneous, and nevertheless there is a system called "the epidemiologist" that comes to have the brain state in which she believes that it is a case of flu. The brain state of the epidemiologist does not measure the systems "out there" since those do not share anything that will make them a kind. What makes them a kind is only the shared brain state of the epidemiologist: If we *extend* the microstates to include not only the systems "out there" but also the brain of the epidemiologist, then the extended systems' microstates *do* have a physical aspect in common, namely, the brain state of the epidemiologist. Here, we stressed, the epidemiologist does not *discover* the special science kind of flu, but *creates* this kind. Option 2, when we consider the extended microstate, is also not one of multiple realization, and is also explained as a case of reductive type-identity physicalism. Cases of Option 2 may *appear* to be ones of multiple realization, if we ignore the epidemiologist and consider only the systems "out there". This is *apparent but non-genuine* multiple realization.

Option 3 is the case of *genuine* multiple realization, and this – we argued – is a case of token dualism.

In these terms, Albert's above example illustrates cases compatible with reductive type-identity physicalism, in which special sciences kinds are *not* multiply realized by physical kinds, and so this example does not give rise to an argument about (or against) multiple realization. It seems that Albert mainly talks about Option 1, in which the epidemiologist detects the physical aspect that is shared by all the tokens in which there is a flu – this is clearly a case of reductive type identity physicalism. He does not point out that in Option 2 there might be an apparent but non-genuine multiple realization, and does not address Option 3 of genuine multiple realization. The argument we outlined against Option 3 (in Sects. 2.5 and 2.6) shows that genuine multiple realization amounts to token dualism, and is therefore not a form of physicalism.

Another argument that Albert brings against the thesis of multiple realizability concerns the law of conservation of energy: here "multiple realization" would mean that different quantities, in worlds with even radically different physical laws than our own, satisfy the same kind of regularity, described and explained by Noether's theorem. Albert writes:

> "We know that the conservation of energy is a law of the actual world – and that is manifestly a substantive and interesting and altogether nontrivial claim. But outside of that, all we seem to know is that the principle is a law in all and only those worlds whose fundamental laws share this particular feature (that is, the feature of entailing the conservation of energy) with the actual one. All we seem to know (to put it slightly differently) is that the principle is a law in just those worlds in which it is a law – which is not to know anything, at least of an empirical kind, at all." (Albert, 2014 p. 14–15)

Notice that this kind of multiple realizability is not the one envisaged by Fodor (1974, 1997), and addressed recently by for example Polger and Shapiro (2016), for they – as most of the contemporary philosophers of science and of cognitive science – seem to have in mind multiple realization in our world. (Putnam, 1975 may have other-worldly cases in mind as well). Here Albert says that the counterfactual cases he imagines share the feature of "satisfying this law", but we don't know why they do; we don't know which features of them make it the case that they do. However, if, in the end, we shall discover (theoretically, of course) which feature those counterfactual worlds share, in virtue of which they all satisfy the law of conservation of energy, then this will be a case of **Option 1** above, in the following sense. This shared feature would be described in terms of some "super-physics", in which all these worlds can be described. (Lacking such a super-physics language we are not sure how to make sense of this example.) Option 1 here would mean that the different worlds share a super-physical aspect.

But it seems that this is not what Albert has in mind, since he writes:

> "[T]he conservation of energy can obviously be *realized*, the conservation of energy can obviously be *underwritten*, by any number of distinct sets of fundamental laws of physics—laws which will in many cases be radically different, in any number of other respects, from our own." (Albert, 2014 p. 14–15)

The other options are the following. If, in the end, we shall discover (theoretically, of course) that those counterfactual worlds do not share anything super-physical, then their shared satisfaction of this law (if it is describable at all!) would be a mere coincidence, and in fact they will not be exhibiting the same law at all: they will be satisfying different laws, with some parallel *structural* features. This might be understood as **Option 1** as well, where the shared feature is the structural one, or it can be understood as **Option 2**, where the shared feature is the theoretician's belief (i.e. brain state). **Option 3** would mean that these counterfactual worlds do not share anything physical, or super-physical, or something in the theoretician's brain. This would mean that we have super-physical token dualism.

In sum, Albert has the intuition that the thesis of *multiple realization* is unacceptable, but his discussion does not offer an argument against this thesis. We offered such an argument in Sects. 2.5 and 2.6.

5 Conclusion

We have argued that Flat Physicalism is a better system of our world in the Lewis-style sense of the term than The Mentaculus, because it is more reductive and more physicalist.

In The Mentaculus, the PH + SP are not derived from the microscopic dynamics of our world given by CF + NM, but are *added* as independent laws about counterfactual worlds. In addition, The Mentaculus is not physicalist, *if* multiple realizability of any of the special sciences by heterogeneous facts is accepted. This holds also with respect to the relation between thermodynamics and fundamental physics (see the example of temperature in Sect. 2.5). As we mentioned, Lewis endorsed multiple realizability as part of his functionalism about the mind. This means that Lewis's view is not physicalism at all, but (as we showed) *token dualism.* We are not sure what are the current positions of Albert and Loewer on this matter.

By contrast, Flat Physicalism is fully reductive and physicalist. It accommodates the fact that given CF + NM, it is possible that PH + SP are false, but if they are true, we have shown how to account for the facts they describe on the basis of CF + NM alone. Likewise, Flat Physicalism accommodates the fact that the Second Law of thermodynamics is *not* a general theorem of mechanics. But: if it is true of our world (i.e. our micro-trajectory), as seems to be suggested by our experience, then this fact can be accounted for on the basis of CF + NM alone. Similarly, the account of all the other special sciences on the basis of CF + NM is fully reductive by Flat Physicalism which is a strict type-type reductive *identity* theory. All the special sciences kinds are strictly identical with physical kinds (aspects of the micro-trajectory) and there is no multiple realization whatsoever. We have shown (see Sects. 2.5 and 2.6; Hemmo & Shenker, 2015a, 2019b, 2021a, b, c) that supervenience of higher-level kinds on physical kinds is not enough, since it is compatible with multiple realization, which in turn *entails* dualism. The laws of the special sciences (if true) are derivable from CF + NM via the probability rule (1) in the same way that the Second Law is.

Let us reiterate that within the Humean "best system" approach token dualism is not to be rejected as such: the only criterion for endorsing or rejecting dualism is whether the laws of nature turn out to be the "best system" in providing the optimal balance of informativeness and simplicity with respect to the Humean base. Still, if The Mentaculus turns out to be a tokan dualism view, it is of interest to know that this is the case.

Finally, the general question of whether the "best system" of our world is fully reductive in the sense that it consists only of a partial description of the Humean base with no additional elements, and in particular whether this partial description is purely physicalist, is ultimately a question of fact. As the Humean base becomes closer to physics (and vice versa), the whole metaphysical framework becomes closer to physicalism, as for example in Flat Physicalism. Will the in-principle difference between these metaphysical frameworks cease to be interesting at some point? We leave this question open.

Acknowledgements We are grateful to Barry Loewer and David Albert for valuable comments in clarifying with us the main points of disagreement. The result of this correspondence as we understand it is summarized in Sect. 1. This research was supported by the *Israel Science Foundation*, grant numbers 1148/2018 and 690/2021.

References

Albert, D. (2000). *Time and chance*. Harvard University Press.

Albert, D. (2014). *After physics*. Harvard University Press.

Arabatzis, T. (2018). Perspectival realism about what? In M. Massimi & C. D. McCoy (Eds.), *Understanding perspectivism*. Routledge. Forthcoming.

Arntzenius, F. (2000). Are there really instantaneous velocities? *The Monist, 83*(2), 187–208.
Bell, J. S. (1982). On the impossible pilot wave, *Foundations of physics, 12*, 989–999. Reprinted in Bell, J. S. 1987. *Speakable and unspeakable in quantum mechanics* (pp. 159–168). Cambridge University Press, .
Callender, C. (1999). Reducing thermodynamics to statistical mechanics: The case of entropy. *Journal of Philosophy, XCVI*, 348–373.
Callender, C. (2007). The emergence and interpretation of probability in Bohmian mechanics. *Studies in the History and Philosophy of Modern Physics, 38*, 351–370.
Chang, H. (2012). *Is water H2O?* Springer.
Chang, H. (2015). "Prospects for an integrated history and philosophy of composition", text of presentation at CLMPS, Helsinki.
Davidson, D. (1970). Mental Events. In D. Davidson (Ed.), *Essays on actions and events* (pp. 207–227). University of California Press.
Davidson, D. (1993). Thinking causes. In J. Heil & A. R. Mele (Eds.), *Mental causation* (pp. 3–17). Clarendon Press/Oxford University Press.
Demarest, H. (2019). Mentaculus laws and metaphysics. *PRINCIPIA* 23(3): 387–399.
Dürr, D. (2001). Bohmian mechanics. In J. Bricmont, D. Dürr, M. C. Galavotti, G. Ghirardi, F. Petruccione, & N. Zanghi (Eds.), *Chance in physics: Foundations and perspectives* (Lecture notes in physics) (pp. 115–132). Springer.
Dürr, D., Goldstein, S., & Zanghi, N. (1992). Quantum equilibrium and the origin of absolute uncertainty. *Journal of Statistical Physics, 67*(5/6), 843–907.
Ehrenfest, P., & Ehrenfest, T. (1912). *The conceptual foundations of the statistical approach in mechanics*. Dover.
Fermi, E. (1936). *Thermodynamics*. Dover. (1956).
Feynman, R. (1965). *The character of physical law*. MIT Press.
Firt, E., Hemmo, M. & Shenker, O. (2020). "Generalizing Hempel's Dilemma", forthcoming.
Fodor, J. (1974). Special sciences: Or the disunity of science as a working hypothesis. *Synthese, 28*, 97–115.
Fodor, J. (1997). Special sciences: Still autonomous after all these years. *Nous, 31*, 149–163.
Frigg, R. (2008). A field guide to recent work on the foundations of statistical mechanics. In D. Rickles (Ed.), *The Ashgate companion to contemporary philosophy of physics* (pp. 99–196). Ashgate.
Frigg, R., & Hoefer, C. (2015). The best Humean system for statistical mechanics. *Erkenntnis, 80*, 551–574.
Ghirardi, G., Alberto R. and Weber, T. (1986). Unified Dynamics for Microscopic and Macroscopic Systems. *Physical Review D*, 34(2): 470–491. https://doi.org/10.1103/PhysRevD.34.470
Gibbs, J. W. (1902). *Elementary principles in statistical mechanics*. Yale University Press.
Goldstein, S. (2012). Typicality and notions of probability in physics. In Y. Ben-Menahem & M. Hemmo (Eds.), *Probability in physics* (The Frontiers collection) (pp. 59–72). Springer.
Goldstein, S., & Lebowitz, J. (2004). On the (Boltzmann) entropy of nonequilibrium systems. *Physica, D, 193*, 53–66.
Goldstein, S., Lebowitz, J., Tumulka, R., & Zanghì, N. (2020). Gibbs and Boltzmann entropy in classical and quantum mechanics. In V. Allori (Ed.), *Statistical mechanics and scientific explanation: Determinism, indeterminism and Laws of nature*. forthcoming. arXiv:1903.11870
Hall, N. (2020). "David Lewis's metaphysics", *The stanford encyclopedia of philosophy* (Spring 2020 Edition), Edward N. Zalta (Ed.), URL https://plato.stanford.edu/archives/spr2020/entries/lewis-metaphysics/
Hemmo, M., & Shenker, O. (2010). Maxwell's demon. *The Journal of Philosophy, 107*, 389–411.
Hemmo, M., & Shenker, O. (2012). *The road to Maxwell's demon: Conceptual foundations of statistical mechanics*. Cambridge University Press.
Hemmo, M., & Shenker, O. (2015a). The emergence of macroscopic regularity. *Mind & Society, 14*(2), 221–244.
Hemmo, M., & Shenker, O. (2015b). Probability and typicality in deterministic physics. *Erkenntnis, 80*, 575–586.

Hemmo, M., & Shenker, O. (2015c). Letter to the editor. *International Studies in the Philosophy of Science, 29*, 91–93. https://doi.org/10.1080/02698595.2015.1079075
Hemmo, M. & Shenker, O. (2016). "Maxwell's Demon", Oxford Handbooks Online. Oxford University Press. DOI: https://doi.org/10.1093/oxfordhb/9780199935314.013.63.
Hemmo, M., & Shenker, O. (2019a). The second law of thermodynamics and the psychological arrow of time. *British Journal for Philosophy of Science*. https://doi.org/10.1093/bjps/axz038
Hemmo, M., & Shenker, O. (2019b). Two kinds of high-level probability. *The Monist, 102*(4), 458–477.
Hemmo, M., & Shenker, O. (2019c). The physics of implementing logic: Landauer's principle and the multiple-computations theorem. *Studies in History and Philosophy of Modern Physics, 68*, 90–105.
Hemmo, M., & Shenker, O. (2020). Maxwell's demon in quantum mechanics. *Entropy, 22*, 269. https://doi.org/10.3390/e22030269
Hemmo, M., & Shenker, O. (2021a). A challenge to the second law of thermodynamics from cognitive science and vice versa. *Synthese*. https://doi.org/10.1007/s11229-020-03008-0
Hemmo, M., & Shenker, O. (2021b). Why functionalism is token-dualism. In G. Vishne, S. Ioannidis, M. Hemmo, & O. Shenker (Eds.), *The multilevel structure of reality*. Springer. forthcoming.
Hemmo, M., & Shenker, O. (2021c). Flat physicalism. *Theoria*. forthcoming.
Hemmo, M. & Shenker, O. (2021d). "Observer-dependent physicalism: A new argument for scientific realism", forthoming In Y. Ben-Menahem & C. Posy (Eds.), in Memory of Mark Steiner, Springer.
Hoefer, C. (2019). *Chance in the world: A Humean guide to objective chance* (Oxford studies in philosophy of science). Oxford University Press.
Hoefer, C., & Marti, G. (2019). Water has a microstructural essence after all. *European Journal for Philosophy of Science, 9*, 12.
Jaynes, E. T. (1965). Gibbs versus Boltzmann entropies. *American Journal of Physics, 33*, 391–398.
Kim, J. (2012). The very idea of token physicalism. In G. Simone & H. Christopher (Eds.), *New perspectives on type-identity* (pp. 167–185). Cambridge University Press.
Kripke, A. S. (1980). *Naming and necessity*. Harvard University Press/Basil Blackwell.
Ladyman, J., & Ross, D. (2007). *Everything must go*. Oxford University Press.
Lebowitz, J. (1993, September). "Boltzmann's entropy and time's arrow", *Physics Today*, pp. 32–38.
Leff, H. & Rex, A. (Eds.) (2003). Maxwell's Demon 2: Entropy, classical and quantum information, computing. Institute of Physics, Bristol.
Lewis, D. (1966). An argument for the identity theory. *Journal of Philosophy, 63*, 17–25.
Lewis, D. (1973). *Counterfactuals*. Cambridge, MA: Harvard University Press.
Lewis, D. (1980a). Mad pain and Martian pain. In N. Block (Ed.), *Readings in the philosophy of psychology, volume I* (pp. 216–222). Harvard University Press. Reprinted in David Lewis, *Philosophical Papers* (1986), Volume I, pp. 122–132, Oxford: Oxford University Press.
Lewis, D. (1980b). A subjectivist's guide to objective chance. In R. C. Jeffrey (Ed.), *Studies in inductive logic and probability, volume II* (pp. 263–293). University of California Press. Reprinted with Postscripts in Lewis, David (1986), *Philosophical Papers*, Volume II, pp. 83–132, Oxford: Oxford University Press.
Lewis, D. (1986). *Philosophical papers, volume II*. Oxford University Press.
Lewis, D. (1994). Reduction of mind. In S. Guttenplan (Ed.), *A companion to the philosophy of mind* (pp. 412–431). Blackwell/Oxford.
Loewer, B. (1996). Humean Supervenience. *Philosophical Topics, 24*(1) Metaphysics, 101–127.
Loewer, B. (2001). Determinism and chance. *Studies in History and Philosophy of Modern Physics, 32*, 609–620.
Loewer, B. (2008). Why there is anything except physics. In J. Hohwy & J. Kallestrup (Eds.), *Being reduced* (pp. 149–163). Oxford University Press.
Loewer, B. (2012). Two accounts of laws and time. *Philosophical Studies, 160*(1), 115–137.

Loewer, B. (2020a). The Mentaculus vision. In V. Allori (Ed.), *Statistical mechanics and scientific explanation: Determinism, indeterminism and Laws of nature*. forthcoming.
Loewer, B. (2020b). "The Package Deal Account of laws and properties (PDA)" forthcoming in *Synthese,* special issue on Humeanism.
Maudlin, T. (2007a). *The metaphysics within physics*. Oxford University Press. 2007.
Maudlin, T. (2007b). What could be objective about probabilities? *Studies in History and Philosophy of Modern Physics, 38*, 275–291.
Polger, T., & Shapiro, L. (2016). *The multiple-realization book*. Oxford University Press.
Putnam, H. (1975). The nature of mental states. In H. Putnam (Ed.), *Mind, language and reality* (pp. 429–440). Cambridge University Press. (1975); Originally published as "Psychological predicates," In: William H. Capitan and Daniel D. Merrill (Eds.), *Art, mind and religion*, pp. 37–48, Pittsburgh, PA: University of Pittsburgh Press (1967).
Shenker, O. (2017a). Flat physicalism: Some consequences. *Iyyun: The Jerusalem Philosophical Quarterly, 66*, 211–225.
Shenker, O. (2017b). Foundations of statistical mechanics: Mechanics by itself. *Philosophy Compass.*https://doi.org/10.1111/phc3.12465
Shenker, O. (2017c). Foundations of statistical mechanics: The auxiliary hypotheses. *Philosophy Compass.*https://doi.org/10.1111/phc3.12464
Shenker, O. (2020a). Foundations of quantum statistical mechanics. In E. Knox & A. Wilson (Eds.), *Routledge companion to the philosophy of physics*. Routledge. Forthcoming.
Shenker, O. (2020b). "Denialism about consciousness", *Iyyun: The Jerusalem Philosophical Quarterly*, forthcoming.
Shenker, O. (2020c). Common sense and scientific realism: Completing the cycle. In J. de Ridder, R. van Woudenberg, & R. Peels (Eds.), *Cambridge companion to common sense philosophy*. Cambridge University Press. Forthcoming, expected.
Sklar, L. (1993). *Physics and chance*. Cambridge University Press.
Uffink, J. (2001). Bluff your way in the second law of thermodynamics. *Studies in History and Philosophy of Modern Physics, 32*, 305–394.
Uffink, J. (2007). Compendium to the foundations of classical statistical physics. In J. Butterfield & J. Earman (Eds.), *Handbook for the philosophy of physics, part B* (pp. 923–1074). Elsevier.
Uffink, J. (2017). Boltzmann's work in statistical physics, *The Stanford Encyclopedia of Philosophy* (Spring 2017 Edition), Edward N. Zalta (Ed.), URL = https://plato.stanford.edu/archives/spr2017/entries/statphys-Boltzmann/
Wallace, D. (2001). Implications of quantum theory in the foundations of statistical mechanics, unpublished manuscript. Preprint available at http://philsci-archive.pitt.edu/410/
Wallace, D. (2015). The quantitative content of statistical mechanics. *Studies in History and Philosophy of Modern Physics, 52*, 285–293.
Werndl, C., & Frigg, R. (2017). Mind the gap: Boltzmannian versus Gibbsian equilibrium. *Philosophy of Science, 82*(5), 1224–1235.
Weatherson, B. (2021). David Lewis: The Stanford Encyclopedia of Philosophy, Edward N. Zalta (ed.), https://plato.stanford.edu/archives/win2021/entries/david-lewis/

How to Make Possibility Safe for Empiricists

John D. Norton

Abstract What is possible, according to the empiricist conception, is what our evidence positively allows; and what is necessary is what it compels. These notions, along with logical possibility, are the only defensible notions of possibility and necessity. In so far as nomic and metaphysical possibilities are defensible, they fall within empirical possibility. These empirical conceptions are incompatible with traditional possible world semantics. Empirically necessary propositions cannot be defined as those true in all possible worlds. There can be empirical possibilities without empirical necessities. The duality of possibility and necessity can be degenerate and can even be falsified.

Keywords Empiricist · Necessity · Possibility

1 Introduction

What are "small-e" empiricist[1] philosophers of science, like me, to make of possibilities? They are central to science. Our best theories of gravity and stellar structure tell us that it is possible for a sufficiently massive star eventually to collapse

My thanks to Yemima Ben-Menahem, Jeremy Butterfield, Patrick Dürr, Carl Hoefer, Edouard Machery, Brian McLoone, James Norton, Matt Parker, Bryan Roberts, Margot Strohminger and Jim Woodward for helpful discussion. They bear no responsibility for the content of this paper other than for failing to talk me out of it.

[1] A "small-e" empiricist holds that we can only learn contingent truths of the world from experiences of the world. A "big-E" Empiricist holds the stronger view that these experiences are all we can learn.

J. D. Norton (✉)
Department of History and Philosophy of Science, University of Pittsburgh, Pittsburgh, PA, USA
e-mail: jdnorton@pitt.edu; https://sites.pitt.edu/~jdnorton/jdnorton.html

Y. Ben-Menahem (ed.), *Rethinking the Concept of Law of Nature*,
Jerusalem Studies in Philosophy and History of Science,
https://doi.org/10.1007/978-3-030-96775-8_5

under its own gravity to form a black hole. These same theories tell us that it is impossible for you or me to escape any of these black holes, once we have passed the event horizon. These claims are truths of our best science, even though almost all of them describe situations that we will never experience and may never be actualized. Since, we can never have direct experience of them, how are empiricists to understand them?

We look for help in answering to the existing philosophical and metaphysical literature on modality. The results are disappointing. From abstract reflections on possibility and necessity, in work extending back to the scholastics, traditional metaphysics calls the extraordinary into existence. There are contingent beings. Their existence is possible, but not necessary. It follows, we are to infer, that some being exists necessarily to bring them about. The existence of this necessary being can in turn be inferred simply from reflection on the concept of perfection itself. The greatest perfection, supposedly, necessitates existence. Whatever notions of possibility and necessity are employed here, they are not amenable to empiricists, who require that contingent truths of the world be derived from experience.

While it is now generally conceded that these traditional metaphysical reflections fail, they retain a place in present philosophy, along with the hope that, somehow, they might be made to work.[2] However, an extraordinary faith in the creative powers of possibility and necessity persists. It has found new life in the talk of "possible worlds" that has recently come to dominate the modality literature. The precise nature and import of these worlds are debated. In the most extreme case, we are assured by David Lewis (1986) that these many possible worlds exist and are as real as our own, even while they may include strange fantasies of the imagination like talking donkeys and magic mountains.

Other avenues now popular are equally puzzling to the empiricist. We are told in an authoritative source[3] that gold must necessarily be the element with atomic number 79 and could not possibly be anything else, such as iron pyrites ("fool's gold"). But these assertions of possibility and necessity merely record our human use of the word "gold." They require no necessities, but merely the empirical discovery of chemists that elements happen to divide by atomic number. Had we decided that "gold" can also designate iron pyrites, our decision would in no way oblige the world to change the possibilities afforded by its chemistry.

To someone with modest empiricist leanings, interested in understanding what is possible in our world, this literature fails to provide an empirically responsible notion of possibility. Indeed, it is both astonishing and alarming for its excesses. It is not a safe place for empiricists.

---

[2] The two arguments are the "cosmological" and the "ontological" arguments for god's existence. Van Inwagen (2009, Ch. 6 and 7) provides a contemporary formulation and discussion. He coincedes (p. 141) that all versions of ontological argument depends on some logical error or unsupported premise.

[3] Kripke (1980), pp. 123–25.

The present paper has two goals. First, it will seek to provide a lean account of possibility that is amenable to an empiricist and avoids the fantasies rife in the present literature. Second it will argue that this sense of possibility is adequate for scientific applications and that proposals to go beyond it cannot be justified. There are just two notions of possibility: logical and empirical possibility. Section 2 below sketches briefly the logical notion. Since the notion of empirical possibility sought is of possibilities in the world, Sect. 3 requires in advance that any definition of it must be free of dependencies on our thoughts, beliefs and accidents of our language. Section 4 specifies the defining properties of empirical possibility. The guiding principle is that what is possible is what our evidence positively allows; and what is necessary is what our evidence compels. Since evidential support comes in varying degrees or varying strengths, empirical possibility inherits corresponding gradations.

Sections 5, 6, 7, and 8 develop further properties of this empirical conception. The present literature flattens this inherently gradational notion into just two conditions, possibility *simpliciter* and necessity *simpliciter*. This flattened representation is poorly adapted to the empirical notion and, as a result, familiar truisms of the present literature fail. There can be cases in which there are empirical possibilities, but no empirical necessities other than logical consequences of the evidence. The familiar duality of possibility and necessity proves fragile. There are natural conceptions of empirical possibility in which some proposition can be empirically necessary and its negation may also be empirically possible. "Necessary" is not always "not-possibly-not." Since empirical necessities need not be true in all possible worlds, as standardly defined, familiar possible world semantics fails always to apply. There seems to be no way to restore it without destroying the simplicity essential to its appeal.

Section 9 begins the defense of the second major claim of this paper: that logical and empirical possibilities are all that can be employed responsibly. It lists the main alternative conceptions, several of which are already subsumed by logical and empirical possibilities. The more complicated cases are addressed in subsequent sections. Section 10 argues that empirical possibility is distinct from epistemic possibility. The latter violates a requirement of Sect. 3: it is defined in terms of beliefs; and it derives possibility from mere ignorance, where empirical possibility requires positive evidence. In Sect. 11, it is argued that nomic possibility and necessity are properly understood as hypothetical or counterfactual versions of empirical possibility and necessity.

Section 12 addresses metaphysical possibility. Much of what is now described as metaphysics is merely the reporting of results of empirical science. Thus the possibilities provided by this empirical metaphysics lie within the compass of empirical possibility. If metaphysics is to supply a notion of possibility that goes beyond it, the notion has to be derived from a non-empirical body of metaphysical knowledge. I recall the familiar empiricist complaint, tracing back to Hume's fork, that contingent facts cannot be known non-empirically. Further, non-empirical metaphysics attempts an impossible task. It seeks to abstract away the particulars of many cases to recover a universal and enduring conception of some metaphysical category. However, since the variety of cases grows as we learn more of the world, the prospects of a single, unifying conception is correspondingly, continually reducing. Non-empirical metaphysics has insufficient resources to deal with this empirically fueled growth, precisely because it must draw its insight from a non-empirical body of knowledge. A concluding Sect. 13 speculates on the origin of the present disarray in the philosophical literature on modality.

While there has been little empiricist reaction to the excesses of the present philosophical literature on modality, a welcome exception is Ismael's (2017) "An Empiricist's Guide to Objective Modality." My approach has close affinities with hers. We both offer an empiricist understanding of modality in terms of inductive inference. The main difference lies in our areas of focus. I am more concerned with broader possibility claims made in the general scientific context on the basis of actual evidence. Ismael's focus is on modalities as they appear in the formulation of scientific theories. She writes (p. 119):

> Scientific models—on the local and global scale—are embodiments of our very best inductive practices. I am suggesting that the modal content of our models—the overlay of laws, dispositions, capacities, and potencies—are to be understood in terms of their role guiding prediction and decision.

This is a welcome perspective which I will use in Sect. 11 below in the analysis of nomic possibility.

2 Logical Possibility

The empiricist account of possibility to be developed here employs just two senses of possibility: logical and empirical. A proposition is logically possible with respect to the propositions of some body of evidence just if a contradiction cannot be deduced from their conjunction. A proposition is logically necessary with respect to some body of evidence just if the propositions of that body of evidence deductively entail the proposition.

Here, deductive entailments derive from the meaning of the terms used in the propositions. The case of connectives is the most familiar. If "*A* and *B*" is true, it follows logically from the meaning of "and" that *A* is true. There is no need to restrict this sense to connectives. It applies more generally. If Atlas is an intact quadruped, it follows from the meaning of "quadruped" and "intact" that Atlas has four legs. This usage leads directly to cases of logical possibility and necessity. It is logically possible that some quadruped is a four-legged dog. However, a legless snake cannot possibly be a quadruped; that is, it is a logical necessity that it is not a quadruped.

This conception of logic loosely follows the Humean tradition that divides propositions into "relations of ideas" and "matters of fact." The first fits with the conception of deductive logic here. Of it, Hume (1777, p. 18) writes: "Propositions of this kind are discoverable by the mere operation of thought, without dependence on what is any where existent in the universe." The latter "matters of fact" will be associated with empirical possibility.

This sense of logical possibility should be distinguished from a narrower sense in which logical possibility is confined to possibilities authorized by some formalized logic. The first difficulty with this conception is its narrowness. Since there is, as far as I know, no formal logic of quadrupeds, the Atlas inference would be excluded. The second problem is that the set of inferences authorized by extant formal logics is not fixed. It changes whenever someone invents a new formal logic. Finally, its content is not univocal. Truth functional logics admit inferences that are precluded by relevance logics. Some formal logics eschew contradictions; paraconsistent logics do not. Most curiously, standard modal logics include the duality "possibly = not-necessarily-not." We shall see that this standard identification can fail in the account of empirical possibility developed below.

3 Requirements for an Empiricist Account of Possibility

What I shall call "empirical possibility," the notion defined in an empiricist account of possibility, must satisfy the following requirements.

Empirical possibility is relational.

Since empirical possibility is to express a contingent fact of the world, an empiricist requires that such facts be supported by experience. (Here and henceforth, by "contingent fact" I mean one that is true, but is not a logical truth; that is, it is not true in virtue of the meaning of the terms employed in it.) Since different experiences would in general support different possibilities, these possibilities cannot be made absolute by detaching them from the evidence that supports them.[4]

Empirical possibility is independent of our thoughts and beliefs.

While empiricists hold that we learn of contingent facts of possibility through experience, the notion of possibility sought is of possibilities in the world, independent of what we think, believe and may or may not know. Consequently, empirical possibility, understood as an objective relation among propositions, must be independent of these thoughts and beliefs. An empiricist does expect such facts of possibility to inform our beliefs, so that empirical possibilities should have a central role in belief formation. However they play that role in a subsequent step in the epistemological analysis and is not engaged here. The present goal is to understand empirical possibilities. They are, like the world, independent of our beliefs and thoughts.

[4] The standard philosopher's understanding of experience as that which we sense directly is wholly inadequate to explicate how evidence from the world is used in science. Boyd (2018) provides a serviceable, updated account.

Empirical possibility is independent of our language.

While we may use language to express empirical possibilities, the contingent facts expressed should not be facts of language or depend essentially on them. For then they are no longer simply facts of the world. Our use of language can require certain facts of the world to obtain. That the element with atomic number 79 has metallic properties follows from the quantum mechanics of its electronic structure. Such a fact is required if the element is to be designated by the word "gold," where gold is presumed to be metallic. But that we use the word this way remains a conventional fact of our language. The view that we cannot separate facts in the world from those of our language is at best a form of epistemic skepticism and at worst a form of idealism. Since both excesses can be avoided, we should.

4 Empirical Possibility

The guiding principle for the notion of empirical possibility is this: what is possible is what our evidence positively allows; and what is necessary is what our evidence compels. A version of this notion, restricted to deductive relations, had already been identified by Lewis and Langford (1959, p. 161) in their work which introduced the modal logic S5. They distinguish the strictly logical notion of possibility in their treatment from the "colloquially more frequent," relative conception:

> In this second sense, "possible" means "consistent with the data" or "consistent with everything known"; "impossible" means "not consistent with the data, or with what is known"; and "necessary" means "implied by what is given or known."

The notion is implemented here as a relation defined on some set of propositions over which there are relations of inductive support. Some subset of the propositions comprises the evidence and other propositions, commonly but not always outside the set, enter into the relation of empirical possibility with them. This relation has the following properties.

> *Empirical possibility is evidential support.* A contingent[5] proposition is a possibility with respect to the propositions of a body of evidence just in case the evidence provides inductive support for the proposition to any extent, or, optionally, sufficient[6] inductive support.

Here inductive support is understood in the general sense of ampliative inference. It is an objective relation over propositions, understood as abstract specifications of states of affairs, and obtains whether humans know of them or not. In this sense it satisfies the conditions of Sect. 3.

[5] A contingent proposition is here taken to be one that may be either true or false according to the state of the world. That is, its truth or falsity is not compelled by the meaning of the terms in the proposition.

[6] This option concerning a sufficient level of inductive support will be elaborated below.

In a common Bayesian account of inductive support, a proposition is supported by a body of evidence to the extent that the evidence increases the probability of the proposition; or alternatively accords non-zero probability to it. This account, I have argued at length elsewhere,[7] is only one of many accounts of inductive support, where the choice of the applicable account is determined by the prevailing background facts of the particular situation. These background facts may not warrant the introduction of probabilities, but some other inductive logic. In general, there is no universally applicable logic of induction. The warrant for an inductive inference or relation of inductive support resides in the facts of the domain of application.

Empirical possibility comes in degrees or varying strengths. This follows since empirical possibility is measured by inductive support, which comes in varying degrees or strengths that need not be numerical.

Relative to the evidence of how stars form, it is possible that two stars in some neighborhood of a galaxy have the same mass, within some narrowly specified window of error. The possibility strengthens successively as we consider larger and larger neighborhoods of the galaxy. This follows since the evidence of how stars form provides successively stronger support for the proposition of equally massive stars as the neighborhood grows.

A notion of empirical necessity can be introduced as:

Empirical necessity is evidential compulsion. A contingent proposition is empirically necessary with respect to the propositions of a body of evidence if its inductive support rises above some high but otherwise arbitrarily chosen threshold.

There is a further natural property of empirical necessity:

Empirical necessity includes logical necessity as a limiting case. If the evidence deductively entails a contingent[8] proposition then, by empiricist lights, it is supported in the strongest way.

However logical compatibility without inductive support is not included as a limiting case of empirical possibility. For a proposition can be logically compatible with a body of evidence while the evidence is irrelevant to the proposition. (This point will be developed further below in Sects. 6 and 10.2.)

Assigning mere empirical possibility to some proposition, without inclusion of the strength of support, discards much of the useful content of empirical possibility. The same label—"empirical possibility"—would be assigned to a proposition that has only the slenderest of support on the evidence; and to one that enjoys strong support. It fails to separate the barely possible from the quite possible. The present

[7] In Norton (2021) and earlier papers cited there.

[8] The restriction to contingent propositions is needed since it is usually held that any body of evidence deductively entails a tautology. Without the restriction, we could then find that, on the evidence of a coin toss, it is empirically necessary that either dinosaurs once lived or dinosaurs once did not live. This awkward result would arise in a set of propositions concerning coin tosses and dinosaurs, but where the evidence consists only of propositions about coin tosses.

literature[9] employs this flattened notion. More precisely, it represents the great variability of possibility inadequately by reducing the variability to two modes only: possibility and necessity. These states are treated as interdefinable, dual quantities, connected by the duality relations for any proposition *P*:

possibly *P* iff not necessarily not *P*
necessarily *P* iff not possibly not *P*

These duality relations need not be satisfied by the notions of empirical possibility and necessity as defined so far. While we shall see that failure below in Sect. 7, in order to maintain continuity with the present literature, we can optionally add further conditions that enable the notions of empirical possibility and empirical necessity to satisfy these duality relations in many cases. This is achieved by invoking the "sufficiency" option offered above. That is, we require a minimum threshold for the inductive support accrued to a proposition before it is declared empirically possible:

Optional possibility threshold. A contingent proposition is empirically possible with respect to some specified body of evidence if its support rises above some low, but otherwise arbitrarily chosen threshold of support.

The duality relations are realized by requiring the threshold for empirical necessity to be adapted to the threshold for empirical possibility in just the way the duality relations require:

Optional threshold matching for duality relations. The thresholds for empirical possibility and empirical necessity are set such that: (i) the negation of any empirically necessary proposition in not empirically possible, and (ii) the negation of any empirically possible proposition is not empirically necessary.

These thresholds are most easily defined when support is probabilistic. However non-numerical senses of strength of support can still be strong enough to sustain this sense of empirical necessity.

In the probabilistic case, we might choose a threshold of 0.001 for probability. Then, given the evidence of ten, independent, fair coin tosses, it is not empirically possible that there are no heads among the ten tosses and it is empirically necessary that there is at least one head.[10] Probabilistic measures for the thresholds are not needed, however. On the collected evidence of the energetics of ordinary systems, it is very likely that a proposed design for perpetual motion machine will fail. That is, the failure is empirically necessary and, under the matching condition, its functioning as designed is impossible. This judgment might be made without the need for any assignment of probabilities by those who find the duality appealing.

It is best not to become too attached to this implementation of the duality. We shall see below in Sect. 7 that this implementation is fragile. The duality is degenerate in cases in which there are possibilities but no necessities. If we drop

[9] Kment (2014, p. 2) is a welcome exception in the literature that allows for degrees of possibility.

[10] The probability of no heads in ten tosses is $(1/2)^{10} = 0.0009766$ and of at least one head in ten such tosses is $1 - (1/2)^{10} = 0.9990234$.

the requirement of a threshold for possibility, then we can have cases in which the duality is contradicted. The better notion is just that of possibility, where possibility comes in varying strengths according to the strengths of support provided by the evidence. Necessity is still the case of very high support, but possibility and necessity need no longer conform with the duality relations.

These difficulties affirm that the adoption of the two optional conditions is a contrived way of trying to force the duality of possibility and necessity, when that duality is not constitutive for the notion of empirical possibility. The contrivance is most clearly seen in the second optional condition: the thresholds for possibility and necessity must be set to match in such a way that the duality is preserved. Without that stipulation, nothing in the prior conditions requires the two thresholds to match in the precise way required by the duality.

This much already provides a serviceable account of possibility. However, it is unable to preclude fantasies like the possibility of magical wizards in other universes with powers of levitation. The proposition of such wizards is possible in relation to a fantasy body of evidence from such universes, replete with reports of caped figures with mysterious powers. To preclude such cases, to put "empirical" firmly into the account, there is a final condition:

> *Evidential actuality*: The relation of empirical possibility only employs bodies of evidence that report actual experiences.

The overall import is that we preclude fantasies like these other-wordly, magical wizards from the propositions over which empirical possibility is defined. For the assumption, at the outset, was that empirical possibility is defined over a set of propositions for which inductive relations are also defined. Our experience fails to provide inductive support for or against proposition asserting the existence of such other-worldly, magical wizards.

The restriction to actual bodies of evidence does not preclude the relation from ascribing empirical possibility to propositions describing situations that will never be actualized. For actual bodies of evidence can provide inductive support for many propositions, most of which will never be actualized. One might, nonetheless, think that actual experiences can only support actualities but not non-actualized possibilities. That thought is mistaken, as has been argued convincingly by Strohminger (2015) and Roca-Royes (2017).[11]

There is a place for hypothetical and counterfactual possibilities in the account if we relax this evidential actuality condition. A *counterfactual possibility* is defined as one that derives inductive support from a body of evidence in contradiction with the evidence provided by experience. A *hypothetical possibility* is defined as one that derives inductive support from a body of evidence not fully recovered from experience but logically compatible with it. These options will prove useful below

[11] Roca-Royes (2017) develops an empiricist epistemology of *de re* possibility (but not necessity). Her project differs from that of this paper in that it is epistemic. It concerns how we come to know some possibilities inductively. Here those possibilities are reduced to the inductive relations themselves.

in Sect. 12 where the modal content of scientific theories is recaptured in inductive terms.

5 The Modest Semantics of Empirical Possibility

This account of empirical possibility is quite modest in the resources needed to give its assertions of possibility a clear meaning. To say that a proposition *P* is possible or necessary, on some body of evidence *E*, just means that it accrues some or very high inductive support from the evidence *E*. To interpret this last assertion, we require only what is already in a philosopher's toolkit. We need to know what it is for a proposition *P* to be true or not in the actual world; and we need some serviceable account of inductive support. The account needs no notion of modality as a primitive.

In this way, the present account escapes the enduring woes faced by the larger modality literature in determining the meaning of an assertion of a possibility. Modality is, we may be asked to believe, some sort of primitive aspect of the world whose reach extends well beyond the actual. We may even be lured to accept that there is more in our basic ontology than the actual. It also contains real possibilities. They hover about in some manner, real but forever invisible, like ghostly spirits in a paranormal fable, and in extravagantly greater number than actualities. These difficulties arise most pointedly through the need to provide some space in our ontology for possibilities that cannot be actualized, but are nonetheless conceived as somehow possible. Here one might think of the possibility of talking donkeys (which is a running example in Lewis (1986)) or, more exotically, of the other-worldly, magical wizards mentioned above.

The present literature in modality has been hard pressed to give meaning to such possibilities. The notorious and extreme case is David Lewis' (1986) modal realism. According to it, other worlds with all manner of fantasy situations are as real as our world. "... absolutely *every* way that a world could possibly be is a way that some world is." (p. 2) Ours is distinguished only by indexicality (pp. 92–93): it is actual only in so far as it happens to be our world.

An empiricist can only stand aghast at this explosion of worlds whose real existence is beyond the reach of any possible experience. Many modal theorists also react with incredulity, generating what Lewis (1986, p. 133) calls the "incredulous stare." A more modest proposal—decried by Lewis (1986, Ch. 3) as "ersatzism"—seems to have become a routine default. A possible world is just an exhaustive, consistent assignment of truth values to some set of propositions.[12] The most minimal conception of possible worlds halts at that bare propositional conception.

[12] "To say that a world *w* is a *complete* way things might have been is to say that for each proposition in the sphere of discourse under discussion, *w* either verifies or falsifies that proposition; ..." From the entry "Possible Worlds," in Kim (2009, p. 502, emphasis in original).

The conception has become more prominent in ordinary philosophical discourse from its role in the semantics of formal modal logic. Hence Kripke, whose (1959) completeness proof reinvigorated work in formal modal logic, is most direct: "A possible world is *given by the descriptive conditions we associate with it*" (1980, p. 44, his emphasis). His disagreement with modal realism is stated tersely: "'Possible worlds' are *stipulated*, not *discovered* by powerful telescopes." (p. 44, his emphasis) We may wish to trim even further. We may drop the possible world talk and just posit that there are possibilities residing in some factual realm that extends beyond the actual. We eschew giving any further account of it.

However even this most minimal position, welcome for its modesty, is troubled. For what in it precludes propositions asserting talking donkeys and magical wizards? In the possible world picture, they are possible in the sense that truth is assigned in some valuation to the propositions asserting their existence. The natural response is that propositions in the realm of the possible should be restricted by some sort of reasonable accommodation to the world. What better restriction than to require the propositions to be within the scope of what our actual experience supports evidentially? Indeed, what *other* restriction is there? If modal theorists agree with this restriction, I welcome them. Their view is functionally equivalent to empirical possibility, as defined here. It has become tacitly relational, since the judgments of possibility are now dependent on some body of evidence. If they wish to use some other restriction or none at all, then I am left to wonder what could sustain their alternative. We shall return to this problem in Sect. 9 and later, where alternative conceptions of possibility are considered.

6 Properties of Empirical Possibility

The evidential actuality condition precludes some judgments of possibility that we would otherwise find natural. To illustrate, take our present body of evidence concerning our solar system. It contains the proposition that the earth has one moon. It is not possible in relation to this body of evidence that our earth could have two moons. This conclusion conflicts with the natural intuition that two moons are, in some sense, a possibility for the earth. That intuition can be accommodated in the present account in two ways. In one, we reduce the body of evidence so that it no longer contains the fact that our earth actually has only one moon (and other associated facts). Then, relative to this reduced body of evidence, an earth with two moons is something that could be the actual case. In another approach, we replace the earth-moon system with a surrogate. We consider some distant sun-like star about which an earth-like planet orbits. We do not know whether the planet has moons at all. On that evidence, we can say that it is empirically possible for this planet to have one moon; or two moons. There is no need to add to the actuality of our earth's single moon, an invisible second moon brought into being by the powers of primitive possibility.

The more striking properties of empirical possibility come from the extreme cases. As the body of evidence grows, many propositions can be judged less and less possible, while support grows for one possibility. Thus, the errant motion of Mercury was over a century ago good evidence for the possibility of another planet orbiting within the orbit of Mercury. That changed when observations failed to detect the planet and the accumulating evidence focused support on the possibility of gravity deviating from Newtonian theory.[13]

Given this dynamic, as it grows, a body of evidence is weakened in the range of possibilities it authorizes. Greater possibility accumulates on fewer propositions, with the remainder accorded less and less possibility. In the limit, a body of evidence becomes modally inert once it contains all the propositions of the set over which the possibility relation is defined. For example, take the set of propositions to consist of all actual facts of our solar system and take the evidence set to be that totality. It follows that there are no unactualized possibilities concerning our solar system. For, by supposition, there are no propositions in the set that are not already contained within the evidence.

A similar effect occurs at the opposite extreme when we have no evidence at all. Empirical possibility is a relation between a proposition and a body of evidence. If there is no evidence, then one of the relata is missing and the relation cannot be defined. There is no empirical possibility.

Closer to this extreme is the case of evidence that is so thin that it provides equal support to all possibilities. An example of this is the "completely neutral support" developed in Norton (2010). It would arise in some novel physical theory with a characteristic parameter "h" as follows. We assume the evidence tells us merely that h has some value greater than zero, but it tells us nothing more. Then I show that we must assign equal support to the parameter lying in each of the intervals (0,1), (1,2), (2,3), ... However, we must also assign the same support to it lying in each of the intervals (0,1), (1,2), (1,3), (1.4), ..., $(1,\infty)$.[14] That is, all contingent, non-empty[15] intervals accrue the same minimal level of support. Assuming that this one level of support exceeds the threshold required for possibility if there is one, all these intervals are empirically possible.

This extreme case of weak evidence still assures us that a parameter value in these intervals is possible. The evidence, while extremely weak, is still relevant to the parameter. This case is distinct from one in which the evidence is irrelevant to the proposition. The conception of empirical possibility applies to possibilities that can be inductively supported to some extent, even if small, from the evidence of experience. If the evidence is irrelevant to the proposition, no relation of empirical possibility or necessity can be established. Thus, as noted in Sect. 4 above, the evidence of some coin toss is irrelevant to whether dinosaurs ever lived on earth.

---

[13] For one account of this episode, see Norton (2021, Ch. 9, Sect. 7).

[14] This is not a typographical error. For justification, see Norton (2010).

[15] An interval (x, y) is non-empty just if $y > x$.

Relative to that evidence, no assertion of past dinosaurs' empirical possibility can be made.

7 Failure of the Duality of Possibility and Necessity

The duality relations for possibility and necessity in Sect. 4 are fragile when applied to the empirical notions. The relations can become degenerate or fail entirely.

In the first case, the relations become degenerate when we have empirical possibilities but no empirical necessities, beyond what is deductively entailed by the evidence. These cases arise whenever the evidence is unable to provide inductive support for any proposition that rises above the necessity threshold, without the proposition being a deductive consequence of the evidence.

The coin toss example of Sect. 4 illustrates how this can happen. On the evidence of ten independent coin tosses and the threshold of necessity of probability 0.999, we saw that it is empirically necessary that there is at least one head. Matters are otherwise in the case of nine independent coin tosses if we use the same thresholds of probabilities 0.001 for empirical possibility and 0.999 for empirical necessity. For then it is empirically possible that we have no heads in nine tosses. But it is not empirically necessary that we have at least one head in the nine tosses.[16] More generally, consider any contingent proposition *P* not entailed by the evidence in the Boolean algebra of propositions of nine-coin toss outcomes. The probability of *P* cannot rise above 0.999.[17] It follows that there are no empirically necessary proposition *P* that are not entailed by the evidence.

In this example, the duality relations can still obtain, but at best in a degenerate or vacuous sense, since there are no empirically necessary propositions not entailed by the evidence. That is, for any contingent proposition *P* not entailed by the evidence, it is always the case that *P* is empirically possible, but never the case that *P* is empirically necessary. That is, we never have "necessarily *P*." However, for each of these *P*, we do have "not necessarily *P*" and "possibly not *P*." The duality relations survive in the specific form "not necessarily iff possibly not."

The duality relations fail directly, however, if we drop the requirement that there is a minimum threshold of inductive support for empirical possibility. It is natural to drop this requirement since empirical possibility is a permissive notion. All that it requires is that the evidence positively favors the proposition, no matter how weak the favoring. This does mean that some propositions with very weak support will be deemed empirically possible. Since the notion is always qualified by its degree, this

[16] The probability of no heads in nine tosses is $(1/2)^9 = 0.001953$; and the probability of at least one head in nine tosses is $1 - (1/2)^9 = 0.998047$.

[17] The atomic propositions specify a specific sequence of nine coin toss outcomes, each with $(1/2)^9 = 0.001953$. So the most probable proposition not entailed by the evidence is just a negation of a single atom, which has probability $1 - (1/2)^9 = 0.998047$.

inclusion is not troublesome. To preserve the duality, the corresponding adjustment to empirical necessity would require that the threshold to be passed is that of complete support. This is clearly too high a standard. For then propositions very strongly supported by the evidence would cease to be empirically necessary. For example, our accumulated evidence of energetic processes very strongly supports the necessary failure of a perpetual motion machine, while not precluding the extremely unlikely possibility of exceptions.

Without the minimum threshold for empirical possibility, we readily arrive at cases in which all contingent propositions are empirically possible, while at the same time there are also empirically necessary propositions not entailed by the evidence. The ten-coin toss case of Sect. 4 illustrates this. The proposition P that there is at least one head in ten-coin tosses is empirically necessary since its probability is greater than the threshold 0.999. However, its negation, not P, is also empirically possible, since it accrues non-zero probability. Thus, we have for P both

necessarily P and possibly not P

in contradiction with the duality relations. A second and more extreme example arises in the case of completely neutral support, just developed. For no contingent proposition in the outcome space of non-empty intervals gains more inductive support than the minimum support accrued just to the narrow outcome of the parameter h lying in (0,1). In these cases, the duality relation can no longer be used to define a possibility as the dual of necessity, as does the axiom system of Kripke (1959, p. 1).

8 Failure of Possible World Semantics

The possible world semantics employed widely in the philosophical literature is admirably simple. It is based on the notion of a possible world, where, minimally, a possible world just is a maximal, consistent set of truth value assignments to some set of propositions. The set of possible worlds is just the set of all such assignments. A proposition is necessary if it is true in all the worlds of the set. A proposition is possible if it is true in at least one of the worlds of the set.

This semantics is ill-suited to empirical necessity, since it is inhospitable to a gradational conception of necessity. Under this definition of possible worlds, there will be many cases of empirically necessary propositions that are not true in all possible worlds. For example, the conservation of energy is very likely for all ordinary systems. However, its failure is not absolutely precluded. The steady-state cosmologists of the mid twentieth century included the failure as an essential element of their theory.[18] In it, mass-energy is continuously created everywhere

[18] For a brief survey of decline of steady-state cosmology, see Norton (2021, Ch. 9, Sect. 8). It has also proved difficult to find a general formulation in general relativity that expresses

in space at just the rate needed to balance the dilution of mass-energy resulting from the expansion of the universe. Steady-state cosmology has fared poorly in its competition with big bang cosmology. Our evidence now strongly favors big bang cosmology. However, steady-state cosmology is not absolutely ruled out. Since the evidence for it is extremely weak, just a sliver of possibility remains for it. Thus, the propositions defining the pertinent set of possible worlds should include the proposition "The cosmos conforms with steady-state cosmology." Since we generate possible worlds by assigning all possible truth values to these propositions, there is a possible world in which steady-state cosmology is true and energy conservation is false, even though the evidence makes it very unlikely. The result is that the empirical necessity of energy conservation is not captured by the truth of the proposition in all possible worlds. It fails in some possible worlds.

This semantics automatically instantiates the duality relations of Sect. 4, which are here a version of the duality of universal and existential quantification in predicate logic. Necessity is instantiated as "For all possible worlds..." and possibility as "There exists a possible world..." This instantiation brings further problems for possible world semantics if we consider cases in which these relations fail. In those cases, as we saw in Sect. 7, we can have an empirically necessary proposition *P* whose negation is also empirically possible. Possible world semantics cannot accommodate this case.

It is disappointing to lose the familiar possible world semantics. For that semantics reduces initially perplexing questions of possibility and necessity to simpler questions of set theory. It may seem that a simple adjustment to possible world semantics adapts it to empirical necessity. In it, we require that necessity not be attributed to propositions true in *all* possible worlds, but only to those true in *most* possible worlds. If we employ a threshold for possibility, then possibility is attributed not just to propositions true in at least one possible world, but to those true in *sufficiently many*. If each specific sequence of ten-coin toss outcomes is a possible world of the coin toss example above, then in over 99.9% of them, there is at least one head. Thus, an outcome of at least one head is empirically necessary. In fewer than 0.1% of them, there are no heads, so the outcome of no heads is not empirically possible.

This stratagem will work sometimes, such as in this coin toss example. What enables it to work is that meanings for "most" and "sufficiently many" are provided by the example's probability measure. It is not merely that at least 99.9% of the toss outcomes have at least one head. It is that each toss outcome has equal probability, so that their combined probability is at least 0.999. This is an exceptional case. As I have argued at length in Norton (2021, Ch. 10–16), probabilities are not in general available. Sometimes a probabilistic meaning can be given to "most"; and sometimes not.

---

the conservation of energy, understood as a conserved quantity extended over space. For a comprehensive survey of these problems, see Duerr (ms).

One might hope that there are surrogates available, such as a mere counting of the number of favorable and unfavorable outcomes. However, that counting can give unexpected results. If the outcome set is countably infinite, cardinalities do not match our normal intuitions about "most." We would want to say that most natural numbers are not powers of ten: {10, 100, 1000, ...}. However, there are in cardinality just as many powers of ten as there are numbers that are not powers of ten.

Even when natural measures are available, they may give results incompatible with empirical necessity. In the example of the unknown positive real-valued parameter *h* above, that *h* lies in the interval $(1,\infty)$ accrues infinite Lebesgue measure, far exceeding the zero measure of (0,1). However, as a case of completely neutral support, each interval (0,1) and $(1,\infty)$ accrues that same measure.[19] Both are merely empirically possible. Neither is empirically necessary.

In sum, the appeal of possible world semantics is lost. Questions of possibility and necessity can no longer be reduced to simple questions in set theory. If we persist in using this amended possible world conception, in each case we will have to ascertain whether some measure is available and, if so, which it is. It is simpler to address the question of empirical necessity within the particular inductive logic that happens to be applicable to the case at hand. We only create more problems if we try to find some set theoretic implementation of that particular logic. There may be no simple one.

9 And Nothing More

This empirical notion of possibility conforms with the requirements laid out in Sect. 3. Together with the logical notion sketched in Sect. 2, it is, I believe, all that an empiricist needs to accommodate possibilities. Moreover, I shall now argue, it is all that anyone needs, in so far as the analysis of possibility is made responsibly. To establish this "and nothing more" claim, the following reviews accounts of possibility that may go beyond the logical and empirical and argues that they are either actually already subsumed by them; or, if not, they are not cogent.

The philosophical literature reports many varieties of possibility. A fairly good list is:

> logical possibility, conceptual possibility, metaphysical possibility, nomic possibility, physical possibility and epistemic possibility.

Several of these need only a brief mention since they are already within the scope of logical and empirical possibility as defined above.

---

[19] How can this be? Very briefly, given the paucity of evidence, we do not know if the correct representation of the parameter is *h* or $1/h$. The conditions $0 < h < 1$ corresponds to $1 < 1/h < \infty$; and $0 < 1/h < 1$ corresponds to $1 < h < \infty$.

Conceptual possibilities are those arising from the meaning of the relevant concepts. That makes them a version of logical possibility. They have no special powers to create possibilities in the world. We may conceive in fiction a Pegasus, a horse that by our conception can fly. That we conceive it makes it no more possible that there are flying horses than if we had never conceived it. Our concepts of perfection similarly have no power to bring things into existence. In the optimistic conception, existence increases perfection. In the pessimistic conception, existence decreases perfection, for the real thing always falls frustratingly short of the perfections we imagined for it. The optimistic conception assures us of the existence of the most perfect and the nonexistence of the least perfect. The pessimistic conception reverses these judgments. Creative powers cannot be attributed to mere conceptions on pain of inconsistency.

Physical possibility refers to mundane possibilities learned through the routine operation of science. They are empirical possibilities. Nomic possibilities, those authorized by scientific laws, are similar in that they are also derived through the routine empirical investigations of scientists. However, since one might imagine that this law-based notion is somehow more elevated and thus distinct, in Sect. 11 below, I argue otherwise. Epistemic possibilities are those left open by what we do not know. While its name suggests that it is a version of empirical possibility, I argue in Sect. 10 below that the notion is distinct from empirical possibility and violates one of the requirements laid out in Sect. 3.

The most troublesome entry to address in the list is metaphysical possibility. It is troublesome since there is an enormous literature declaring all manner of possibilities and necessities on its authority. Yet there is little agreement on exactly what it is. In Sect. 12 below, I will separate notions of metaphysical possibility into two types. First are possibilities authorized by empirical science. While they may be called "metaphysical," they conform with empirical possibility precisely because of their empirical foundations. Second are notions of metaphysical possibility that go beyond logical and empirical possibility. Precisely because they lack proper empirical foundation, they are all "sophistry and illusion," or so I will argue.

10 Empirical Possibility Is Not Epistemic Possibility

Among existing accounts of possibility in the philosophical literature, epistemic possibility comes closest to the presently defined notion of empirical possibility. For both are motivated by the question of what we can know to be possibly the case. However, the similarity stops there. The two notions differ greatly in how the question is answered. In brief, first, epistemic possibility is defined in terms of the knowledge of agents. It requires there to be agents with thoughts and beliefs to know things or not to know them. Empirical possibility is defined as an inductive-logical relation over propositions, independent of agents with thoughts. Second, epistemic possibilities are defined negatively, in terms of ignorance, as scenarios that the agent's knowledge cannot rule out. Empirical possibilities are defined positively, as propositions accruing some inductive support, even if small, from a designated body of propositional evidence.

In both aspects, epistemic possibility fails to satisfy the requirement of Sect. 3 that its judgments are independent of our beliefs and knowledge[20] and thus fails as a conception of possibility in the world.

10.1 Dependence on Beliefs and Knowledge of Agents

In their synoptic introduction to the volume *Epistemic Modality*, Weatherson and Egan (2011, p. 1) give this formulation as a starting point that requires many qualifications:

> A possibility is an epistemic possibility if we do not know that it does not obtain.

The characterization seems fairly stable in the literature. For example, Chalmers (2011, p. 60, his emphasis) has a similar starting point: "We normally say that it is *epistemically possible* for a subject that *p*, when it might be that *p* for all the subject knows." These opening definitions make clear that epistemic possibility is *defined* in terms of the beliefs and knowledge of specific agents, in contradiction with the requirement of Sect. 3. While many refinements of the conception follow, the dependence on an agent's knowledge remains at the core of the notion. That fact produces problems that occupy the literature. Weatherson and Egan (2011, pp. 4–17), for example, note that the starting formulation makes no explicit reference to relata that must tacitly be there. Their list of candidate relata is "a person, group, evidence set, or information state." They then offer three options as expansions of the opening definition: the context in which the agent makes the assertion ("contextualism"); variations in how different agents use terms to refer ("relativism"); and that assertions of possibility are simply direct expressions by agents of uncertainty ("expressivism"). These sorts of worries are long-standing. Writing decades earlier, DeRose (1991, p. 584) sought to determine whether *S* can assert some epistemic possibility just if *S* does not know of its falsity; or whether the assertion requires that no one in the relevant community knows of its falsity. Then we must deal with the awkward problem of determining just what counts as the relevant community.

Recall the question asked: what possibilities can we know? My concern has been to find an answer that is independent of the vagaries of our thoughts and beliefs and how we may happen to use language. Of course, these vagaries will be a part of a full account of how some particular agent or some community comes to know some possibility. But whether the evidence is known by a single agent, or a relevant community, or expressed in language understood differently by different agents in different contexts—none of these vagaries are constitutive of what possibilities in the world we can know. Unfortunately, the epistemic possibility literature has allowed itself to become dominated by these vagaries.

[20] Of course, theorists of epistemic possibility know this, but that does not ameliorate the failures.

The notion of empirical possibility arises when we take a different course and strip away all these distractions. We just ask a question in inductive logic: given this evidence, what is possible? How some agent or some community or some idiosyncratic language user might exploit the answer goes well beyond the original question.

10.2 Dependence on Ignorance

These last remarks suggest that epistemic possibility might become empirical possibility if we replace the real, cognitively limited agents of the standard account by ideal reasoners (such as is included in the definition by Kment (2017, § 1)). That is not the case, for the epistemic possibility literature has been diverted from answering the question of what possibilities we can know. Epistemic possibility is defined in an obliquely negative manner. The starting definition above says "we do *not* know that it does *not* obtain." That is, the epistemically possible is delimited by what we cannot positively rule out.

Unless one has exaggerated confidence in the creative powers of our ignorance, this definition is vastly too permissive. Its embrace includes an unimaginable plethora of situations that we do not now know and could never know to be the case. By deriving possibility from ignorance, it allows the assertion of possibilities where prudence would dictate silence. For example, nothing that we know rules out levitating wizards living in magic mountains in other universes, disconnected fully from ours. Their possibility is allowed by this notion of epistemic possibility; and they are given the same label—"epistemically possible"—as is the supposition (prior to the Apollo missions) that there are mountains on the other, hidden side of the moon.

The notion of empirical possibility handles these cases very differently. Its concern is what we can positively come to know. As empiricism dictates, it delimits the possible according to whichever body of evidence is at hand. In the absence of any evidence pertaining to magic mountains in other universes, it allows no assertion of empirical possibility. However, we have a substantial body of evidence concerning our moon and the formation of mountains on its visible surface. That body of evidence allows the assertion of the empirical possibility of mountains on the hidden side of the moon.

11 Nomic Possibility

Empirical possibility, as defined here, is a relation between our actual evidence and a proposition. Theories in science routinely make claims of possibility and necessity that are detached from any particular body of evidence. According to general relativity, it is possible for black holes to form; and it is impossible to escape them once the event horizon is passed. According to thermodynamics, it is possible to convert heat into work, but perpetual motion machines fail of necessity. These are law-based—nomic—notions of possibility and necessity. Since they are detached from evidence, are they a non-empirical notion of possibility? Does that mean scientific theories employ modality as an irreducible primitive? They are not, I now argue.

First, I do not accord scientific laws any special status qualitatively in comparison with other contingent facts. They are not some higher order of truth beyond ordinary contingent truths. They are merely contingent truths of very broad scope. None that we know has universal application.[21] Thus they can afford us no notion of possibility and necessity that is distinct from that afforded by ordinary contingent truths. They merely do so with greater scope.

After that qualification, I follow Ismael's (2017) analysis. The modal aspect of scientific theories is, when examined more closely, captured fully with inductive notions. Ismael (2017, pp. 119–20) describes that content as "partially prepared solutions to frequently encountered problems." In my rendition, a scientific theory is simply a large collection of possibility and necessity claims, often encoded elegantly in quite compact statements. In their domain of application, they assert propositions of the form "If this happens, then that may possibly ensure, or that other may necessarily ensue, but that other again necessarily cannot ensue." Propositions of this form are inductive statements. They can be restated as "If this happens, then there is some evidence for that; and compelling evidence for that other, etc." That is, they are large collections of propositions that lie within the compass of empirical or logical possibility. Since most of the "this's" will not describe our actual evidence, these propositions almost all assert hypothetical or counterfactual possibilities (as defined in Sect. 4 above).[22] There is nothing primitively modal about them. To imagine that there is anything more in them is superfluous to their function. There is no need to declare their primitive modality, that is, irreducible, or worse to reify their possibilities.

I cannot resist likening a theory to a restaurant menu. It is a collection of "if . . . then . . . " propositions: "if you order the Reuben sandwich, then this rye bread sandwich will be prepared and brought to you." The assertion made does not require a distinct notion of culinary modality to give it meaning. And it certainly does not require that every meal on the menu is already prepared in advance and that all will be discarded, save the one that you happen to order.

---

[21] In physics, where we might hope to find universal truths, we know of none. General relativity is our best account of space, time and gravitation, but it fails in the very small. Quantum theory is our best account of matter in the very small, but it fails where general relativity thrives, in domains of very strong gravity.

[22] This holds of a case examined in more detail by Ismael (2017, p. 116): theories based on physical chances, formulated in terms of probabilities. For they just tell us that the evidence of the antecedent supports various consequents according to the inductive strengths of the associated probabilities.

12 Metaphysical Possibility

12.1 The Grand Promise

The promises made on behalf of metaphysical possibility are grand, even grandiose. Gendler and Hawthorne (2002b, pp. 4–5) put it this way:

> The notion of metaphysical possibility, meanwhile, is standardly taken to be primitive.[23] It is taken as the most basic conception of "how things might have been"—gestured at by talk of how "God might have made things" or "ways it is possible for things to be."

Fine (2002, p. 259) continues the theological metaphor:

> There appears to be an intuitive difference to the kind of necessity attaching to metaphysical and natural necessities (granted that some natural necessities are not metaphysical). The former is somehow "harder" or "stricter" than the latter.[24] If we were to suppose that a God were capable of breaking necessary connections, then it would take more of a God to break a connection that was metaphysically necessary than one that was naturally necessary.

Our expectations are now set very high as we seek distinctive examples of metaphysical impossibilities challenging even to the gods. Instead, aside from empirically dubious cases like the necessary existence of the most perfect being, we find (Sider, 2003, pp. 181–82):

> Examples [of metaphysical impossibilities] might include the existence of a round square, someone's being taller than himself, someone's being in two places at once, George W. Bush being a donkey, there existing no numbers, and there existing some water that is not made of H_2O. Exactly what is metaphysically impossible beyond logical and analytic contradictions is unclear; this unclarity is what makes the analysis of metaphysical possibility and necessity so difficult.

If these are the best illustrative examples of distinctively metaphysical impossibilities, we must begin to suspect that it is an inchoate idea that can admit no good examples. All of these examples are problematic. If the definition of a square includes the requirement that it is a figure with four pointed corners, then round squares are merely a logical impossibility. Their definition precludes them. If we use a definition that does not preclude them logically, then it turns out that round squares are only impossible contingently if, for example, space happens to be Euclidean. Other geometries admit round squares. If we define a square as a figure bounded by four straight lines and symmetric under rotation by a right angle about its center, then, in the spherical geometry of the surface of a sphere, the equator ABCD in Fig. 1 is both a square and a circle with center O. Round squares might even be more than mere possibilities. If our three-dimensional space has a spherical geometry, as present cosmology might allow, then there are round squares in our space.

---

[23] Their footnote: "In contemporary discussions, at any rate."

[24] Fine's footnote: "It seems to be something like this that [Kripke (1980, p. 99)] has in mind when he talks of necessity 'in the highest degree.'"

Fig. 1 A round square

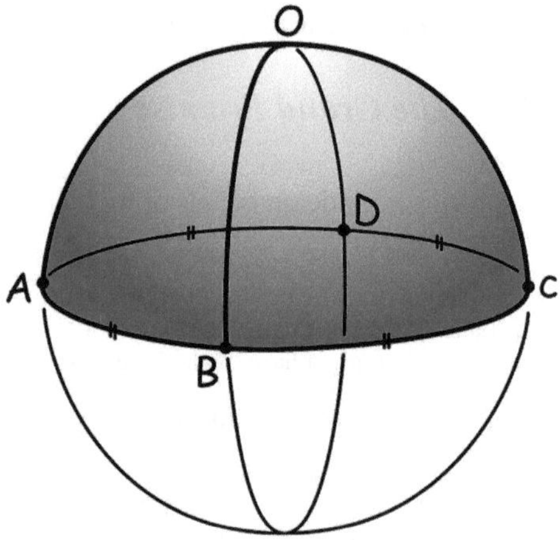

Being taller than oneself is a logical contradiction following from the meaning of "taller." The relation is irreflexive. If we replace "taller" by the reflexive relation "no taller than," then it is trivial: we are no taller than ourselves. The world has a habit of supplying actualities that contradict the impossibilities of metaphysics. Electrons in quantum theory can be in two places at once,[25] as can larger configurations of particles like people, absent quantum measurement. The being designated by "George W. Bush" is a convention of our language. Plausibly, someone, somewhere has so named their cat or even their donkey. Or perhaps the idea is more subtle. Is it that whichever being is designated by "George W. Bush" *could not* possibly be a donkey? That this being *is not* a donkey is a mundane fact. What added facts in the world, as opposed to conventions of language, allow us to replace "is not" with "could not be"? Whether numbers exist is something that philosophers of mathematics continue to debate. The question is unhelpful as an instructive example. Finally, that water is necessarily H_2O alludes to Kripke's celebrated analysis of necessity (as does the proper name, George W. Bush).

Kripke's analysis has an outsized role in the present literature and many favorite examples of metaphysical necessity derive from it. The difficulty is that Kripke's analysis treats the terms appearing in science on the model of proper names.[26] This is a poor starting place and was fated not to end well. The reference of proper names depends essentially on the conventions of a group of language users. Whatever

[25] An electron in a bimodal well can be present equally in both, even if the wells are separated by great distances. This impossibility required no god to break it, but merely Erwin Schroedinger.

[26] Kripke (1980, p. 127) "According to the view I advocate, then, terms for natural kinds are much closer to proper names than is ordinarily supposed."

necessities they bring are not facts in the world, independent of us and our use of language.[27] The requirements of Sect. 3 are violated.

The most familiar necessities (e.g. Kripke, 1980, p. 116) in this literature are that water is necessarily H_2O[28]; and that gold is necessarily the element with atomic number 79.[29] There are two sorts of facts here. First are the empirical facts discovered by chemists that substances are naturally divided best according to atomic numbers for elements and molecular formulae for compounds. Second is a convention of language concerning the words we choose to use to label the divisions found empirically. There are no necessities in these two facts that outstrip empirical and logical necessities. Nothing in our society obliges a community of language users to use the words in this way. Nothing compels nature to provide us substances formed from elements with atomic natures and discrete atomic numbers. Nature may not oblige. The latter happens with another of Kripke's (1980, p. 99) necessities:

> ...characteristic theoretical identifications like "Heat is the motion of molecules", are not contingent truths but necessary truths, and here of course I don't mean just physically necessary, but necessary in the highest degree-whatever that means.

Heat is simply the motion of molecules only in the special case that most occupied the nineteenth century physicists, ideal gases. In other cases, heat is not simply molecular motion, but the energy of that motion plus other forms of energy. In a classically modeled crystal, heat resides in the kinetic and potential energy of the vibrating atoms bound in the lattice. The dominant form of heat in the early universe resides in the energy of radiation. This necessity "in the highest degree" is not broken by the exertions of a powerful god, but merely by consulting a physics textbook. Perhaps a more charitable reading is needed here. The exact nature of heat is a topic for specialists in thermal and statistical physics. Kripke is not one.[30] Might the unfortunate remark be more properly rendered as "Heat is necessarily whatever our best science tells us it is (and I am told that it is molecular motion)." The reformulation does not save this notion of necessity. It now just gives us an empirical necessity, for science determines the nature of heat by empirical investigations. Unless the added qualification "necessarily" is merely an empty flourish, it asserts an unrevisability of scientific findings incompatible with the history of science. If

[27] A growing body of empirical evidence, such as in Machery et al. (2004), shows that there is nothing fixed and thus nothing necessary about these conventions. They vary with the culture of the language users.

[28] Putnam is also celebrated for the claim that "water" rigidly refers to H_2O. He, however, seems to defer to Kripke for the claim that there is a metaphysical necessity to the rigidity. See Putnam (1975, p. 151).

[29] Kripke (1980, pp. 35–36) identifies the sense of necessity explicated as metaphysical: "The second concept which is in question is that of necessity.... what I am concerned with here is a notion which is not a notion of epistemology but of metaphysics, in some (I hope) nonpejorative sense."

[30] The admission is explicit for a neighboring science: Kripke (1980, p. 117) writes "... I don't know too much chemistry."

we once thought that heat is necessarily identical with the motion of molecules, we have since learned that it is not necessarily so.

12.2 Empirical Metaphysics

These last examples show that the literature has faced great difficulties in providing cogent examples of possibilities and necessities that outstrip the logical and the empirical. Are these failures merely failures of exposition? Perhaps eager authors are oversimplifying their examples to connect more easily with novice readers? Whether such failures must happen depends on the deeper question of whether metaphysics has within its scope the capacity to provide modalities that go beyond the logical and the empirical.

To answer, we ask "what is metaphysics?" A clear answer proves elusive. Discourses on metaphysics routinely begin by admitting the difficulty of identifying just what metaphysics is.[31] This has not discouraged metaphysicians from writing treatises on the subject. A standard starting point in historical surveys, such as Simmons (2009, p. 3), is Aristotle's definition of metaphysics as the study of "being *qua* being." It agrees in general import with what van Inwagen (2009, p. 1) calls "the best definition of metaphysics I have seen," that is, "metaphysics is the study of ultimate reality." This goal of metaphysics then aligns well with that of my small-e empiricism—to learn about reality—unless the qualification "ultimate" hides some evasion.

For me, empirical science is our best attempt at learning about reality. For the nature of space and time, we look to special and general relativity. For the nature of matter, we look to quantum theory. For the nature of life, we look to biology. For the nature of mind and thought, we look to cognitive science and neuroscience. A great deal of what passes as metaphysics in the present literature is simply reporting what our best science tells us. Since all these metaphysical investigations fall within ordinary empirical science, the associated notions of metaphysical possibility and necessity fall within the empirical conception developed here.

This empirically grounded metaphysics is diffused throughout the metaphysics literature, as a scan of collections like Loux and Zimmerman (2003), Kim et al. (2009) and Le Poidevin et al. (2009, Part III) shows. Finding clean examples, however, is difficult, since metaphysics with and without empirical foundations are routinely mingled. For example, the present literature on the metaphysics of time makes heavy use of empirical science in its reliance on special relativity, its representation in a four-dimensional Minkowski spacetime and its extension to gravity in general relativity. See for example Loux and Zimmerman (2003, Part IV), Hawley (2009) and Ney (2014, Ch. 5). But these ideas are freely intermingled

[31] Van Inwagen and Sullivan (2020)'s first sentence is: "It is not easy to say what metaphysics is."; and the first section begins: "The word 'metaphysics' is notoriously hard to define."

with conceptions with no empirical manifestations, such the distinction between eternalism and presentism; and between perdurantism and endurantism.[32]

12.3 *Non-empirical Metaphysics*

Our present concern, however, is to see if there is a defensible notion of metaphysical possibility that extends beyond the logical and the empirical. If there is such a thing, it must be supplied by a body of metaphysical knowledge that is not derived from empirical sources: "non-empirical metaphysics." That such a non-empirical body of knowledge is viable has been subjected to an enduring critique.

Hume's fork (Hume, 1777, Section IV), as it is commonly called,[33] tells us that all we can know derives either from logical inference ("relations of ideas") or from experience ("matters of fact"); and nothing more. It supports his celebrated, closing riposte against any work that consists of propositions of neither kind (Hume, 1777, p. 120): "Commit it then to the flames: For it can contain nothing but sophistry and illusion." The fork has been a fixture, in one form or another, in empiricist thinking to the present day. It was foundational for the logical empiricists. They took it to a dangerous extreme as giving the only basis on which propositions could be meaningful. Hempel's (1965, p. 101) treatment of cognitive significance begins with a version of the fork. It allows sentences to be cognitively significant only if they are logical truths or empirically testable. A continuation of this literature now distinguishes empirical metaphysics from non-empirical metaphysics. Ladyman and Ross (2007, pp. 7–10) denounce the latter as "Neo-Scholastic metaphysics." Bryant (2020) calls the latter "free range metaphysics" and finds it epistemically inadequate.[34]

12.4 *Fatal Abstraction*

Here I sketch a version of this traditional empiricist critique of non-empirical metaphysics that focusses on a foundational assumption of metaphysics. That assumption was in place at the outset. Simons' historical survey (2009, p. 4) reports of Aristotle's conception of metaphysics:

> Unlike other branches of knowledge, which concern themselves with part of what there is, metaphysics is universal: it is about absolutely everything, not with every detail, but only those matters which all things share.

[32] For the complaint that these distinctions lack empirical manifestations, see Norton (2015).

[33] Editor's notes, p. xxxv in Hume (1777).

[34] For a more equivocal appraisal of the two approaches, see Guay and Pradeu (2020).

The same assumption persists in modern accounts of metaphysics. Le Poidevin (2009, p. xx) gives it as[35]:

> So far, then, we have characterized metaphysics as concerned with what it is to be or be real, with what things there are, with the way that they are, and with the connection between the way things are and what things there are. And all this is pursued at a higher level of abstraction than typifies any of the special sciences like physics, geology or chemistry.

The assumption is that it is possible to abstract away the specifics of the treatments of the various sciences and be left with non-trivial content of enduring, universal applicability. For example, each of the sciences gives an account of how the entities in its domain interact. Abstract away the specifics of the individual sciences and what should remain is a universal metaphysics of causation. It was, perhaps, once a reasonable conjecture that there might be non-trivial content at this general level of abstraction. It had to be a conjecture since there is no prior guarantee that such content is there to be found. It might fail. The developments recorded in the history of science point to failure. They show that the many aspects of the world come in a great variety. The more we learn of them, the greater becomes the variety and less commonality can be found within it.

Developments in our theories of matter illustrate this continually growing variety. In the seventeenth century, matter came to be regarded as consisting merely of corpuscles in space. The problematic question was whether they could be attributed further properties beyond their occupation of some small volume of space, such as an active power of gravitational attraction. In the course of the nineteenth century, this corpuscular conception of matter was supplemented by, and even sometimes fully replaced by, a field conception. This additional form of matter is not localized in the manner of corpuscles. The electric field of a single charge can pervade all of infinite space. In the twentieth century, in Einstein's general theory of relativity, the strict division between space the container and matter the contained began to break down. The gravitational field, with its material energy-momentum, was now merged into the geometrical structure of spacetime itself (and not without some complications. See Dürr (ms)). Einstein's unified field theory program sought a theory in which this breakdown would be complete, embracing all matter. There would be no distinction between matter and spacetime. At the same time, the new quantum theory brought different conceptions of matter that we still find problematic today: matter is not particle-like or wave-like, but both at the same time. Its matter is non-local in the sense that each particle may have no definite position and that entangled particles remain so even when separated by astronomical distances. Quantum field theory then breaks down the idea of individual particles. They are merely excitations in a quantum field and we can have states with particles but no definite number of them. These results seem extraordinary on first

---

[35] Ney (2014, p. xiii) gives a similar characterization: "Unlike the natural and social sciences that seek to describe some special class of entities and what they are like—the physical things or the living things, particular civilizations or cultures—metaphysicians ask the most general questions about how things are, what our universe is like."

acquaintance, but they soon become everyday facts of familiar physics. Theories of quantum gravity promise that this parade of the extraordinary is far from complete.

This growing variety is typical in science. As we learn of the ever-greater extent of it, the scope for commonalities reduces. Where does it end? We must take seriously the possibility that there is no non-trivial commonality among all the instances that do now and will in future fall under some metaphysical category. They may just be individual cases bundled together by us since they seem similar to us. If those similarities are merely superficial, then abstracting away the specifics of the individual cases will leave nothing. There will be no non-trivial content in common and no metaphysics. The abstraction would be fatal.

One can only feel sympathy for a metaphysics trying to recover an enduring notion for some metaphysical category, such as substance, when faced with this history. It is forever chasing a moving target. New science repeatedly produces new results that outstrip our old imaginings. Empiricist metaphysics has some chance of coping. It can accommodate this endlessly growing variety by restricting itself to what can be learned responsibly from the empirical evidence and by making no immutable pronouncements. Non-empirical metaphysics, however, cannot do this. By definition, it cannot draw on this new empirical evidence. It is in an impossible position. It must somehow have already divined a place for the latest novelty in its conceptions from a fixed body of metaphysics that cannot be informed by new experiences. It is trapped in an ill-fated attempt to do a priori science.

The history of these efforts in non-empirical metaphysics has been one of persistent failure, as one would surely expect for any attempt at a priori science. I have investigated one case in greater detail and believe the mode of failure I found there will be found throughout non-empirical metaphysics. Norton (2003) traces the long history of the failure of metaphysics to identify non-trivial, universal truths of causation. Time and again, it would pronounce with great certainty some universal, causal truth, adapted to the then present understanding of processes. Subsequent developments in science would overturn it.[36] Since causal relations underwrite many claims of possibility and necessity, this failure of causal metaphysics is automatically also a failure of modal metaphysics.

Another ready candidate for fatal abstraction is the current literature on "grounding." It seeks to abstract away all the specifics of particular cases in which something depends upon something else. The goal is to find a universal, non-trivial notion of grounding. However, what is debated is whether there even is just one notion of grounding.[37]

This analysis of this section indicates that a non-trivial, non-empirical metaphysics is impossible. If this is so, it follows that metaphysics can provide no

[36] These continuing corrections to causal metaphysics are unlike the continuing corrections within science. For the corrections in science derive from the science itself through the epistemic power of its empirical methods. Causal metaphysics has no corresponding power and no means for finding its own errors. The corrections are imposed from the outside.

[37] For a sketch of the debate, see Bliss and Trogdon (2016, Sect. 1).

support for metaphysical possibilities and necessities that outstrip the logical and the empirical. In this conclusion, I concur with Callendar's (2014, p. 44, his emphasis) remark[38]:

> ... there is no interesting species of metaphysical modality that is largely immune to science. Our modal intuitions are historically conditioned and possibly unreliable and inconsistent. The only way to weed out the good from the bad is to see what results from a comprehensive theory that seriously attempts to model some or all of the actual world. If the intuitions are merely 'stray' ones, then they are not ones to heed in ontology. *In metaphysics we should take possibilities and necessities only as seriously as the theories that generate them.*

13 Conclusion: How Did This Happen?

To an empiricist, the convolutions of the philosophical literature on modality are bewildering. The ideas are, for the empiricist, so simple. What is possible is what our evidence positively allows. What is necessary is what our evidence compels. How much more can there be to say?

Let me raise a conjecture on how the modality literature arrived at its present disarray. It is a broth touched by too many cooks. First, there are the traditional metaphysicians, long criticized by empiricists as far back as Hume. These metaphysicians mistakenly but tenaciously believe it is possible to know profound, contingent truths of the world through means other than experience. By mere reflection, we can know that there is a necessary being in the world. They offer us an epistemic free lunch.

This old-fashioned metaphysics has been overlaid by philosophers who take the reach of formal, modal logics to extend too far beyond the refined realm of abstract logic. In their work in which the popular modal logic S5 was introduced, Lewis and Langford (1959, p. 160) went to some pains to emphasize that their notion of possibility was narrowly logical: "Thus [possibly] *p* means '*p* is self-consistent' or '*p* does not imply its own negation.' " They distinguished it from the more colloquial notion that is related to data or what is known (quoted in Sect. 4 above).

The explicit narrowness of their conception is prudent. The world is greatly varied in its states and properties. A notion of possibility that can go beyond pure logic and capture this variety must itself be greatly varied. In this regard, formal modal logics are a meager resource. The major claims of a modal logic reside in a few postulates, such as the few axioms of the logic S5. The expectation has now become that this one, logically motivated size will somehow fit all cases. The characterization of possible worlds in the later formal semantics is, by design, as lean as possible, so as to ensure the widest application. These logics are at best a

[38] For a further critique, see Clarke-Doane (2019) who argues that metaphysical possibility lacks the absoluteness routinely claimed for it.

convenient starting point for a richer notion of possibility, while in need of extensive supplementation; or at worst, a misdirection at the outset.

In autobiographical reflections, Kripke (1980, pp. 3–6) recalled doubting his claims on the necessities associated with rigid designators in natural language. His confidence in the claims was restored by "(self-evident) theses of philosophical logic independent of natural language."[39] They were, it now seems, an unreliable guide concerning the necessities of the physical world.

The mathematical demands of formal logic encourage oversimplification. Possibilities range from the wildly unlikely to the expected; and the gradations in between really do matter. A mere assertion of possibility crams them all together. Why use it? It is easy to formalize. The duality of possibly as not-necessarily-not mimics the duality of existence as not-all-not that is so familiar and comfortable in ordinary predicate logic. The result is that the subsequent modality literature largely limits itself to these flattened notions of possibility and necessity.

These flattened notions of possibility and necessity also support a simple semantics in the formal logic: *P* is possible if it is true in at least one possible world, where a "possible world" is an abstract entity within the structures of formal modal logic. Of them, in his treatment of the modal logic modal logic $S5^{*=}$, Kripke (1959, p. 2) writes:

> The basis of the informal analysis which motivated these definitions is that a proposition is necessary if and only if it is true in all "possible worlds." (It is not necessary for our present purposes to analyze the concept of a "possible world" any further.)

But now we need to give some meaning to natural language assertions of possibility and necessity and we do need to analyze the concept of a possible world. It is all too tempting to remove the talk of "all possible worlds" from the artificial environment of a formal logic and, without much further analysis, treat them as a notion fully intelligible in a more general discourse. We may even succumb to the temptation to reify them. They are all real, so now the semantic manipulations of the formal logic can be carried over to natural language as well. This, in spite of Kripke's (1980, p. 48, fn) admonition:

> The apparatus of possible words has (I hope) been very useful as far as the set-theoretic model-theory of quantified modal logic is concerned, but has encouraged philosophical pseudo-problems and misleading pictures.

Conveniences designed to ease the construction of a formal logic have become foundational philosophical principles. The final spice added to the broth comes from a sound literature that simply reports what our latest science tells us about foundational matters. The well-meaning philosophers who undertake this last project have started to call their efforts "metaphysics." Yet as long as they draw

[39] In his demonstration of the completeness of the modal logic $S5^{*=}$, Kripke (1959, p. 2) defines a subformula *B* as necessary if it is assigned **T** by every member of some set of interpretive assignments **K**. Crucially, if *B* contains free individual variables $x_1, \ldots, x_n$, each must be assigned the same individual in the domain **D** by every assignment of **K**. Williamson (2013, p. 120) describes this as "treating variables as rigid designators."

the authority of their results from the idea that our best science is so because it has the strongest empirical support, their project is distinct from the traditional metaphysician's and the modal logician's.

These three approaches are mixed indiscriminately in the literature. The result is bewilderment when an empiricist opens a new work proclaiming metaphysical possibilities in some area. Of what sort are they? Are they more *a priori* muddles of the traditional analysis? Are they misidentifications of conventions of language or formal conveniences as metaphysical necessities? Or are they merely a benign empiricist analysis of our current science? And do any of the claims, if made responsibly, require anything more that the two notions of logical and empirical possibility articulated here?

References

Bliss, R., & Trogdon, K. (2016). Metaphysical grounding. In E. N. Zalta (Ed.), *The Stanford encyclopedia of philosophy* (winter 2016 ed.). https://plato.stanford.edu/archives/win2016/entries/grounding/

Boyd, N. (2018). Evidence enriched. *Philosophy of Science, 85*, 403–421.

Bryant, A. (2020). Keep the chickens cooped: The epistemic inadequacy of free range metaphysics. *Synthese, 197*, 1867–1887.

Callender, C. (2014). Philosophy of science and metaphysics. In S. French & J. Saatsi (Eds.), *The Bloomsbury companion to the philosophy of science* (pp. 33–54). Bloomsbury.

Chalmers, D. J. (2011). The nature of epistemic space, in Egan & Weatherson (2011, pp. 60–107).

Clarke-Doane, J. (2019). Metaphysical and absolute possibility. *Synthese*, published online. https://link.springer.com/article/10.1007/s11229-019-02093-0

DeRose, K. (1991). Epistemic possibilities. *The Philosophical Review, 100*, 581–605.

Dürr, P. (ms), *Gravitational energy and energy conservation in general relativity and other theories of gravity*. DPhil Thesis, University of Oxford.

Egan, A., & Weatherson, B. (Eds.). (2011). *Epistemic modality*. Oxford University Press.

Fine, K. (2002). *Varieties of necessity*, Ch. 7 in Modality and Tense: Philosophical Papers. Oxford: Clarendon Press.

Gendler, T. S., & Hawthorne, J. (2002a). *Conceivability and possibility*. Clarendon Press.

Gendler, T. S., & Hawthorne, J. (2002b). Introduction: Conceivability and possibility, pp. 1–70 in Gendler & Hawthorne (2002a).

Guay, A., & Pradeu, T. (2020). Right out of the box: How to situate metaphysics of science in relation to other metaphysical approaches. *Synthese, 197*, 1847–1866.

Hawley, K. (2009). Metaphysics and relativity, Ch. 47 in Le Poidevin et al. (2009).

Hempel, C. (1965). Empiricist criteria of cognitive significance: Problems and changes. In *Aspects of scientific explanation: Essays in the philosophy of science* (pp. 101–119). The Free Press.

Hume, D. (1777). *An enquiry concerning human understanding* (P. Mullican, Ed.). Oxford University Press, 2007.

Ismael, J. (2017). An empiricist's guide to objective modality. In M. Slater & Z. Yudell (Eds.), *Metaphysics and the philosophy of science: New essays* (pp. 109–125). Oxford University Press.

Kim, J., Sosa, E., & Rosenkrantz, G. (Eds.). (2009). *A companion to metaphysics* (2nd ed.). Wiley Blackwell.

Kment, B. (2014). *Modality and explanatory reasoning*. Oxford University Press.

Kment, B. (2017). Varieties of modality. In E. N. Zalta (Ed.), *The Stanford encyclopedia of philosophy* (spring 2017 ed.). https://plato.stanford.edu/archives/spr2017/entries/modality-varieties/
Kripke, S. A. (1959). A completeness theorem in modal logic. *Journal of Symbolic Logic, 24*, 1–14.
Kripke, S. A. (1980). *Naming and necessity*. Harvard University Press.
Ladyman, J., & Ross, D. (2007). *Everything must go: Metaphysics naturalized*. Oxford University Press.
Le Poidevin, R. (2009). General introduction: What is metaphysics? pp. xviii–xxii in Le Poidevin et al. (2009).
Le Poidevin, R., et al. (Eds.). (2009). *The Routledge companion to metaphysics*. Routledge.
Lewis, D. (1986). *On the plurality of worlds*. Blackwells.
Lewis, C. I., & Langford, C. H. (1959). *Symbolic logic* (2nd ed.). Dover.
Loux, M. J., & Zimmerman, D. W. (Eds.). (2003). *The Oxford handbook of metaphysics*. Oxford University Press.
Machery, E., et al. (2004). Semantics, cross-cultural style. *Cognition, 92*, B1–B12.
Ney, A. (2014). *Metaphysics: An introduction*. Routledge.
Norton, J. D. (2003). Causation as folk science. *Philosophers' Imprint*, *3*(4), reprinted in pp. 11–44, Price, H., & Corry, R., *Causation, physics and the constitution of reality*. Oxford University Press.
Norton, J. D. (2010). Cosmic confusions: Not supporting versus supporting not-. *Philosophy of Science*, *77*, 501–23.
Norton, J. D. (2015). The burning fuse model of unbecoming in time. *Studies in History and Philosophy of Modern Physics., 52*, 103–105.
Norton, J. D. (2021) *The material theory of induction*. BSPSOpen/University of Calgary Press.
Putnam, H. (1975). The meaning of meaning. In *Language, mind, and knowledge* (*Minnesota studies in the philosophy of science*) (Vol. 7, pp. 131–193).
Roca-Royes, S. (2017). Similarity and possibility: An epistemology of de re possibility for concrete entities. In B. Fischer & F. Leon (Eds.), *Modal epistemology after rationalism* (pp. 221–245). Synthese Library.
Sider, T. (2003). Reductive theories of modality, pp. 180–208 in Loux & Zimmerman (2003).
Simons, P. (2009). Millennia of metaphysics, pp. 3–7 in Le Poidevin et al. (2009).
Strohminger, M. (2015). Perceptual knowledge of nonactual possibilities. *Philosophical Perspectives, 29*, 363–375.
van Inwagen, P. (2009). *Metaphysics* (3rd ed.). Westview.
van Inwagen, P., & Sullivan, M. (2020). Metaphysics. In E. N. Zalta (Ed.), *The Stanford encyclopedia of philosophy* (Spring 2020 ed.). https://plato.stanford.edu/archives/spr2020/entries/metaphysics/>
Weatherson, B., & Egan, A. (2011). Introduction: Epistemic modals and epistemic modality, in Egan & Weatherson (2011, pp. 1–18).
Williamson, T. (2013). *Modal logic as metaphysics*. Oxford University Press.

Laws of Nature as Epistemic Infrastructure Not Metaphysical Superstructure

Richard Healey

> *"We need to rethink the idea of a scientific law pragmatically or functionally, that is, in terms of what scientific laws let us do rather than in terms of some ideal of a law . . . "*
>
> *Mitchell (2002).*

Abstract The status of laws of nature has been the locus of a lively debate in recent philosophy. Most participants have assumed laws play an important role in science and metaphysics while seeking their objective ground in the natural world, though some skeptics (Giere, van Fraassen, Cartwright) have questioned this assumption. So-called Humeans look to base laws on actual, particular facts such as those specified in David Lewis's Humean mosaic. Their opponents (including Maudlin) argue that such a basis is neither necessary nor sufficient to support the independent existence of scientific laws.

This essentially metaphysical debate has paid scant attention to the details of scientific practice. It has mostly focused on so-called fundamental laws, assumed to take a particular form (such as Maudlin's FLOTEs). I propose a pragmatist alternative—not as another position in the debate but as an alternative to the debate itself. This pragmatist alternative offers a view that questions the representational conception of truth presupposed by participants to the debate as well as the metaphysical import of fundamental laws.

Statements of law serve many different purposes in science: I'll look at some. But their central role is in inference, primarily to improve the epistemic state of a scientist with limited access to information. To play this role, a scientific law statement need be neither necessary, unconditionally universal nor even true. It must merely be sufficiently reliable within its domain of application. The use of laws in astrophysics and metrology will help to illustrate these points.

R. Healey (✉)
University of Arizona, Tucson, AZ, USA
e-mail: rhealey@email.arizona.edu

Y. Ben-Menahem (ed.), *Rethinking the Concept of Law of Nature*,
Jerusalem Studies in Philosophy and History of Science,
https://doi.org/10.1007/978-3-030-96775-8_6

Keywords Laws of nature · Humeanism · Metaphysics · Pragmatism · Inferentialism · Truth · Necessity · Scientific knowledge

1 Introduction

Philosophers have long been puzzled by laws of nature. They strike us as more than brute regularities, and scientists tell us that they have discovered laws that lie behind and perhaps even control the patterns of events. But any such laws are hidden from us: we observe only the patterns, and scientific instruments merely reveal events exhibiting deeper regularities. Taking scientists at their word, contemporary metaphysicians have offered rival analyses of the concept of a natural law. But other philosophers see this analytic project as misguided since scientific practice makes no serious use of the concept. While my sympathies lie with the latter group, I think laws do play important roles in science. But when we see how they play these roles it will become clear that laws of nature are not what metaphysicians have taken them to be.

Laws are important in science even though we know of no empirical laws that apply everywhere, are necessary, or are even true. A law does not need to be any of these things to contribute to scientific knowledge. What it does need to do is to support reliable inferences.

After a preliminary survey of our target in the next section, Sect. 3 distinguishes two rival views of scientific laws—as representations of an objective reality that many philosophers take to consist of more than actual (non-modal), particular facts; or as what Ryle called inference tickets that make explicit those inferences we find to be reliable but not deductively valid. Taking the latter view helps illuminate the general function of laws in what I call the epistemic infrastructure of science. But a closer examination reveals several ways in which scientists use laws to gain knowledge.

Section 4 gives examples of ways in which laws have been used vastly to increase our knowledge of what there is in the universe and how it behaves. Section 5 exhibits the crucial role of laws in guiding our construction of theoretical models—in many ways the work-horses of contemporary science, whose successful application has given us so much scientific knowledge. Section 6 is devoted to the underappreciated value of scientific laws in metrology. It shows how laws have contributed to what I call the social epistemology of practical knowledge by permitting scientists and technologists to coordinate their actions by calibrating their instruments to shared standards.

While accepting that what pass for scientific laws have functioned to back inferences in all these ways, a proponent of the representational view may nevertheless maintain that an actual scientific law offers a more secure backing because it truly represents a deep feature of objective reality, where depth is measured by the modal strength associated with natural or metaphysical necessity, objective chance, or the logical necessity by which a theorem follows from laws taken as axioms. A law statement could offer such backing only if it were true. But law statements like those used in previous sections to illustrate the varied inferential functions of laws in science are not true. Either we know no scientific laws because none of our law statements faithfully correspond to objective reality, or truth is not a necessary condition on a statement of a scientific law. In Sect. 7 I grasp the second horn of this dilemma.

Section 8 briefly considers rival accounts of laws of nature offered recently in what some call the metaphysics of science. All these accounts take a representational view of laws though they differ as to what exactly law statements represent. All are thereby committed to the truth of any statement at least of a fundamental scientific law (less fundamental laws may then be considered strictly false though approximately true in certain circumstances). But the scientific laws we know admirably serve their general and specific epistemic functions independent of their truth. So all these rival metaphysical accounts impose a requirement on laws of nature that our scientific laws don't meet now, and perhaps never could meet. Some accounts add further superfluous modal requirements.

In conclusion I argue for a turn away from metaphysics and toward pragmatism in our attempts to understand scientific laws. Ramsey took this turn already in 1929 when he denied that law statements are evaluable either as true or as false. Later, Ryle (1949) said that a law statement is true or false. But Ramsey had argued that their role in supporting reliable inferences showed that law statements were neither. Examination of the scientific function of laws reveals only shadows of the metaphysician's requirements of unconditional universality, necessity and even truth. Scientists seek maximal reliability (the shadow of necessity), maximal applicability (the shadow of unconditional universality) and maximal endorsability (the shadow of truth).

2 What Is a Law of Nature?

When he called his four key statements about motion and gravitation laws of nature, Newton (1687) enshrined a usage that is now the common currency of scientists, philosophers and the general public. But the modern concept of a law of nature is now largely free of the theological associations it had for Newton. As a deist, Newton thought God created the world so that it would develop in accordance with these laws, permitting rare violations only in the form of its creator's miraculous interventions. But we still hear echoes of Newton's concept in metaphors like natural laws *governing* what happens, or of natural laws' *dominion*, or of events as *subject to* laws of nature. Such metaphors suggest that events manifest laws that are embedded in or even in some sense prior to any patterns they display.

With the twentieth century's linguistic turn, philosophers came to focus on law statements rather than anything in the world they may purport to describe or represent. Many have joined Carnap (1966) in identifying scientific laws with statements expressed by a class of general sentences: Ramsey (1929) had called these variable hypotheticals. But some have more recently returned laws to the world. Maudlin (2007) takes them as ontological primitives, while Weinberg (1996) took them to be as real as the rocks in the fields.

By beginning with language one can avoid premature discussion of what would make a law statement true and whether law statements are even truth-evaluable. Accordingly I'll take scientific laws to be statements expressed by sentences of natural or scientific language, usually if not always universal in form. Newton's laws are of this form, whether stated in a natural language (like Newton's Latin) or mathematics (e.g. $\boldsymbol{F} = d(m\boldsymbol{v})dt$).

Many laws are usually called something else; equations (like the Schrödinger equation), principles (such as Archimedes principle and the superposition principle), constraints (including the Gauss and Hamiltonian constraints), hypotheses (Avogadro's hypothesis) and rules (the Born rule, Chargaff's rules). These alternative terms often mark the history or some special use of the law in question. In science some things are called laws although they are stated or used rather differently than typical scientific laws (Bode's law, Hubble's law, Moore's law, Zipf's law). Such variations in terminology should not distract attention from something important: that a typical scientific law has a characteristic general function in science. But what is that function?

3 Laws as Inference Tickets

By assuming that our statements about the laws of physics are in a one to-one correspondence with aspects of objective reality, Weinberg (1996) joined a venerable philosophical tradition. This tradition takes it to be the general function of a scientific law to truly represent a widespread if not universal regularity in nature that is more than just a pattern of coincidences. This then explains why by applying such laws we can better predict, explain and control natural events.

But a rival tradition maintains instead that the general function of a scientific law is simply to license inferences, including those required to predict, explain and control natural events. This tradition is skeptical of the idea that a scientific law truly corresponds to any modal aspect of the world like natural necessity, tendency or chance. And it takes improvement in our ability to predict, explain and control events to follow just from successful exercise of this general function in the absence of any such correspondence.

In his (1929, [1990, p. 149]) Ramsey took his place in this rival tradition when he said "variable hypotheticals are not judgments but rules for judging". He maintained further that these sentences (including those used to express causal laws and laws of nature) express cognitive attitudes without [what they express?] being propositions. By doing so he distanced himself from the view sketched in his (1928) that warranted his name's inclusion with J.S. Mill and David Lewis as the second initial of the acronymic MRL view of laws.

His *Concept of Mind* (1948) made Gilbert Ryle a more widely known exponent of an inferential view of laws of nature, especially through his description of them as inference-tickets.

> Law-statements are true or false but they do not state truths or falsehoods of the same type as those asserted by the statements of fact to which they apply or are supposed to apply. They have different jobs. The crucial difference can be brought out in this way. At least part of the point of trying to establish laws is to find out how to infer from particular matters of fact to other particular matters of fact, how to explain particular matters of fact by reference to other matters of fact, and how to bring about or prevent particular states of affairs. A law is used as, so to speak, an inference-ticket (a season ticket) which licenses its possessors to move from asserting factual statements to asserting other factual statements. It also licenses them to provide explanations of given facts and to bring about desired states of affairs by manipulating what is found existing or happening. (p. 105)

Ryle's distinction between two types of truth (statements of fact and the true law statements to which they may apply) is similar to Carnap's (1966). In Carnap's view the distinction is made already at the syntactic level, to be backed up by application of the standard semantics of first order logic. Ryle on the other hand locates "the crucial difference" at a pragmatic level in terms of their distinct functions: unlike statements of fact, the function of law statements is to license inferences. While sharing Ryle's view of the function of law statements, Ramsey takes a further, more radical step. In his (1929) he appears to accept Carnap's view of the logical form of sentences used to express law statements. But by denying these the title of propositions he declines to evaluate them as true or false in accordance with the standard semantics of first order logic. Rather than presenting an alternative semantics of variable hypotheticals he offers a subtle account of a speaker or hearer's differing cognitive attitudes toward an utterance or inscription of a variable hypothetical that may lead him to judge it not true or false but right or wrong.

The dispute within the inferentialist tradition of how law statements are best understood has significant implications for their representational status, so it bears on the metaphysical import of laws of nature. I'll return to this issue after describing and illustrating several ways in which the inferential power of laws contributes to scientific knowledge. The illustrations are of independent interest. But I have deeper motives for offering them. One is to develop the schematic characterization of the goals of science as prediction, explanation and control by exhibiting other ways in which laws undergird the epistemic and practical activities of scientists. I see this as a way of combating skeptics such as van Fraassen (1989), Giere (1999) and Cartwright (2016) who have minimized or dismissed the importance of laws in understanding the scientific enterprise. Another motive is to show how each of these epistemic contributions supports an inferentialist approach by explaining why scientific laws lack many of the features attributed to them by contemporary metaphysicians as well as previous generations of philosophers of science.

4 Laws Extend Our Knowledge

A few people have left the earth to visit just a tiny portion of the solar system, and we have sent machines barely outside of it. Yet science has given us a great deal of knowledge of the universe. How has this been possible? The construction of instruments such as telescopes and now gravitational observatories like LIGO and VIRGO has certainly played an essential role. But without scientific laws these would be unable to deliver the knowledge we get by using them. Laws fuel such epistemic engines: laws form an essential part of the epistemic infrastructure of the scientific enterprise.

Helium provides a striking example. This element was first found not on earth but on the sun from which it derives its name. Spectroscopic analysis of telescopically collected light from the sun revealed lines that did not correspond to those characteristic of any known terrestrial element. The laws of spectroscopy that underlay this characterization were known well enough to justify scientists in adding a new element characterized by the newly observed spectral lines, even though it was only 27 years later that a sample of an element with this spectrum was isolated on earth.[1] Of course the justification of the claim that a hitherto unknown element was present on the sun depended on much more than just the laws of spectroscopy. But they played an essential role in establishing this as scientific knowledge.

Astrophysics is a rich source of examples of the use of scientific laws to transcend our local situation to vastly expand our knowledge of the universe. We can see laws in action in the current controversy over the value of the Hubble constant H. Recall that Edwin Hubble is credited with the discovery that on a large scale the universe is expanding in such a way that, on average, the relative velocity at which a distant galaxy is receding from the Earth is proportional to its distance from us: $v = Hd$, where H is now known as the Hubble constant.

A chain of inferences in conventional observational astronomy based on the so-called cosmic distance ladder led to the 2017 estimate that $H = 73.2$ km/sec/megaparsec. But, when applied to observations of anisotropies in the cosmic microwave background radiation, laws used to construct current cosmological models license the inference to the value $H = 67.8$ km/sec/megaparsec. This may not look like a big discrepancy. But careful analysis of the observational errors in each case indicated that the two estimates are separated at the 3σ level, meaning

[1] The discovery of helium is a bit like that of oxygen (Kuhn (1962, pp. 56–56)). In 1868 Lockyer observed a line in the spectrum of light from the solar corona that he had not seen before in any spectrum, although Janssen had seen this line during a solar eclipse earlier that year. Lockyer first took the line to be produced by an unknown element present in the solar corona which he went on to call helium, but he later came to doubt that a new element gave rise to this line. The spectral line was observed in gases emitted by Vesuvius in 1882, but only in 1895 was a sample of an inert gas producing this line isolated in the laboratory. After obtaining a sample himself and observing its spectrum, Lockyer identified the gas with the new element helium he had initially taken to be responsible for the new spectral line he had observed in light from the solar corona in 1868.

that there is a probability of only 0.13% that the conclusions of both inferential chains are correct.

The cosmic distance ladder is based on observation of light emitted by distant stars. Geometric parallax measurements yield good estimates of the distance only to nearby stars in our galaxy. But these include so-called Cepheid variables, whose analysis by Henrietta Leavitt and Harlow Shapley proved crucial to Hubble's work by providing the next rung in the ladder, permitting inference to the distance of other galaxies—indeed, to the *existence* of other galaxies besides our own Milky Way.

In 1912 Leavitt noted a rough empirical regularity between the apparent brightness and period of variation of a number of stars whose brightness varied regularly in the same way over a period of from 1 to several hundred days. Since some she studied seemed to be located at roughly the same distance from the earth (in the small Magellanic cloud) this indicated a corresponding relation between period and absolute magnitude. At this stage no-one knew *why* these stars varied in brightness. It was simply assumed that the regularity was reliable enough to be extrapolated to all similar stars, wherever they were: the regularity was treated as a law licensing subsequent inferences from the apparent brightness of any such variable star to its distance, basing its absolute magnitude on its period and assuming another better-established law—that apparent brightness is inversely proportional to distance squared.

Hubble's first crucial move was to identify individual Cepheid variable stars in "nebulae" and, using the period/magnitude relationship, to estimate their distance from the earth, thereby concluding that these "nebulae" were actually entire galaxies of stars far beyond our own galaxy.[2]

His second crucial move was to use other laws to estimate the velocitie*s* of many of these galaxies.

Two kinds of laws were needed here, both on much firmer epistemic ground than the Leavitt-Shapley relation for Cepheid variables. A law associated with the Doppler effect permitted an inference from the observed frequency of a moving source of radiation to its velocity of recession from the observer. And laws of spectroscopy correlated frequencies of particular spectral lines with the element in whose spectrum they lay, so that even a "red-shifted" line in the spectrum of a particular element in a moving source could still be uniquely correlated with the same line for an atom of the element at rest.

After using all these laws to estimate both the distance to and the velocity of several galaxies, Hubble noticed a regular but far from perfect correlation between these two magnitudes. This came to be known as Hubble's law describing the expansion of the universe. A refined version came to stand on firmer ground after further observational and theoretical developments.

[2] Hubble also used two other plausible principles in a complex interlocking set of inferences to estimate distances. I thank John Norton for alerting me to the finer structure of Hubble's reasoning that he explores in Norton (unpublished), as well as his suggestion to look more closely into the history of the "discovery" of helium on the sun.

The mechanism of Cepheid variability came to be understood, first in terms of thermodynamics and later in terms of the difference between how singly- and doubly-charged helium ions in its outer layers absorb the radiation emitted by a certain kind of large star. These more fundamental laws improved the epistemic status of the Leavitt-Shapley law, while also showing why other kinds of variable stars do not conform to it. This is a nice example of how something that began as a somewhat shaky empirical regularity can be refined as well as strengthened using inferences licensed by other laws, thereby improving the connections in the inferential web constituting what I called the epistemic infrastructure of science.

Many galaxies are still too far away for us to identify Cepheid variable stars in them using even the best telescopes—they are just not bright enough. But more recently astronomers have found a much brighter "standard candle" among supernovae—exploding stars that can briefly outshine their entire galaxy. The mechanism of a particular kind of supernova is believed to be well enough understood for astrophysicists to feel confident that at peak they are all equally bright. (A Type IA supernova is believed to occur when, through accretion from its companion, a white dwarf star orbiting a larger star approaches the Chandrasekhar limit of 1.44 solar masses beyond which it cannot support itself by degeneracy pressure. This can initiate an uncontrolled fusion reaction in the white dwarf that blows it apart in a few seconds, releasing an enormous amount of energy as radiation.) The peak brightness is constant insofar as it always arises from the same mechanism in stars of the same mass.

It is because the laws used to infer this conclusion are quite fundamental that the observed brightness/actual brightness correlation is taken to permit a reliable inference to the distance of the galaxy in which the supernova occurred. Observations of Type IA supernovae since 1998 provide our strongest evidence that the expansion of the universe is actually accelerating: our distance from a typical distant galaxy is increasing faster than linearly with time. But the Hubble "constant" is not constant but *decreases* with time: a galaxy at a certain large distance from us is typically receding from us slower than another galaxy at that distance in the past. In this case, refinement of a law has shown that it does not hold at all times, contrary to what many philosophers have considered essential to laws of nature. Nevertheless, the inferential procedures I've just sketched seem reliable, and by 2017 they had yielded a value of $H = 73.2$ km/sec/megaparsec for the current value of the Hubble "constant".

Now rungs of the cosmic distance ladder have given way in the past: Using a more primitive form of the same basic methodology, Hubble himself estimated the value of H as 500! And we now have a very different chain of inference to the current value of H that doesn't even involve observing stars and figuring their distances from the Earth. Instead it appeals to a careful analysis of anisotropies in the cosmic microwave background radiation (CMB) that provides some of the most conclusive evidence for the "Big Bang" that initiated the universe's current expansion.

My treatment of this alternative inferential chain will be brief and unoriginal I'll simply quote Joshua Sokol's account from a (March 10, 2017) issue of *Science*.

> In 2003, the Wilkinson Microwave Anisotropy Probe (WMAP) published its first map showing the speckles of temperature variations on the CMB. The maps provided not a standard candle but a standard yardstick: a pattern of hotter and colder spots in the primordial soup created by sound waves rippling through the newborn universe.
>
> With a few assumptions about the ingredients in that soup—familiar particles like atoms and photons, some extra invisible stuff called dark matter, and dark energy—the WMAP team could calculate the physical size of those primordial sound waves. That could be compared to the apparent size of the sound waves as recorded in the CMB speckles. The comparison gave the distance to the microwave background, and a value for the expansion rate of the universe at that primordial moment. By making assumptions about how regular particles, dark energy, and dark matter have altered the expansion since then, the WMAP team could tune the constant to its current rate of swelling. Initially they came up with a value $H = 72$. [But more recently] Planck, a European Space Agency mission that has now imaged the CMB at higher resolution and greater temperature sensitivity, has settled on the value for H of 67.8.

In this episode we see a problem in the epistemic infrastructure of science. It's as if two teams, using the most advanced surveying and engineering techniques available, began tunneling from different sides of the Alps to build a road or rail link between Switzerland and Italy, but found that their tunnels failed to meet in the same place under the mountains.

Since 2017 several other relatively independent methods of measuring the Hubble constant have been developed, resting on other observations, laws and astrophysical models. As I write these have still not resolved the discrepancy between estimates based on observations of supernovae and observations of the cosmic microwave background. But the different methods may be reconciled either by locating systematic errors of observation or by suitably modifying astrophysical or cosmological models. The use of laws in constructing these and other scientific models is the topic of the next section.

5 Laws Guide Theoretical Modeling

According to Giere (1999, p. 90)

> . . . the whole notion of "laws of nature" is very likely an artefact of circumstances obtaining in the seventeenth century. To understand contemporary science we need not a proper analysis of the concept of a law of nature, but a way of understanding the practice of science that does not simply presuppose that such a concept plays any important role whatsoever.

Giere proposed instead that we should understand science as essentially a modeling activity. He allows that there are principles that guide scientists in constructing the models they use to represent natural systems and he includes Newton's "principles" of mechanics in this category. Newton's three key "principles" of mechanics continue to serve as exemplary scientific laws whatever he may have believed about their pedigree. So while Giere was right to stress the importance of modeling as a scientific activity it is no abuse of usage to acknowledge the vital role of laws in guiding the practice of theoretical modeling in science.

The previous section gave several examples in which laws have been used in constructing models of natural systems: the laws of thermodynamics and laws of the emission, absorption and propagation of radiation are involved in modeling the behavior of Cepheid variables; Pauli's exclusion principle and laws from the theory of relativity are involved in modeling type 1A supernovae; laws of general relativity and equations of state for different forms of matter are involved in cosmological models. But laws are used to guide modeling in quite different sciences.

The Hodgkin-Huxley model of the creation and propagation of action potentials in the giant axon of the squid was a seminal achievement in neuroscience. The mathematical model is a set of nonlinear differential equations representing currents through and voltages across channels in the axon through which may flow sodium and potassium ions. Simple laws of electric circuits form the basis of the model. The model equations were constructed by inserting variables and constants into these laws to represent the conductances of channels through which ions pass or the level of a channel activation. The way these additional variables varied with voltage differences and time was modeled by independently postulated equations containing parameters that could be estimated experimentally by fitting the model to experimental data.

While used by Hodgkin and Huxley to construct a model in neuroscience, Ohm's law, the capacitance equation and Kirchoff's laws of electric circuits are laws of physics. But laws of other sciences are also used in model construction. The Hardy-Weinberg law is used as a starting point for constructing models in population genetics, typically by modifying it to represent the effects of factors it does not represent, such as migration, mutation or selection. The Lotka-Volterra equations form a basis for constructing more realistic models of predator-prey dynamics.

Many laws are used in constructing climate models of the earth. These include the laws of thermodynamics, the Navier-Stokes equations for fluid flow, the Stefan-Boltzmann law, the Clausius-Clapeyron equation, and the laws of conservation of mass and momentum. Key inputs to these models include generalizations about the absorption by various gases (including carbon dioxide, methane and water vapor) of electromagnetic radiation at different wavelengths. These are sufficiently insensitive to environmental conditions to warrant classification as scientific laws.[3]

It is common for scientists to respond to skepticism about the use of climate models by appeal to the epistemic pedigree of the laws used in their construction. They say that most if not all of these laws have been thoroughly tested and used for a very long time successfully to predict, explain and control a wide range of phenomena otherwise unrelated to the climate.

> Each of the components (submodels) and the individual processes must obey the basic physics and chemical laws of the world around us. An important overlooked fact is that the fundamental principles of climate modeling are not new. Simulating the earth system relies on principles of physics and chemistry that have been known for 100–300 years. (Gettelman and Rood (2016, p. 38))

[3] They are strong, stable and relatively abstract in Mitchell's (2000) dimensions of scientific law.

These laws are fundamental to the construction of successful models of the climate and many other phenomena. But as Gettelman and Rood (*ibid*) go on effectively to acknowledge, few if any of them are universally applicable.

> The existence of a new subatomic particle does not require us to change our climate models. They contain no complex physics (like presumptions of warping space-time).

All the laws mentioned in the previous paragraph fail to apply to phenomena that are believed to occur in extreme conditions, some of which can be created in the laboratory while others obtain at other space-time locations in the universe. Most physicists would be willing to call a law fundamental only if it applied even to such phenomena.

6 Laws Align Individual Actions

Section 4 showed how scientific laws may be used to extend our knowledge of the universe. By using them we first learned that helium exists on the sun and now know that it composes 24.7% of its mass. Section 5 illustrated the use of laws to construct models. By using law-based models of nuclear physics, gravity and thermodynamics we now understand how the sun is able to continue radiating vast amounts of radiant energy, primarily through the so-called *p-p* fusion process that produces helium from the hydrogen that the sun is mostly made of.

This knowledge is all theoretical. But more practical knowledge also depends on the use of scientific laws. Geometry (earth-measurement in the original Greek) has been a source of practical knowledge at least since the Babylonians used Pythagorean triples of numbers to form right angles when dividing up land. But since then empirical laws of Euclidean geometry (including Pythagoras's theorem) have been successfully used by surveyors, architects, builders and mechanics of all kinds as a basis for their practical knowledge. Today our knowledge of where we are and how to get where we are going increasingly depends on the laws of general relativity incorporated in the software of the satellites whose precisely timed signals form the basis of the global positioning system. Warped space-time is not just a presumption: even practical knowledge depends on its laws when these are applied in situations in which Euclidean geometry is no longer applicable.

We measure a lot more than just the earth. My high-school physics teacher offered a definition of physics as the science of measurement. But that title has already been bestowed on metrology, the science which, "embrac[es] both experimental and theoretical determinations at any level of uncertainty in any field of science and technology" to quote the International Bureau of Weights and

Measures (BIPM, 2004). Scientific laws are basic to metrology. Metrology, in turn, is basic to practical knowledge.[4] This is made manifest by a recent episode.

As of May 20th, 2019 the basic units of measurement of the international SI system have been redefined. No longer is the unit of mass defined as that of the standard kilogram—a particular platinum-iridium cylinder stored in a vault near Paris. Just as the meter is now defined in terms of a fundamental constant c (the speed of light in a vacuum) the kilogram is now also defined in terms of another fundamental constant—Planck's constant h—which has also been assigned a fixed value by definition, as has the unit of electrical current, the ampere, which is defined in terms of the now fixed value of a third fundamental constant, the charge e of the electron. As before, the second is not defined in terms of a fundamental physical constant, but nor is it defined in terms of any *particular* object: one second is defined as the duration of 9 192 631 770 periods of the radiation corresponding to the transition between the two hyperfine levels of the ground state of any cesium 133 atom. The fundamental constants c, h, and e all figure in fundamental physical laws. The dimensionless fine structure constant $2\pi e^2/hc$ is a fundamental number in physics representing the strength of the electromagnetic interaction between two elementary charges

The place of fundamental constants in fundamental laws suggests that the idea behind these changes in standards is that a standard unit should be accurately reproducible by anyone anywhere, given enough effort. This implies that a fundamental standard should not depend on the properties of any particular object, and preferably not on those of a particular kind of object. Realizing the standard should be possible "whatever objects are to hand": and its realization should involve the most accurate[5] measurements that are available.

But there is a problem with this idea. The constraint of accurate measurement is in severe tension with the constraint of universal availability of materials. A more adequate justification will illustrate several important contributions scientific laws make to what we might call the social epistemology of practical knowledge.

Our science and technology increasingly depend on the coordination of actions throughout a network of diverse and spatiotemporally separated groups and individuals. This is true whether the goal is to build and operate a commercial jet or a new astronomical telescope. Since many of these actions depend on locally performed measurements, their outcomes must have the same significance throughout the network, lest (to take an extreme case) a tire measured in one country last year to have internal diameter 36 cm in one factory fail to fit the rim of a wheel measured in another today to have 36 cm external diameter. Epistemic as well as practical

[4] An engineer once offered this alternative definition: "Metrology is the science of measurement and the technology behind the quality assurance processes in manufacturing that ensure a car runs the way it should, the computer's processor works properly, aircraft stay in the air and more aspects of daily life that we take for granted – until something goes wrong." (https://www.ee.co.za/article/metrology-the-science-of-measurement.html online 4th February, 2019).

[5] Here the term 'accurate' may be taken in any of the five senses distinguished by Tal (2011, pp. 1084-5).

considerations are best served by locating epistemic authority for standardization of measurement units with a group (BIPM) delegated with the task of defining those units and certifying methods for their accurate material realization and global dissemination. Currently the most accurate realization of a standard kilogram is provided by a watt balance.

A measurement with a watt balance involves accurate measurements of a current and a voltage. The connection to Planck's constant and the elementary charge comes in because two quantum effects permit the extremely precise calibration of this current and voltage against standards directly related to h and e.

A law associated with the Josephson effect connects the voltage drop V across a junction in a superconducting circuit to Planck's constant h and the magnitude of the charge e of an electron by $V = nhf/2e = nf/K_J$, where f is the (accurately measurable) microwave frequency with which the junction is illuminated, n is an integral number of voltage steps, and K_J is known as the Josephson constant. A law associated with the quantum Hall effect connects quantized values of the "resistance" R exhibited in a special kind of electrical circuit in a strong magnetic field at low temperatures to the same magnitudes by $1/R = I/V = ne^2/h = 1/R_K$ where n is an integer or simple fraction and R_K is known as the von Klitzing constant. Together these laws imply $h = 4/K_J^2 R_K$, so precise determinations of Josephson and von Klitzing constants formerly permitted a precise determination of h and e by reference to accurately measured currents and voltages. But since the values of h and e have been fixed *by definition* measurements in the Josephson and quantum Hall effects may be taken instead to permit the accurate reproduction of standards of voltage V and resistance R, hence current $I = V/R$ and power $W = VI$.

Determining the mass of an object by a Kibble balance requires measurements in two stages. In weighing mode the object is placed in one pan of the balance. The downward force on the object due to gravity is balanced by an upward force whose origin is electromagnetic. This force is exerted on a coil connected to the side of the balance containing the object because a current is passed through the coil as it sits in a strong magnetic field. When the forces are equal

$$Mg = IBl \tag{1}$$

where M is the mass of the object to be measured, g is the local acceleration due to gravity, l is the length of the coil through which current I is passing, and B is the magnetic field strength where the coil is. The stationary balance point is located using laser interferometry.

In moving mode the object is removed and the same side of the balance is slowly moved vertically at a constant speed u through the magnetic field B, inducing a voltage V in the coil, where

$$V = uBl \tag{2}$$

Putting (1) and (2) together we get

$$Mgu = VI \tag{3}$$

The watt balance gets its name from Eq. (3), equating (virtual) mechanical and electrical power. A laser interferometer is used to accurately measure u, and g can also be measured extremely accurately, essentially by dropping the mirror of an interferometer and applying Galileo's law!

Many different laws are used to model a watt balance, from very different areas of physics and with very different pedigrees, from Galileo's law of free fall through basic laws of classical mechanics and electromagnetic theory to quantum theoretic laws of the Josephson and Hall effects. Many other laws are used to correct for potential errors, including laws of the relativistic 2nd order Doppler effect; and laws of classical rigid body mechanics.

Note that the watt balance does not realize a mass unit by comparison to the mass of a real or ideal mass: this is no Reichenbachian coordinative definition. By contrast, realizations of the second are made by comparison to oscillation periods of real or ideal Caesium atoms. Instead the watt balance realizes a mass unit by relating mass to other quantities whose values are now fixed by definition: Planck's constant, the elementary charge, the speed of light, and a hyperfine frequency of an ideal caesium atom. The law stated by Eq. (3) is key to establishing these relations. So this law plays a role in coordinating the quantitative mass concept to objects and processes in the physical world. But other laws are also involved in the coordination, such as the laws of the Josephson and quantum Hall effects and some form of Galileo's law.

By fixing the values of fundamental constants, the new definitions of SI units leave open the possibility for improved realizations of a mass unit that relate mass to quantities that are more accurately measurable, *via* other laws involving these constants. This is one way in which the discovery of scientific laws like those of the Josephson and quantum Hall effects has contributed (and they or other laws may in future contribute again) to improvement of scientific concepts.

Tal (2017) has offered another justification for the 2019 change in SI units that assigns a further role to laws. A theoretical model makes it possible to estimate the systematic errors involved in the readings of any instrument by comparison to the outcomes of a second instrument whose measurement outcomes are taken as reference values of the measured quantity. Multiple such pairwise comparisons can secure coherence within a group of instruments: they are coherent only if the deviations of each instrument's outcomes from the outcomes of any other instrument lie within that instrument's range of uncertainty. But such pairwise comparisons are impracticable in a large network of instruments, and coherence within each small part of the network does not guarantee global coherence.

By adopting a measurement standard scientists can secure coherence across an epistemic network. The assumptions used to model a standard measurement process then *prescribe* how the measured magnitude is to be applied in that process by regulating the correction of systematic errors across the network: the standard is taken to be subject to no systematic errors. Since this involves the laws used to construct the model, those laws become more than just empirical generalizations. When a standard so modeled is declared free of systematic errors the laws used to model it acquire normative authority over the network.

For a while before the 2019 redefinition, the Josephson and quantum Hall effects realized working standards of electrical measurements and so regulated the treatment of systematic errors in electrical measurements. Since the definitions of basic SI units already had standard realizations regulating systematic errors, electrical measurements drifted free from the rest of the network. That epistemic electrical "island" has now been reconnected to the main network by giving the working electrical standards legislative authority as a measurement standard also for mass and electric current.

The new SI definitions effect an improvement in the social epistemology of science by enabling the consistent application of measurement standards for magnitudes represented in the SI system by all scientists everywhere. The uniform calibration of instruments under the ultimate authority of these standards is important practically, not only in science but in any activity that involves accurate measurement. A great deal of practical knowledge is required to realize the new standard of mass, since a watt balance is very difficult to construct and operate successfully. But, somewhat paradoxically, adoption of the new standard has had only minimal impact on the practical activities of most "end users" in science and technology, very few of whose instruments are sensitive enough to require recalibration.

Each of the different laws involved in modeling the operation of a watt balance played its role in building the tight inferential web supporting the final inference to the mass of an object weighed by the balance. Once built, this web better coordinates our quantitative mass concept with the physical world. Adopting the watt balance as a standard realization of this concept endows the laws involved in modeling its operation with epistemic normative force in regulating systematic errors involved in mass measurements over a global social network. But each of these laws was applied only in a domain in which it could be relied on, despite the fact that the conjunction of all these laws would quite likely be inconsistent. Here, as elsewhere, the reliable inferential application of a law in science does not require either its truth or its unconditional universality, let alone its necessity. Scientific practice reveals these requirements as philosophers' myths, even when laws are applied in pursuit of a goal requiring maximum overall accuracy.

7 Are Laws True?

If it were the general function of a scientific law truly to represent a universal empirical regularity that is more than just a pattern of coincidences then every law mentioned in this paper falls short, since none of them are applicable to all physical phenomena. Considered as universal empirical generalizations, most or all of them are known to be subject to exceptions and hence not to be true. Even the law of conservation of mass-energy is problematic in its application to general relativistic phenomena where the concept itself is difficult to define.[6]

It is common to locate the best current candidates for fundamental laws of nature in the physics of the Standard Model of so-called elementary particle physics and in the general theory of relativity. But despite its unblemished record of success, the Standard Model is now regarded as merely an effective theory—a low energy limit of some unknown (quantum field?) theory that is still applicable at the higher energies where the Standard Model breaks down. And there are believed to be extreme phenomena involving black holes and the early universe that cannot be modeled by general relativity, whose laws do not encompass quantum effects. If unconditional truth were a requirement for a law of nature then no laws of nature are known to current science.

Undeterred by this absence of examples, a proponent of the representational conception of scientific laws as unconditionally (i.e. not *ceteris paribus*) universal, non-accidental, empirical truths may hold up this conception as a regulative ideal of science—an ideal that science has been approaching by way of progressive approximation. But the radical conceptual changes revealed by the history of fundamental science render this notion of approximation dubious at best. More importantly, what science has come up with by way of scientific laws have admirably performed epistemic and practical functions including those described earlier in this paper. Our scientific knowledge rests on the epistemic infrastructure they constitute.

In general epistemology a statement is usually considered to be known only if it is true. If a scientific law is not true then how can it contribute to scientific knowledge? The inferential conception of scientific laws offers an answer to this question: whether or not it is true, a scientific law contributes to knowledge by licensing inferences that are generally reliable in those contexts in which they are licensed. Physicists often call such a context a regime, as when they say that Newton's law of gravity is applicable in a regime of low relative velocities and small space-time curvature. If several laws are used in constructing a model of an object or process, it becomes important to judge that all of them together permit reliable inferences in that context. Physicists prioritize the search for a theory of quantum gravity because laws of currently fundamental theories of quantum theory and general relativity do not permit reliable inferences in regions in the interiors of black holes and in the early universe. So we don't know what happens there. Some may share Weinberg's (1992) dream of a final theory whose laws would be universally applicable and truly

[6] Carroll (2010) argues that it is less misleading to call this conservation law false in general relativity than to say that mass-energy cannot there be defined. But he acknowledges that "plenty of experts in cosmology or GR would not put it in these terms. We all agree on the science; there are just divergent views on what words to attach to the science. In particular, a lot of folks would want to say 'energy is conserved in general relativity, it's just that you have to include the energy of the gravitational field along with the energy of matter and radiation and so on.'"

correspond to objective reality at the deepest level. But during their waking hours scientists have more modest goals whose attainment they could recognize.

While agreeing about the inferential function of laws, Ramsey and Ryle disagreed about their truth status. Ryle (1949) claimed that law-statements are true or false, while Ramsey (1929) claimed that they are not propositions capable of the two cases of truth and falsity: as he put it, they are not judgments but rules for judging. Ramsey here embraced the instrumentalist view that a law-statement (a "variable hypothetical") lacks a truth-value.

A standard objection to such instrumentalism is to point to examples in which a scientist makes an inference from the conjunction of two or more scientific laws that is not a logical consequence of any single one of them (as in the derivation of Galileo's law of free fall from Newton's laws of motion and gravity). Such an inference may be deductively valid, but it is sound only if the premise laws are true.

What the objection fails to recognize is that scientific inferences from laws need not be sound but merely reliable in the context in which they are made. Hubble's inference to the existence of other galaxies was based on Leavitt's generalization about Cepheid variables as well as the inverse square law for light intensity. But Leavitt's law was not even true of the observed instances on which it was based: she arrived at it only by fitting a curve to data points that clearly failed to lie on it. When a scientist makes an inference from a law she does not assume its truth: she assumes only that she is warranted in using it to infer her conclusion in this specific context. Further investigation may undermine her assumption or bolster it while more precisely delineating the scope of the law's reliable application. This is an important example of the fallibility of scientific knowledge.

Even if scientific laws are not true, it seems harsh to call them all false. Elgin (2017) calls them (along with much else in science and art) true enough, thereby nicely alluding to their function in serving our purposes. Ramsey's alternative was to take laws' constitutive role in inference as a reason to refuse to evaluate them in terms of truth or falsity. How or whether we choose to apply the term 'true' to scientific laws is not of great importance in understanding science. I think we should take our cue from scientists themselves who rarely use the term in this context, talking instead of a law's scope, limits and reliability. When a scientist talks of known and unknown laws this conception of knowledge rarely carries a commitment to truth.

8 Laws and Naturalistic Metaphysics

According to a contribution (Dorato and Esfeld (2015)) to a recent collection of essays on metaphysics in contemporary physics, there are three main stances with respect to laws of nature in current philosophy of science: Humeanism, primitivism and dispositionalism. I stand for none of these. As a philosopher of science focusing on physics I have illustrated my general arguments here mostly by examples chosen from that science. But the arguments are meant to apply anywhere scientific laws are used to gain knowledge of the world: and their conclusions are about science, not about metaphysics. They seek not only to undermine all "three main stances" but also to motivate a turn away from metaphysics in the philosophy of science and toward pragmatism.

Humeans consider the world to be a mosaic of matters of actual, particular fact, and laws to be the axioms of the description of this mosaic that achieve the best balance between simplicity and strength. Lewis (1994) gave a seminal presentation of this view, and others have offered modifications, following in his footsteps (e.g. Cohen and Callender (2009), Hall (2015), Loewer (2021)). Lewis credits Ramsey with a forerunner of this view in a paper (1928) whose main idea Ramsey summarized later (1929): "... laws [are] a consequence of those propositions which we should take as axioms if we knew everything and organized it as simply as possible in a deductive system".

But Ramsey here summarized his earlier view only immediately to completely reject it on the grounds that it is impossible to know everything and organize it in a deductive system. Doubtless Lewis was well aware of Ramsey's change of mind, but he offered no good objection to Ramsey's reason for rejecting his former view, despite the fact that it would apply equally to Lewis's own account of laws of nature. Scientists rarely try to formulate even a single theory as an axiomatic system; the scientific laws we have are not even true, let alone universally applicable; and we have excellent reasons to doubt that the progress of science will converge on any body of knowledge that could be organized as a single axiomatic system.

According to primitivism, true scientific law-statements correspond to modal facts about the world over and above all (or even any) matters of actual, particular fact. In Maudlin's (2007) version, fundamental laws of temporal evolution (FLOTEs) do not organize but *generate* actual, particular facts as well as whatever consequent patterns these display. I find talk of a generation relation among particulars as obscure as Ramsey found talk of a real relation among universals:

> ... may there not be something which might be called real connections of universals? I cannot deny it, for I can understand nothing by such a phrase; what we call causal laws I find to be nothing of the sort. (1929; [1990, p. 160]).

What about dispositionalism? On this view (Dorato and Esfeld (2015)) local matters of particular fact instantiate one or more properties that fix the behavior of these local happenings, for example the temporal development of an initial configuration of particles. Such properties are dispositions or powers that ground a law in the sense that the latter is made true by the former. So in this view laws are not primitive but nor do they offer a simple yet strong summary of non-modal facts. Instead laws inherit their modal content from that of modal properties instantiated by local matters of particular fact.

But as we have seen, whether a law fulfills its function in science is largely independent of whether it is true or even truth-evaluable. Hence a scientific law requires no ground of truth in dispositions, powers or other modal properties. This is fortunate because we have no epistemic access to any such modal properties even if we can convince ourselves that we mean something when we talk about them.

While doing science, scientists seldom evaluate the laws or theories they use in terms of truth or falsity. But when reflecting on what they have done they are prone to join some philosophers and other commentators in applying those terms quite casually. Praising the achievements of fundamental physics while commenting on its currently fluid state, an August 2021 issue of *The Economist* included the following passage.

> Physicists have got used to the idea that mathematical theories can be turned into reliable representations of reality, thus producing understanding. And one discovery based on maths that physicists were pretty confident of making was of a phenomenon called Supersymmetry, which gives coherence to the current, rather ad hoc explanation of the menagerie of fundamental particles that has been collected since the 1890s. Supersymmetry is a stalking horse for a yet-deeper idea, string theory, which posits that everything is ultimately made of infinitesimally small objects that are most easily conceptualised by those without the maths to understand them properly as taut, vibrating strings.
>
> So sure were most physicists that these ideas would turn out to be true that they were prepared to move hubristically forward with their theorising without experimental backup—because, for the first decades of Supersymmetry's existence, no machine powerful enough to test its predictions existed. But now, in the form of the Large Hadron Collider, near Geneva, one does. And hubris is turning rapidly to nemesis, for of the particles predicted by Supersymmetry there is no sign.

The passage never mentions laws, and it applies the term 'true' directly not to the theories it does mention but to the ideas behind them. But it is not unreasonable to assume that its authors share the popular view that a theory contains laws, and that the threatened nemesis would involve rejection as false of some law or laws of supersymmetry.

It is noteworthy that the quoted passage also assigns to fundamental physics the role of reliably representing reality. The passage suggests (without quite endorsing) the view that it is a distinctive goal of fundamental physics to contribute to metaphysics by faithfully representing reality at the deepest level. The contrast is with other parts of physics and the rest of science which may achieve their goals by deploying idealized models and laws that are at best only approximately true in certain circumstances. In this view, whatever the function of laws in other parts of science, fundamental physics seeks laws whose function is to offer objectively true representations of reality.

Weinberg (1998) expressed this view as follows

> We hope that in the next great step forward in physics we shall see the theory of gravitation and all of the different branches of elementary particle physics flow together into a single unified theory. This is what we are working for and what we spend the taxpayers' money for. And when we have discovered this theory, it will be part of a true description of reality.

Earlier in the same essay he had defended such uses of the terms 'true' and 'real'

> I know that it is terribly hard to say precisely what we mean when we use words like "real" and "true." That is why, when I said that the laws of nature and the rocks on the ground are real in the same sense, I added in parentheses "whatever that is." I respect the efforts of philosophers to clarify these concepts, but I'm sure that even Kuhn and Rorty have used words like "truth" and "reality" in everyday life, and had no trouble with them. I don't see

> any reason why we cannot also use them in some of our statements about the history of science.

But in his (1996) Weinberg had acknowledged that he was "making an implicit assumption (which in everyday life we all make about rocks) that our statements about the laws of physics are in a one-to-one correspondence with aspects of objective reality". Ramsey (1929) would have rejected this assumption, while Rorty exposed its vacuity in his essay "Thomas Kuhn, Rocks and the Laws of Physics" (1999).

If fundamental physics were to offer a true description of reality including laws in a one-to-one correspondence with basic aspects of objective reality, then this could justifiably be considered a contribution to naturalistic metaphysics. Maudlin (2007) took physics to have even greater metaphysical content when he said

> I am a physicalist about laws: the only objectively primitive laws I believe in are the laws of physics. Speaking picturesquely, all God did was to fix the physical laws and the initial physical state of the universe, and the rest of the state of the universe has evolved ... from that. Once the total physical state of the universe and the laws of physics are fixed, every other fact ... supervenes. (p. 158)

But even Weinberg took the search for a theory whose laws offer true descriptions of objective reality as a work in progress, motivated by the hope that we will eventually succeed in finding a true, powerful, unified theory in terms of which we will be able to understand the world at the deepest level. Even if there is such a theory, we have no reason to believe that it will include any of the empirical laws that physicists have formulated so far. For we have good reasons to doubt that any of these is both universally applicable and true.

I have argued that a scientific law can admirably serve its purposes without being true. But a law that is not true, or not a proposition capable of the two cases of truth and falsity, is not something to support a metaphysics.

A naturalist metaphysician may well object to this quick argument by making the following analogy between the methods of science and metaphysics. Scientific progress validates the worth of an older scientific theory by showing it to be derivable as a limiting case of current theories in certain regimes, so we can regard it as a useful approximation. Arguably, this permits one to claim that the earlier theory was approximately true, at least in some of its key claims. Similarly, we can make progress in the metaphysics of science by basing metaphysics on the best current science, confident that our metaphysical conclusions will also be approximately true, at least in part.

But as McKenzie (2020) has pointed out, whatever one thinks of attempts to portray science as progressing toward the truth by successive approximation, metaphysics does not deal in approximations. Either laws of nature are primitive or they are not: laws of nature either have or lack modal content. Metaphysicians' shared assumption that there are true laws of nature seems non-negotiable: and a metaphysics of the non-true laws that science continues to produce is of questionable value.

9 Conclusion

Natural laws function in science as inference-tickets or rules for judging rather than judgments. In so doing they provide science with an epistemic infrastructure that helps us expand our knowledge of the universe far beyond its starting point here and now. Laws guide scientists in constructing models and coordinating our interventions. We use them to predict, explain and control the world. We can use laws successfully to do all these things even while none of them live up to their philosophical billing as necessarily true, unconditionally universal generalizations.

Metaphysicians have proposed analyses of laws of nature that take them to be something more than reliable empirical generalizations about actual, particular facts. For Humeans they are not much more since they are determined by such facts, given our proclivity to axiomatize as well as our standards of simplicity and strength. Primitivists and dispositionalists add more by way of objective modal content, whether inherent in the laws themselves or inherited from their ground in particular modal facts about the world. All these views saddle laws of nature with excess metaphysical baggage by adding a metaphysical superstructure to the world that science investigates—a world consisting of a rich tapestry of actual physical events.

But behind the use of laws in science we can see shadows of the traditional philosophical conception of laws of nature as necessarily true, unconditionally universal generalizations of unlimited scope.

. Scientists seek laws of maximal applicability because a law of wider applicability permits reliable inferences in more domains of inquiry and thereby forges more epistemic connections between otherwise disparate fields of knowledge. The application of laws of particle physics to cosmology is one salient example. We can see another in the current drive for unification in fundamental physics. The search for laws of maximal applicability is the epistemic shadow of the metaphysical condition of unconditional universality.

Hume was right to relocate necessity from the world to us. But we see the shadow of natural laws' modal content not in an individual mind's felt necessity when confronted with a causal sequence, but in the demand that a scientific law support *reliable* inferences. Ryle's analogy of laws to season-tickets works insofar as both can be relied on to get you where you want to go, in science as in life. The demand for reliability is so important in science that, on finding a scientific law, it is hard to resist the temptation to project necessity onto the law itself rather than resting content with the reliable inferences it permits. Dispositions lie not in the states but in ourselves. Having learnt them, we are disposed to apply terms like 'red', 'flat', 'mass' and 'force' to particulars in the same way as each other. One can reasonably infer from one's correct application of such a term not only that others would generally apply it to the same things; but also that they, too will bleed red and stop at red lights, rest wine glasses only on flat surfaces, and see forces at work whenever a massive body accelerates.

Truth may be the hardest condition to abandon in an analysis of scientific laws. Earlier we saw Ryle still clinging to its possibility, even while advocating an inferentialist conception of laws. Did Ramsey give up too much by denying that laws of nature are even truth-evaluable? The fate of superseded theories provides a reason to think he did. After the general theory of relativity had passed its three classic tests, Popper was by no means alone in taking Newton's law of universal gravitation to have been falsified. Since then general relativity has successfully predicted many other phenomena like black holes whose occurrence seems clearly incompatible with Newton's law.[7] Surely we can now evaluate Newton's law of universal gravitation as false rather than true, even while continuing to use it to make reliable inferences in many applications?

A good inferentialist response to this rhetorical question is beyond the scope of this paper since it would require a defense of a non-representationalist account of the function of truth like that proposed by Price (2003). But Ramsey (1929) took the first steps toward such an account by considering various possible attitudes one could take toward a present annunciation of a law such as Newton's law of universal gravitation (not his example).

Suppose that someone who knows nothing of twentieth or twenty-first century science has learned Newton's laws of motion and the law of universal gravitation and has acquired some skill in using these to make reliable inferences about the behavior of various bodies in the laboratory as well as the solar system (someone like Halley or Wren, perhaps). All his experience (including the testimony of his teachers) supports Newton's law. So his unprovoked annunciation of that law seems clearly to take the form of an assertion: he is neither play-acting nor acting as historian of science. In the light of contemporary science, what is the appropriate attitude to what he says?

Clearly not to believe his assertion. But also not to disbelieve it on the grounds that there is a pair of Newtonian masses separated by a Euclidean distance exerting on each other gravitational forces that differ from those predicted by Newton's law. According to general relativity there can be no such counter-instances of the law since there is no force of gravity or Newtonian masses and no enduring Euclidean space. No force of gravity accelerates a falling apple—the apple does not accelerate since its world-line traces out a straightest path in a space-time curved by the nearby earth. One can say that the attitude he has expressed toward Newton's law is not right but wrong; but one's own attitude should not be to disbelieve his assertion as false because one believes the negation of Newton's law.

The key question is not whether a law of nature is evaluable for truth but whether the sincere annunciation of a law of nature should be counted as an assertion of that law. As Ramsey said elsewhere (1927, [1990, p. 38]) "there is really no separate problem of truth, but merely a linguistic muddle"; "the problem is not as to the

[7] Although an appendix in Hawking and Ellis (1973) includes an English translation of Laplace's striking 1798 demonstration of how a large body could prevent all the massive light corpuscles it emits from escaping from its Newtonian gravitational attraction.

nature of truth and falsehood, but as to the nature of judgment or assertion" [1990, p. 39]. Only if an annunciation counts as an assertion should it be considered evaluable as either true or false. That in turn depends on what social commitment accompanies the annunciation. If Price (2003) is right, the notion of truth has the social function of spurring rational debate as to whether everyone should accept the commitments undertaken by treating the annunciated sentence as an assertion, and therefore subject to criticism if and when these commitments come to be judged as untenable. In their work, scientists rarely if ever undertake such commitments in their use of laws, and so rarely assert those laws. Only when reflecting on their scientific activities do some adopt an uncritical view of truth as something a law either has or fails to have whether or not anyone should or even could assert it, but simply according to whether or not it in itself corresponds to objective reality.

In her (2016), Misak considers Ramsey and Price to be practitioners of what she calls Cambridge pragmatism. Here she means the English Cambridge, not Cambridge, Massachusetts where pragmatism originated in the work of Peirce and James. All these philosophers have rejected correspondence to reality as anything more than an unhelpful metaphor when trying to understand the concept of truth. James is notorious for his view of truth as what works, especially after its uncharitably hostile treatment at the hands of analytic philosophers like the (Cambridge!) philosophers Moore and Russell. But the Peircean strand of pragmatism took root in the fertile mind of Ramsey, who took a very different view of truth. As true pragmatists, Ramsey and Price have asked not what truth represents or corresponds to, but what function the concept of truth serves in our intellectual lives. I take a consequence of their answers to be that there is little or no point in applying the concept to scientific laws.

My treatment of scientific laws here has been in the same pragmatist spirit. By examining several important functions of laws in science I have tried to show that these can all be understood as ways in which laws support scientists' epistemic and practical activities. They do so without being necessary, unconditionally universal or true. All a scientific law needs to do is to support reliable inferences in each domain in which it can be responsibly applied—a domain whose extent science itself explores and strives to maximize.

References

BIPM. (2004). "Worldwide metrology", *Bureau Internationale de Poids et Mesures*. Retrieved September 1st, 2021, Available at https://web.archive.org/web/20170324081343/http://www.bipm.org/en/worldwide-metrology/

Carnap, R. (1966). *Philosophical foundations of physics*. Basic Books.

Carroll, S. (2010). "Energy is not conserved". Retrieved September 1st, 2021, Available at https://www.preposterousuniverse.com/blog/2010/02/22/energy-is-not-conserved/

Cartwright, N. (2016). The dethronement of laws in science. In N. Cartwright & K. Ward (Eds.), *Rethinking order after the laws of nature* (pp. 25–52). Bloomsbury Academic.

Cohen, J., & Callender, C. (2009). A better best system account of laws. *Philosophical Studies, 145*(1), 1–34.
Dorato, M., & Esfeld, M. (2015). The metaphysics of laws: Dispositionalism *vs*. primitivism. In T. Bigaj & C. Wuthrich (Eds.), *Metaphysics in contemporary physics* (pp. 403–424). Rodopi | Brill.
Elgin, C. (2017). *True enough*. MIT Press.
Gettelman, A., & Rood, R. (2016). *Demystifying climate models*. Springer Nature.
Giere, R. (1999). *Science without laws*. University of Chicago Press.
Hall, E. (2015). Humean reductionism about laws of nature. In B. Loewer & J. Schaffer (Eds.), *A companion to David Lewis* (pp. 262–277). Wiley.
Hawking, S., & Ellis, G. (1973). *The large scale structure of space-time*. Cambridge University Press.
Kuhn, T. (1962). *The structure of scientific revolutions*. University of Chicago Press.
Lewis, D. K. (1994). Humean supervenience debugged. *Mind, 103*(412), 473–490.
Loewer, B. (2021). "What breathes fire into the equations?" Consulted September 1st, 2021, Available at http://philsci-archive.pitt.edu/19446/
Maudlin, T. (2007). *The metaphysics within physics*. Oxford University Press.
McKenzie, K. (2020). A curse on both houses: Naturalistic versus *a priori* metaphysics and the problem of progress. *Res Philosophica, 97*(1), 1–29.
Misak, C. (2016). *Cambridge Pragmatism*. Oxford University Press.
Mitchell, S. (2000). Dimensions of scientific law. *Philosophy of Science, 67*, 242–265.
Mitchell, S. (2002). *Ceteris Paribus*: An inadequate representation for biological contingency. *Erkenntnis, 57*(3), 329–350.
Newton, I. (1687). *Philosophiae Naturalis Principia Mathematica*. Royal Society.
Norton, J. (unpublished). "The recession of the nebulae", draft of proposed book chapter.
Price, H. (2003). Truth as convenient friction. *The Journal of Philosophy, 100*(4), 167–190.
Ramsey, F. P. (1927). "Facts and propositions". In Ramsey, F. P. [1990], pp. 34–41.
Ramsey, F. P. (1928). "Universals of law and of fact". In Ramsey, F. P. [1990], pp. 140–44.
Ramsey, F. P. (1929). "General propositions and causality". In Ramsey, F. P. [1990], pp. 145–163.
Ramsey, F. P. (1990). In D. H. Mellor (Ed.), *Philosophical papers*. Cambridge University Press.
Rorty, R. (1999). Thomas Kuhn, rocks and the laws of physics. In *Philosophy and social hope* (pp. 175–189). Penguin Books).
Ryle, G. (1949). *The concept of mind*. Hutchinson.
Tal, E. (2011). How accurate is the standard second? *Philosophy of Science, 78*, 1082–1096.
Tal, E. (2017). Calibration: Modelling the measurement process. *Studies in History and Philosophy of Science, 65–66*, 33–45.
Van Fraassen, B. (1989). *Laws and symmetry*. Oxford University Press.
Weinberg, S. (1992). *Dreams of a final theory*. Pantheon.
Weinberg, S. (1996). Sokal's hoax. *The New York Review of Books, XLIII*(13), 11–15.
Weinberg, S. (1998). The revolution that didn't happen. *The New York Review of Books, XLV*(15), 48–52.

The Babylonian Conception and Conventionalism About Laws in Physics

Mathias Frisch

Abstract In this paper I discuss two features of laws in physics and ask to what extent these features are compatible with different philosophical accounts of laws of nature. These features are (i) that laws in physics fit what Richard Feynman has dubbed the "Babylonian conception" of physics, according to which laws in physics form an interlocking set of 'theorems'; and (ii) that the distinction between dynamics and kinematics is to some extent contextual. These features, I argue, put pressure on any philosophical account of laws that presupposes that the laws of physics have a unique quasi-axiomatic structure, such as the Mill-Ramsey-Lewis account of laws and metaphysical accounts of laws that assume that there is a privileged explanatory nomological hierarchy.

Keywords Laws · Humeanism · Best-System-Account · Conventionalism · Dispositionalism · Instrumentalism · Babylonian conception of laws · Dynamics-kinematics distinction

1 Introduction

During much of the eighteenth and nineteenth centuries Newton's laws of motion were taken to be the paradigm of scientific laws, thought to constitute universal and necessary eternal truths. But since the turn of the twentieth century, we know that Newton's laws are not universally valid. Does this mean that their status as laws of physics has changed? Have we discovered that the principles, which were once thought to be laws of nature, are not in fact laws? Or have we merely learned that the domain of application of Newton's laws is more restricted than we once thought, while the laws' role within their proper domain is unaffected by this discovery? What is more, with the demise of Newtonian physics as universal

M. Frisch (✉)
Leibniz University Hannover, Hannover, Germany

Y. Ben-Menahem (ed.), *Rethinking the Concept of Law of Nature*,
Jerusalem Studies in Philosophy and History of Science,
https://doi.org/10.1007/978-3-030-96775-8_7

theory the entire paradigm of lawful predictability appears to have reached its limits—not only with the discovery of quantum probabilities but also with the emergence of the physics of nonlinear and complex systems, for which the notions of determinism and predictability come apart. How do these developments affect our philosophical conception of laws of nature? Are laws of nature the (perhaps probabilistic) determinants of how the world evolves? Or are they rather our best predictive tools, where what is best does not only depend on the ways the world is but also on who uses the laws in what context and for what purpose? That is, are the laws of nature fully objective and context-independent fundamental features of the world to be discovered by us, or is what the laws are at least partly also a reflection of our human cognitive capacities and of particular contexts of investigation? In this paper I want to make progress towards a philosophical understanding of the role of laws in physics by closely examining a number of features of how putative laws are treated in physical theorizing.

In the next section I will distinguish three broad philosophical conceptions of laws that play a particularly prominent role in contemporary philosophical discussions of laws of nature. These are, first, views that take nomological necessities to be metaphysically basic, including 'governing conceptions' that treat laws as metaphysically fundamental drivers of the temporal evolution of the world; second, Humean views, that take laws to be reducible to patterns of instantiations of non-modal properties; and, third, a cluster of conventionalist or even instrumentalist views that understand laws at least partly as a reflection of a particular perspective of investigation and as part of the epistemic toolkit for building models, which represent aspects of the world from a particular and partial perspective, relative to certain goals and interests.

Sometimes philosophical discussions of laws of nature are situated at several removes from examinations of the actual use of putative laws in physics, taking as their starting points purely philosophical reflections on the nature of laws, with only tenuous support from actual scientific practice. By contrast, I here want to approach the issue from the other end, as it were, by first examining several features that I take to be characteristic of laws in physics (taking methodological reflections of eminent physicists as a cue) and then asking to what extent these features are compatible with various philosophical conceptions of laws. The two features of physical theorizing on which I will focus in particular concern the place of axiomatization in physics (in Sects. 3 and 4) and the relation between dynamics, kinematics, and initial conditions (in Sects. 5 and 6). To anticipate somewhat, there is a conventional and context-dependent element in the way in which laws are treated in physics. Different sets of laws provide us with different perspectives from which to model or represent the phenomena, partly serving different purposes or answering different questions. And while laws at different levels are related in many and intricate ways, and hence there is little evidence for an image of physics as consisting in completely balkanized, disjoint groups of research activities, there is equally little evidence for a conception of physics as providing us with a single hierarchical and uniquely axiomatized structure. Thus, I will argue that even within physics there is support for a view of laws that is more frequently associated with the so-called 'special sciences', according to which laws function as domain- and context-specific generalizations (see, e.g., Mitchell, 2009).

2 The Metaphysics of Laws

Laws of nature, according to one common philosophical understanding of that notion, are universal and necessary truths. The necessity in question is taken to be distinct from logical necessity and sometimes also from (other brands of) metaphysical necessity. The various ways of fleshing out the view that laws are universal truths can be distinguished along different dimensions. First, we can distinguish views that take modalities to be a non-reducible part of the basic inventory of the world from views that understand nomological necessity to be ultimately reducible to non-modal features of the world. And, second, we can divide the former further into views that take laws to be reducible to necessities that reside in the fundamental properties that objects instantiate and views that take laws to be primitive or basic (see Loewer, 2012; Chen & Goldstein, Chapter "Governing Without a Fundamental Direction of Time: Minimal Primitivism About Laws of Nature", this volume).

Dispositional essentialist, such as Alexander Bird (see Bird, 2007), are among those who maintain that laws are reducible to or supervene on necessities that reside in fundamental dispositional properties of objects. For example, on this view massive objects possess the disposition or power to attract other masses and it is this disposition that grounds the law of gravitational attraction. On Bird's view, dispositional essences underwrite counterfactual conditionals, to which the laws are reducible. On Heather Demarest's more recent view (Demarest, 2017), the initial distribution of particles and their dispositions (or potencies) ground the subsequent behavior of the particles and the instantiation of dispositions, which determine the laws. One should note, however, that there are also defenders of a dispositional fundamental ontology, like Nancy Cartwright, who deny that these dispositions ground universal laws.

Other proponents of the view that take modalities to be irreducible to non-modal features of the world take laws themselves to be metaphysically fundamental and to constrain or *govern* the patterns of instantiations of accidental or categorical properties in the world. The laws are primitive. On one such 'primitivism' about laws, fundamental dynamical laws govern how spatiotemporal objects evolve from one moment to another (see Maudlin, 2007). Tim Maudlin's conception of primitivism includes a commitment to a primitive direction of time. The fundamental dynamical laws—the fundamental laws of temporal evolution—on his view, generate later states of the universe from earlier ones. Eddy Chen and Sheldon Goldstein (Chapter "Governing Without a Fundamental Direction of Time: Minimal Primitivism About Laws of Nature", this volume), by contrast, defend a view they call 'minimal primitivism', according to which the laws govern not by producing or generating later states from earlier ones but merely by constraining the space of physical possibilities. Nomological constraints may take the form of differential equations that allow us to determine future states from past ones. But the constraints can have other forms as well. As they put it "laws govern by constraining and constraining is what they do." (Chen & Goldstein, Chapter "Governing Without a Fundamental Direction of Time: Minimal Primitivism About Laws of Nature", this volume) And even if the laws in question are dynamical laws, the laws govern by constraining metaphysical possibilities and not by producing later states from earlier states.

All metaphysically 'weighty' accounts of laws stress the role of modalities, and in particular nomological necessity, in explanation. For the dispositional essentialist explanations ultimately bottom out with an appeal to the dispositions, powers, or potencies of objects. The world evolves the way it does, on these accounts, in virtue of objects realizing their dispositions. The laws supervene on these dispositions and their realizations. On primitivist governing conceptions, laws play a more central role in explanation. On Maudlin's primitivism, laws govern the evolution of the world from its initial state and in virtue of their generative or productive role explain how later states evolve from earlier states. Minimal primitivism holds that laws explain by constraining possibilities.

Primitivists are in agreement on a hierarchical view of laws of nature. According to both Maudlin's account and minimal primitivism, we can distinguish metaphysically fundamental laws that provide ultimate explanations of the world evolves from derived or less fundamental laws that are grounded in the fundamental laws. Crucially for the primitivist, the truth of the fundamental laws, as well as their lawhood and their fundamentality, are independent of our beliefs, our psychologies, or our practices.

Views that nomological necessities are part of the basic inventory of the world contrast with *Humean conceptions* of laws according to which laws and nomological necessity supervene on the totality of particular non-modal matters of fact. Nomological claims, according to Humeans, are true in virtue of certain patterns of fundamental properties being instantiated. One way to spell this out further is the account associated with John Stuart Mill, Frank Ramsey and David Lewis (MRL) (Lewis, 2001). According to the MRL-account, laws provide us with that summary of the mosaic of particular matters of fact that best balances simplicity and strength. The MRL-account asks us to consider different ways of organizing statements about particular matters of fact into deductive systems. The strength of a system is a measure of how many truths the deductive system allows us to derive, while a system's simplicity is both a measure of the system's number of axioms and of the axioms' syntactic complexity. The laws are the axioms of that deductive system that best combines simplicity and strength.

While the traditional Humean account, which identifies laws with the regularities exhibited by the 'Humean mosaic,' appears to offer a reductionist but fully objective account of laws, the MRL-account arguably introduces a pragmatic element. As defenders of the view themselves emphasize, *both* the criterion of simplicity itself *and* what the proper relative weighting of the criteria of simplicity and strength is cannot be spelled out in a purely objective, mind-independent manner: overall goodness ultimately is goodness *for us*. Thus, according to the MRL account what the laws are depends at least partly on us and our cognitive capacities. While this does not mean that what the laws are is arbitrary, it does contrast with primitivists' accounts, which emphasize that which laws govern the world is a "fundamental, objective, and mind-independent fact" (Chen & Goldstein, Chapter "Governing Without a Fundamental Direction of Time: Minimal Primitivism About Laws of Nature", this volume).

Ron Giere, among others, has traced the history of the notion of scientific laws as universal and necessary truths to the idea of God as divine law-giver—an idea one can find for example in Newton (Giere, 1999). The universality of the laws is a consequence of God's supreme power, which ensures that nature has to obey Her laws everywhere and everywhen. Independently of whether Giere's account accurately captures the history of the concept of law of nature, the image of God as law-giver can provide a useful illustration of the contrast between the governing and Humean conception of laws (see also Cartwright, 1999). The difference between the governing-law conception and the Humean conception concerns the question whether laws are metaphysically fundamental or whether what the laws are is reducible to—or supervenes on—the totality of particular matters of fact. According to the former conception, when God created the world, She had two tasks: She had to decree what the fundamental laws of the world would be; and She had to specify what the initial state of the universe shall be. Nothing more was required of Her. Once the initial state was created, the universe evolved forward in time governed by the laws of nature. At least this would be so, if we assumed that the laws are deterministic and contain no 'gaps.'

On the minimal primitivist's conception God's tasks similarly only involves the task of specifying the laws plus appropriate conditions on some initial value surface. But on this conception, God's act of creation is an atemporal act: By specifying the fundamental nomological constraints plus data on some appropriate surface (which need not be a surface at the beginning of the universe) the entire history of the universe is given. Contrary to Maudlin's primitivist conception, the laws do not temporally asymmetrically govern how the universe evolves.

By contrast with both of these conceptions, the Humean God is not a law-giver but rather a pointillist painter, who set Herself the rather more involved task of sprinkling the universe's four-dimensional space-time canvas with particular matters of fact. The Humean God does not merely have to decree what the universe's initial state should be but has to paint all of space-time—that is, all of space throughout all times—with a mosaic of events. What the laws of nature are, is then given by the totality of particular matters of fact. According to the traditional Humean, the laws strictly supervene on the pattern of particular matters of fact. According to the MRL view, what the best way of summarizing the mosaic created by God might be *for us* also partly depends on our cognitive capacities.

The image of God as pointillist painter does not yet tell us how easily the patterns on the canvas can be summarized. Lewis's God and the God of the standard MRL account of laws appears to have been kind to physicists searching for universal regularities on Her canvas: Her painting contains global patterns, is systematic and highly organized so that we can justifiably hope to be able to summarize the goings-on on the entire canvas with the help of a relatively small number of relatively simple fundamental physical laws. But we can also imagine a less systematic painter—a messy and mad genius who painted different parts of the canvas in ways that resist being unified within a single simple framework. Any description of the goings-on on the mad genius's canvas that is both true and universal might be absurdly complex—so complex that our best strategy might be to hunt for small pockets of order that permit being summarized in terms of simple 'lawish' statements rather than aim for a single unified description. If God was messy and the world was 'dappled', we would have to lower our ambitions: instead of aiming for a single best system, we might have to rest content with a multiplicity of domain-specific and perhaps even interest-specific systems. Instead of hoping to discover universal laws, we would search for ways to summarize small pockets of regularities. And instead of hoping to discover laws that are strictly true, we might have to make due with 'lawish' statements that fit the patterns we observe reasonably well to a certain degree of approximation, and aim for domain-specific systems of laws that best combine simplicity, strength and fit and serve our context-dependent purposes as well as possible.

The metaphor of God as a mad genius, who paints a dappled world on Her canvas, represents a third conception of laws—a conception that gives up the idea of laws as universal, context-independent truths and instead takes what the laws are to depend both on the ways the world is and on our interests and purposes—which I have here presented as a variant of the MRL-account (see Cohen & Callender, 2009; Callender & Cohen, 2010; Frisch, 2011a, b). The conception replaces the idea of a single best system with that of a multiplicity of systems, reflecting different perspectives and interests, and instead of demanding that laws be true takes goodness of fit to be just one criterion by which to evaluate a system, to be jointly maximized together with the competing criteria of simplicity and strength. A closely related view insists in addition that the relationship between laws and the world is more indirect than the accounts surveyed so far would have it: the primary role of laws, on this view, is to aid us in the construction of models of the phenomena. The representational role of science, on this view, is ascribed not (or at least not primarily) to the laws but rather to the models constructed with the help of laws: laws are tools for model-building (Cartwright, 1983; Giere, 1999). Instead of providing us with the uniquely correct fundamental underpinnings of how the world evolves, laws are taken to be part of "expanded understandings of both the world and our representations of it as a rich, variegated, interdependent fabric of many levels and kinds of explanations that are integrated with one another to ground effective prediction and action," (p. 19) as Mitchell puts it.

Just as Humeanism can come in different flavors, from a fully objective account of laws to the MRL account with its irreducibly pragmatic element to Callender and Cohen's "better best systems" account allowing for a multiplicity of systems, so can dispositionalist accounts. In fact, dispositionalism seems to allow for the same range of views as Humeanism does, with the main difference between it and Humeanism, of course, concerning the ontological basis the two different types of view postulate: a mosaic of localized non-modal matters of fact as opposed to a fundamental dispositional properties, potencies or powers.

This concludes my brief sketch of the conceptual landscape concerning laws of nature. Which of these three conceptions should we adopt? In what follows I want to describe two features of laws of physics, which, I want to submit, can provide us with some clues that allow us to assess different answers to this question.

3 The 'Babylonian' Conception of Laws

The first putative feature of laws in physics I want to discuss is what Richard Feynman calls the "Babylonian" character of laws in physics. In *The Character of Physical Law* Feynman discusses the mathematized statements at the core of physical theories and asks whether there is "one set of statements that is more fundamental and one set of statements that is more consequential" (Feynman, 2001, 46). According to what Feynman calls the "Greek" or "Euclidean tradition," physics and the mathematized sciences should be thought of as having an axiomatic structure, similar to that of Euclidean geometry, in which the entire content of a theory is derivable from a simple set of fundamental axioms. That is, the Euclidean tradition takes the laws of physics to exhibit a unique and context-independent hierarchical structure with a set of simple, fundamental laws as axioms from which all other laws can be derived. Feynman contrasts the Euclidean view with what he calls the "Babylonian tradition." Feynman says that in the Babylonian tradition general rules were arrived at by working through a large number of examples: rules are arrived at inductively rather than through derivations from more fundamental axioms. But more important for our purposes here is Feynman's suggestion that even once we know all the rules, theorems, or laws, the Babylonian is not interested in the question whether these laws can be arranged into an axiomatic structure. Rather, the Babylonian's main focus is on "calculating things out." And, as Feynman says, the most "efficient way of getting around in the territory" is not necessarily always to start from axioms. This does not mean that he takes the theorems of physics to be disconnected from one another. Rather, the laws of physics provide us with an interconnected (and over-connected) structure, which, however, Feynman stresses, allow no unique and context-independent way of singling out certain of its parts as the most fundamental principles. Thus, Feynman says, we could "start with some particular ideas which are chosen by some kind of convention to be axioms" (47), but we could have chosen a different starting place as well. By contrast with the Euclidean conception, the Babylonian conception of science is non-hierarchical and less tightly organized: "I happen to know this and I happen to know that, and maybe I happen to know that, and I work everything out from there. Tomorrow I may forget that this is true, but remember that something else is true, so I can reconstruct it all again. I am never quite sure of where I am supposed to begin or where I am supposed to end. I just remember enough all the time so that as the memory fades and some of the pieces fall out I can put the thing back together again every day." (*Ibid.*)

Feynman's evocative presentation of the conception might suggest a view that is concerned either with the question of how laws are discovered or with the purely epistemic question how we might recall or rediscover what the laws on the books are. But I take it the view is meant to be more general than that and is, as the title of Feynman's book suggest, concerned with the character or nature of physical laws: the laws of physics, according to the Babylonian conception do not exhibit a uniquely axiomatizable structure.

Feynman's own illustration of the Babylonian conception is the relation between Newton's three laws together with the law of gravity, on the one hand, and Kepler's second law and angular momentum conservation, on the other. As Newton showed, we can derive Kepler's law that equal areas are swept out by a planet in equal times from Newton's laws, which might suggest that the latter are more fundamental than the former. However, Kepler's area law can be thought of as a special case of the principle of angular momentum conservation and the latter applies much more broadly than just to gravitational forces. In fact, according to Noether's theorem, which establishes a connection between symmetries and conserved quantities, angular momentum conservation follows from the fact that the Lagrangian of a system is rotationally symmetric. Thus, one might think that the symmetry principle and angular momentum conservation is the more fundamental principle and should be an axiom instead of the gravitational law. Yet while this principle provides a constraint on the possible form forces between bodies might take, it does not entail Newton's inverse-square law: we cannot make due with the symmetry principle alone.

Another example of the phenomenon to which Feynman points is the relation between the Lorentz force law and the principle of energy-momentum conservation in classical electrodynamics. We can begin by postulating the Maxwell equations and the expression of the principle of energy-momentum conservation for electromagnetic systems and from these together derive the Lorentz force law. Alternatively, we can take the Maxwell equations and the Lorentz force law for continuous charge distributions as starting points to derive that electromagnetic energy and momentum are conserved, given the standard expression for the field energy and momentum. Finally, there is a third view point: We can posit the Maxwell equations together with the Lorentz force law and demand that energy-momentum ought to be conserved in electromagnetic interactions and use these assumptions together to define the electromagnetic field energy in terms of the electromagnetic field vectors. On the first two approaches we assume that we already know what the expression for the energy of an electromagnetic field is, while the third approach uses the demand of energy conservation to find what combination of electric and magnetic field strengths represents the energy contained in the field. Analogously to Feynman's own example, we are faced with the choice of either taking a general principle, such as the principle of energy conservation, as fundamental starting point or beginning with the specific dynamical laws governing the phenomena, such as the Maxwell-Lorentz equations.

Feynman concludes from his discussion that there is no unique answer to the question which of the different starting points is "more important, more basic". Rather in order to understand physics "one must always have a neat balance" (50) between the different possible starting points—in particular, we must balance approaches that begin with broad, general principles with ones that focus on the particulars of the dynamics for a system. Feynman's view echoes a view expressed by Hendrik A. Lorentz, more than half a century earlier, who maintained that approaching physical phenomena *both* from the perspective of broad and general principles, such as the principle of energy conservation, *and* from what Lorentz calls 'the mechanism of the appearances'—that is the particular dynamical laws, can be fruitful (see Frisch, 2005, 2011a, b). Like Feynman after him, Lorentz maintained that neither of the two approaches is privileged and stressed a pragmatic dimension to any preference for one approach over the other: "there are multiple ways by which we try to understand natural phenomena [. . .] Individual characteristics and inclinations determine the choice for each scientist" (Lorentz, 1900, 333).

Feynman also allows for the possibility that "some day when physics is complete and we know all the laws," there may be a privileged set of axioms (p. 49). Thus, Feynman does not want to conclude from the fact that present-day physics does not consist in a uniquely axiomatizable interconnected structure that a final physics would also not allow for a privileged Euclidean axiomatization. But if Feynman is correct in believing that current physics does not fit the Euclidean conception, this raises the question as to what our grounds for assuming that a final physics might fit the conception are. Moreover, Feynman's argument centrally appeals to the use to which laws of physics are put: we are interested in using laws primarily to account for and explain physical phenomena—and do so efficiently. And if Feynman is right that these interests are best served by adopting a Babylonian conception, then it is difficult to see how this situation might change if some day physics were complete and all the laws were known. Even within current physics it is possible to take various starting points as more fundamental (even though Feynman would deny that any such starting point is uniquely privileged). Thus, Feynman's suggestion seems to be that simply in virtue of the fact that a collection of physical laws couldn't be given a yet more fundamental foundation it follows that this collection exhibits a uniquely privileged axiomatic structure.

Conversely, one might try to argue, *pace* Feynman, that we do not have to wait for the completion of physics and that physics already has a hierarchical, quasi-axiomatic structure. Such an argument might appeal to the role of symmetry and conservation principles and to the connection between the two, to argue for the following hierarchical picture: we begin with certain symmetry principles, such as time-translation invariance as meta-laws. These principles explain, via Noether's theorem, which posits a connection between symmetries and conserved quantities, the "great conservation principles", as Feynman calls them. Since the symmetry principles have the status of meta-laws, every particular dynamical law has to satisfy the conservation laws. Thus, we have an explanatory hierarchical scheme beginning from symmetries to conservation laws to particular fundamental dynamical laws.

Yet, even if it were possible to organize theories hierarchically in this manner—from symmetry principles to conservation laws to dynamical laws—the 'anti-Babylonian' needs to establish that this scheme is uniquely privileged and is not merely one among several, and it is not clear that this can be done. For, first, Noether's theorem presupposes a Lagrangian formulation, but not all dynamical systems can be represented in this manner (see Brown & Holland, 2004; Wigner, 1954). Second, where Noether's theorem does apply, it establishes an equivalence and not an explanatory priority of the symmetry principles. That is, in perfect Babylonian fashion one can take either symmetry or conservation principles as starting point.[1] Third, we can in principle take two different viewpoints on the relation between conservation and symmetry principles, on the one hand, and particular dynamical laws, on the other. We could take the general principles to be more fundamental and provide a meta-nomological constraint on the form any dynamical law has to take. But it also possible to take the opposite viewpoint, as Harvey Brown and Peter Holland suggest: "the real physics is in the Euler-Lagrange equations of motion for the fields, from which the existence of dynamical symmetries and conservation principles, if any, jointly spring." (Brown & Holland, 2004) It is compatible with this second viewpoint that we adopt the general principles as heuristic guides in our search for new dynamical laws, without treating them as meta-nomological constraints. Again, the coexistence of these two viewpoints (even if it is not always perfectly peaceful) supports Feynman's Babylonian conception.

I want to end this discussion by pointing to an area of physics that provides perhaps particularly vivid support for a Babylonian conception: the physics of non-linear and complex systems. The physics of complex systems concerns the aggregation of smaller objects into higher-level systems with a very large number of degrees of freedom that can exhibit new types of behavior not present at the lower level. Complex systems exhibit a certain robust organization that arises out of the interplay of underlying randomness and interactions of many elements that result in the higher-order structure (see, e.g., Anderson, 1972; Ladyman et al., 2013) The emergence of new properties in these macroscopically ordered structures involves the breaking of symmetries of the underlying micro-dynamics, which can be characterized as a phase change.

According to one notion, complex systems are systems whose behavior cannot be algorithmically predicted from the micro-dynamical equations governing the system. An example of this is fluid turbulence, which can be modeled at a higher level of description by the so-called 'Burgers equation', an equation that can be solved exactly for arbitrary initial condition. At a more fundamental level fluids are

[1] Marc Lange (2007; Chapter "How to Make the World Safe for Autonomy; or, How to Fodor-Kitcher an Albert-Loewer", this volume) proposes a formal framework which allows the symmetry principles to be meta-laws, which explain the conservation principles. But Lange does not provide an argument for the explanatory priority of the symmetry principles but only shows how, once we *presuppose* an explanatory asymmetry between symmetry and conservation principles, we can spell out the meta-nomological character of the symmetry principles.

governed by the Navier-Stokes equation. Yet turbulent behavior cannot be predicted by integrating the Navier-Stokes equation, since turbulence involves the interactions of different features at different length scales—of big whirls, little whirls and lesser whirls, as the applied mathematician Lewis Richardson has put it—and the locations and times at which these features emerge is sensitive to the precise microscopic initial conditions (see Scott, 2007). That is, complex chaotic phenomena, such as fluid turbulence, pose an entirely new challenge for Laplace's demon in his attempt to predict the universe's evolution: not only does he need to have unlimited computational power, but he also would have to know *exactly* what the universe's *precise* micro-state at some time was, since approximate knowledge of the state, however ever close to the actual state, could result in arbitrarily large divergences in his predictions for the future.

This means that independently of the question whether complex systems might be *ontologically* reducible to their constituents, they are in general not *epistemically* reducible and a description of the higher-level patterns exhibited by the system in terms of higher-level theories is essential. We can be committed to the view that complex systems are in principle governed by the fundamental micro-dynamical laws, yet due to non-linearity, sensitivity to initial conditions, the existence of feedback effects, and the overall computational complexity of the problem, we cannot derive higher-level patterns and laws from the micro-dynamics.

From the perspective of an underlying micro-theory, higher-level descriptions will involve abstractions and idealizations. Nevertheless, to the extent that higher-level features and patterns are robust, have explanatory power and are predictive reliable, they can be thought of as real (see Dennett, 1991; Wallace, 2003). Even if nature were hierarchically organized, this does not entail an explanatory reductionism and a uniquely hierarchical structure of the laws of nature. The physicist Philip Anderson, echoing Feynman's Babylonian conception, expresses this hierarchical yet non-reductionist view as follows:

> The ability to reduce everything to simple fundamental laws does not imply the ability to start from those laws and reconstruct the universe. [...] The constructionist hypothesis breaks down when confronted with the twin difficulties of scale and complexity. The behavior of large and complex aggregates of elementary particles, it turns out, is not to be understood in terms of simple extrapolations of the properties of a few particles. Instead, at each level of complexity entirely new properties appear, and the understanding of the new behaviors requires research which I think is as fundamental in its nature as any other. (Anderson, 1972, 393)

4 The Babylonian Conception and Philosophical Accounts of Laws

I now want to ask what consequences a Babylonian character of physics has for the different philosophical accounts of laws we distinguished above. If we assume that Feynman's account is correct, what would this imply for what account of laws we could adopt? Feynman's view is obviously incompatible with any account that posits the existence of a uniquely privileged axiomatization of physical knowledge. Thus, the Babylonian conception is in tension with the standard MRL view. If Feynman's depiction of physics is correct, then there is no unique context- and interest-independent answer as to what the axioms of the Best System and, thus, what the fundamental laws are. Lewis thought that the world would be such that there would be a uniquely privileged deductive system whose axioms were the laws. The Babylonian conception denies this.

Yet Feynman's motivation of the Babylonian conception actually bears some similarities to the manner in which Lewis motivates his account of laws. A single axiomatization, according to Feynman, does not offer the most efficient way of accounting for getting around in the territory or for accounting for physical phenomena. One way to understand this, is that a single axiomatization is less simple, since it requires overly lengthy and cumbersome derivations. The loss in simplicity in one sense, by allowing for multiple starting points and in effect positing a not tightly axiomatized structure of laws, is more than made up for, or so Feynman can be understood as arguing, by the gain in simplicity in another sense, by cutting down on the length of derivations needed in different contexts. Lewis had hoped that God were kind and that there would be a single deductive system that would clearly and uncontroversially come out ahead as far as a balance in simplicity and strength are concerned. Feynman's discussion suggests that this hope might have been in vain.[2]

There is a further problem for the standard MRL view: it is not clear how one might reconstruct a nomic structure consisting of putative meta-laws, such as symmetry or conservation principles, together with dynamical laws within the traditional MRL account. Intuitively, it would seem that such principles should be among the laws of physics. Yet it is not clear how they could be, according to the account. On the one hand, a set of axioms consisting *only* of these principles would be too weak, since it would not allow us to derive any specific dynamical laws and, a fortiori, no particular matters of fact about the mosaic. On the other hand, if we enlarged our set of axioms to include the dynamical laws, then the general principles would become superfluous: a system including both symmetry principles and dynamical laws would be less simple without any increase in deductive strength compared to a system consisting of the dynamical laws alone.

The Babylonian conception is also in tension both with the view that there exist fundamental laws that ultimately govern how the world evolves and with the view that the laws are consequences of the physical essences possessed by things. The problem for the latter view is that according to the Babylonian conception it is unclear which set of laws directly reflects the underlying essences. Is it, for example, part of the essence of massive objects to respond to forces, or is it rather in their

---

[2] This Lewisian motivation for Feynman's conception echoes the manner in which advocates of a "Better Best System's view motivate their departure from a Lewisian single best system (See Cohen & Callender, 2009; Callender & Cohen, 2010; Frisch, 2011a, b).

essence to conserve energy and momentum or move along a path that minimizes the action (as a Lagrangian picture suggests)? The problem for Maudlin's or Chen's primitivism is that both accounts posit a unique hierarchical conception of laws, according to which there are truly fundamental laws, which ground all other laws.

Now one might reply that this incompatibility is so much the worse for the Babylonian conception. Chen and Goldstein, in particular, appeal to an account of explanation in support of their hierarchical conception of laws. If laws play a central role in explanation in physics and if explanations require a unique hierarchy of laws bottoming out in truly fundamental laws, then the Babylonian conception of false. While a general defense of the Babylonian conception is beyond the scope of this paper, I want to argue that the core of Chen and Goldstein's own account of explanation does not in fact require a commitment to truly fundamental laws. Laws explain, according to Chen and Goldstein, "by constraining the physical possibilities in an illuminating manner." (Chapter "Governing Without a Fundamental Direction of Time: Minimal Primitivism About Laws of Nature", this volume). So-called "constraint explanations" (see Ben-Menahem, 2018) are one type of broadly counterfactual explanations that explain by embedding a phenomenon into a pattern of counterfactual dependencies and by answering "what-if-things-had-been-different questions" (Woodward, 2005). Chen and Goldstein argue that constraint explanations require a metaphysically and modally richer notion of law than a Humean account can provide. But one may concede *this* point without also accepting that modal, nomological constraints need to be hierarchically structured and that there exist unique *fundamental* laws. Indeed, the two additional demands Chen and Goldstein place on explanations in addition to that of appealing to nomic constraints—metaphysical fundamentality and simplicity—may pull in different directions and may even be incompatible with each other. Arbitrary constraints, Chen and Goldstein say, "do not always provide satisfying explanations", since "many constraints are complicated and thus insufficient for understanding nature. What we look for in the final theory of physics is not just any constraint but simple, compelling ones" (Chapter "Governing Without a Fundamental Direction of Time: Minimal Primitivism About Laws of Nature", this volume). But if genuine explanations need to invoke laws that are truly metaphysically fundamental, we don't get to choose: we will have to make do with whatever laws turn out to be truly fundamental, no matter how complicated any explanations invoking them may turn out to be. By contrast, considerations of simplicity (as well as the additional desideratum mentioned by Chen of Goldstein that explanations be unexpected and striking) are quite congenial to Feynman's Babylonian conception: depending on context and on our interests, we might find different explanations of a phenomenon, starting from different laws particularly illuminating, since they might be particularly simple or may be particularly striking in light of what else we assume in a given context. This tension disappears if we adopt Feynman's Babylonian conception: This conception allows us to retain a constraint or counterfactual account of explanation and also allows us to (if we are so inclined) to assume a metaphysically richer notions of modality than Humeans would allow. But at the same the Babylonian conception allows us to retain what

Chen and Goldstein identify as the "epistemic ingredient" in their own account—a commitment to simplicity-while giving up a notion of metaphysical fundamentality. Thus, Feynman's Babylonian conception, it seems to me, does better justice to Chen and Goldstein's view on explanation than their own minimal primitivism does.

Due to their flexibility and their lowered ambitions, the revised MRL-view and the 'laws-as-tools'-view fare best in accommodating the Babylonian conception. The role of general principles, whose usefulness in model-building extends far beyond the domain for which they were originally proposed presents a challenge, however, for more extreme patchwork accounts of laws. What is needed, in order to do justice to the Babylonian character of physics, is an account that can provide a role *both* for general, overarching principles *and* for domain specific modeling assumptions. The Babylonian conception does not posit different relatively isolated 'best systems' that might even make incompatible predictions. After all, the structure Feynman describes encompasses areas interconnected and, indeed, overconnected through multiple bridges (47). The particular matters of fact that are derivable from different starting points in the structure are overlapping and compatible with one another, the laws posited are compatible, there only are different ways of organizing and structuring the set of laws.

5 Dynamics, Kinematics, and Initial Conditions

In the previous section I discussed Feynman's Babylonian conception of physics, according to which the distinction between fundamental and derived laws in physics is to some extent conventional or context-dependent. In this section I will discuss another area in which some physicists and philosophers have argued to have detected conventionalist elements: the distinction between dynamics, kinematics, and initial conditions. In a series of lectures the physicist David Gross presented 25 questions on the future of physics. Question #13 was the question whether "dynamics, kinematics and initial conditions can be separated". (Gross, 2005) Gross's answer was that perhaps they cannot be disentangled. As Gross characterizes the distinction, kinematics constitutes "the framework for physics and its interpretation" while dynamics provides the "specific laws of nature." Kinematics, that is, specifies the formal framework within which specific dynamical interactions take place. The initial conditions are the values of those quantities specified in the kinematical framework that need to be given as input to obtain solutions to the dynamical laws.

Gross suggests that the distinction comes under pressure in contemporary physics and says: "I suspect that, as we learn to understand string theory and explore the nature of space-time, the distinction between kinematics and dynamics will be blurred." Yet several decades before Gross, the physicist Eugene Wigner suggested that the distinction between laws and initial conditions is, if perhaps not blurred, nevertheless not a fully objective distinction even for well-established physical theories and is a human artifact that we make use of in our representations of physical phenomena in order to distinguish a realm of regularities from one of randomness (Wigner, 1979).

Consider the different frameworks for representing the physics of Newtonian interacting particles. In addition to representing the evolution of a mechanical system of Newtonian particles in terms of Newton's laws and the forces acting between the particles one can also represent the system within either a Lagrangian or a Hamiltonian framework (see, e.g., Belot, 2006). The basic variables in the *Hamiltonian* framework are generalized position and momenta coordinates. Position and momentum, are thus treated on an equal footing and by introducing generalized coordinates the framework can be extended even to systems that cannot readily be represented in terms ordinary space-time coordinates. The state of a system consisting of n particles is represented in a $6n$-dimensional phase space, where a system's instantaneous state is given by the values of $6n$ variables representing the 6 linearly independent position and momentum coordinates for each of the n particles. We can think of phase space as the framework for representing the space of possible position-momentum pairs or the associated with a theory. This space determines the space of initial data for a system representing the possible instantaneous states allowed by the theory, which when plugged into Hamilton's equations for the system, nomically determine the evolution of a system. That is, in the Hamiltonian framework, the space of possibilities for a system is given by the space of possible instantaneous position-momenta states in phase space.[3] Hamilton's equations are equivalent to Newton's equations, but while Newton's equations represent the evolution of a system as the result of forces acting between particles, the Hamiltonian framework represents the system in terms of potentials to which a particle's motion responds.

The arena for *Lagrangian* mechanics is the $3n$-dimensional configuration space, corresponding to the $3n$ degrees of freedom of a system of n particles (assuming that there are no constraints reducing the number of independent variables). The space of possible states is the space of possible particle positions. Any arbitrary curve through this space is kinematically possible. The dynamically or nomically possible curves are stationary curves satisfying a variational principle. In the Lagrangian framework, the evolution of a system is derived from the state at two times and a constraint the system satisfies at times in between.

Thus, in the Lagrangian formulation the kinematics—that is the arena in which the nomological constraints operate—is the space of possible positions, while in the Hamiltonian formulation this arena consists in the space of possible position-momentum pairs. The difference in kinematics is compensated by a difference in the dynamics—Hamilton's equations provide first-order constraints on phase space, while the Lagrange equations provide second-order constraints—thereby making it possible for the Lagrangian, Hamiltonian (and Newtonian) frameworks to make equivalent predictions for systems of gravitating particles. Differences between the

[3] Each state can further be assigned a kinetic energy T and potential energy V, which together encode all the physics of the problem.

frameworks arise for possible extensions to other domains and here it is unclear that one of the frameworks is clearly overall superior to the other two. Thus, comparing Lagrangian with Hamiltonian formulations of physics appears to support Wigner's view and the claim that there is a conventional element in how we want to carve up the territory between kinematics and dynamics.

The three different frameworks also may seem to suggest different causal structures. While the Newtonian formulation naturally suggests a causal picture of particles being pushed and pulled around by locally acting forces, and the Hamiltonian formulation is one of particles responding to local values of a potential, the Lagrangian formulation might seem to suggest a teleological explanation of a system's evolution, since the path of a system is represented as being determined not only by its initial state but by the conjunction of its initial and final states: as Feynman puts it, the system's particles "in some grand fashion [smell] all the curves, all the possibilities, and decide which one to take." (p. 52).

I want to briefly mention three other examples of the conventionality of the choice of background framework and of dynamical laws (see Spekkens, 2012). First, gravitational effects are classically modeled in terms of forces, as in Newtonian gravitational theory. Alternatively, one can treat gravity as a geometric effect even non-relativistically, as is done in the Newton-Cartan theory (see Weatherall, 2011). Second, spin degrees of freedom are treated differently in different versions of Bohm's quantum theory. According to Bohmian mechanics, particles follow classical trajectories, guided by a non-classical 'pilot wave' that satisfies the Schrödinger equation. Spin degrees of freedom can be introduced into Bohm's theory either by modeling the particle as having additional internal degrees of freedom or by taking the wave function to be a multiple-component spinor wave-function. And, third, systems governed by a stochastic dynamics can be alternatively represented in terms of a deterministic dynamics, where the system is modeled as interacting with an additional (hidden) noisy source.[4]

6 The Distinction Between Laws and Initial Conditions and Philosophical Accounts of Laws

While the Babylonian conception puts pressure on an account of laws that posits a hierarchy of laws bottoming out in truly fundamental laws, Wigner's suggestion that the distinction between dynamical laws and the kinematic framework within which the dynamics takes place is a matter of convention or choice, puts pressure on any account of laws that sees the distinction between nomological necessity and contingency as metaphysical basic. For a given set of phenomena there can be different, yet equivalent ways to characterize the set of possible states on

[4] Nicholas Harrigan and Rob Spekkens discuss this equivalence in the context of quantum theories with hidden variables (Harrigan & Spekkens, 2010).

which nomological constraints are imposed, with corresponding differences in the nomological constraints. This does not mean that we would have to deny that there are irreducible modalities in the world, but it follows from Wigner's suggestion that there is no unique way to carve up physical goings-on into nomological constraints and the space of possibilities on which these operate.

Moreover, Wigner's view has the consequence that attempts to read off the essences of objects from different dynamical frameworks within which the evolution of these objects may be represented could result in conflicting answers. Is it in the nature of quantum particles with spin to possess additional internal degrees of freedom or is it part of their essence to be guided by a spinor-valued wave function? Is it part of the essence of material particles to follow geodesics in the absence of non-gravitational forces or do material particles essentially exert gravitational forces? And, to return to our main example, is it in the nature of particles to respond to local forces or to local potentials? Or is it in the essence of a system to satisfy a global variational principle? It is unclear whether physics itself can provide unambiguous answers to these questions.[5]

There is an additional problem for versions of a metaphysical primitivism about laws that take laws as being responsible for, or as producing, the temporal evolution of the state of a system. These conceptions may have difficulties in accommodating a Lagrangian framework, since initial states in configuration space are simply not rich enough for the laws to uniquely determine the state's future evolution. Of course, it is open to defenders of a productive account of laws to argue that the Lagrangian variational principle is a consequence of more fundamental forces acting along infinitesimal parts of a system's path (see Lange, 2009). But such an argument in favor of the fundamentality of a Newtonian framework would presumably have to invoke extra-physical, metaphysical reasons for the priority of that framework.

As is the case for the Babylonian conception, a traditional Humeanism that posits the existence of a uniquely best system has difficulties accommodate Wigner's conventionalism, but Humeans who allow for multiple 'best systems' can readily accept that the distinction between kinematics and dynamics is to some extent conventional and that there might be multiple ways to carve out regularities and initial states in the Humean mosaic. Which framework is the most appropriate to choose can then be allowed to be a context-dependent and pragmatic matter, depending not only on the particular phenomenon we are modeling but also on possible extensions of the framework to other phenomena in which we are interested.

[5] Wolfgang Pietsch (2013) argues on the basis of a case study concerning recent developments in metrology that there is an additional element of conventionalism concerning laws: how units and fundamental constants are defined changes and with it the distinction between a theory's definitional presupposition and its empirical predictions.

7 Conclusion

My aim in this paper was to survey two features of laws in physics and ask to what extent these features are compatible with different philosophical accounts of laws of nature. These features are the following: (i) laws in physics form a 'Babylonian' interlocking set of 'theorems' among which we can choose different theorems as starting points or 'axioms', depending on the context, and which are connected by an interlocking web of more or less tight derivations; (ii) there are mathematically equivalent ways of carving up the phenomena into kinematic background framework and dynamical laws. These features put pressure on any philosophical account of laws that presupposes that the laws of physics have a unique quasi-axiomatic structure, such as the traditional MRL account and metaphysical accounts that assume that there is a privileged metaphysical and explanatory nomological hierarchy.

Due to its flexibility, a pragmatic revised MRL-account that allows for context- or perspective- dependent multiple best systems fares best in accounting for the features we have discussed. This, however, does not mean that the practice of physics supports a subjectivism about laws. What the laws are, within a certain context of investigation and for a certain domain of phenomena, is not up to us to decide.

Defenders of a 'better best systems' account, such as (Callender & Cohen, 2010) tend to focus on the relation between the special sciences and physics (and the apparent independence of the special sciences from physics) as central motivation for their account. These authors allow that different 'best systems' may even resist being integrated with one another or may, in some sense, be in tension with one another. By contrast, I have here argued that there is a role for multiple 'best systems' within physics capturing the same range of phenomena. But these different systems allow us to derive compatible and strongly overlapping sets of facts about the world. They don't disagree about the phenomena but only about which laws are taken to be as starting points in derivations and how to carve up the distinction between kinematics and dynamics. Thus, even within physics the case for a strict and unique hierarchical conception of laws—be it a metaphysical primitivist conception or a traditional Humean account—is weaker than it may seem.[6]

References

Anderson, P. W. (1972). More is different. *Science, 177*(4047), 393–396.

Belot, G. (2006). The representation of time and change in mechanic. In J. Butterfield & J. Earman (Eds.), *Philosophy of physics* (pp. 133–227). Elsevier.

Ben-Menahem, Y. (2018). *Causation in science*. Princeton University Press.

Bird, A. (2007). *Nature's metaphysics: Laws and properties*. Oxford University Press.

[6] This paper is a descendent of (Frisch, 2014).

Brown, H., & Holland, P. (2004). Dynamical vs. variational symmetries: Understanding Noether's first theorem. *Molecular Physics, 102*, 1133–1139.
Callender, C., & Cohen, C. (2010). Special sciences, conspiracy and the better best system account of Lawhood. *Erkenntnis, 73*, 427–447.
Cartwright, N. (1983). *How the laws of physics lie*. Oxford University Press.
Cartwright, N. (1999). *The dappled world: A study of the boundaries of science*. Cambridge University Press.
Chen, E. K., & Goldstein, S. (Forthcoming). Goverig without a fundamental direction of time: Minimal primitivism about laws of nature. In *Rethinking the concept of laws of nature*. Springer.
Cohen, J., & Callender, C. (2009). A better best system account of Lawhood. *Philosophical Studies, 145*(1), 1–34.
Demarest, H. (2017). Powerful properties, powerless laws. In J. D. Jacobs, (Ed.), *Causal powers* (pp. 38–53). Oxford University Press.
Dennett, D. C. (1991). Real patterns. *The Journal of Philosophy, 88*(1), 27–51.
Feynman, R. (2001). *The character of physical law*. MIT Press.
Frisch, M. (2005). Mechanisms, principles, and Lorentz's cautious realism. *Studies in History and Philosophy of Modern Physics, 36*(4), 659–679.
Frisch, M. (2011a). From Arbuthnot to Boltzmann: The past hypothesis, the best system, and the special sciences. *Philosophy of Science, 78*(5), 1001–1011.
Frisch, M. (2011b). Principle or constructive relativity. *Studies in History and Philosophy of Modern Physics, 42*(3), 176–183.
Frisch, M. (2014, May). Laws in physics. *European Review, 22*(Suppl. S1), S33–S49.
Giere, R. (1999). *Science without laws* (1st ed.). University of Chicago Press.
Gross, D. (2005) *The future of physics*. http://indico.cern.ch/conferenceDisplay.py?confId=a05302#2005-01-26. Last accessed 12 Oct 2021.
Harrigan, N., & Spekkens, R. (2010). Einstein, incompleteness, and the epistemic view of quantum states. *Foundations of Physics, 40*(2), 125–157.
Ladyman, J., Lambert, J., & Wiesner, K. (2013). What is a complex system? *European Journal for Philosophy of Science, 3*(1), 33–67.
Lange, M. (2007). Laws and meta-laws of nature: Conservation laws and symmetries. *Studies in History and Philosophy of Modern Physics, 38*(3), 457–481.
Lange, M. (2009). Causation in classical mechanics. In H. Beebee, C. Hitchcock, & P. Menzies (Eds.), *The Oxford handbook of causation* (pp. 649–661). Oxford University Press.
Lewis, D. K. (2001). *Counterfactuals* (2nd ed.). Wiley-Blackwell.
Loewer, B. (2012). Two accounts of laws and time. *Philosophical Studies, 160*(1), 115–137.
Lorentz, H. A. (1900). Elektromagnetische Theorien Physikalischer Erscheinungen. In *Collected papers* (Vol. VIII, p. 333). Martin Nijhoff.
Maudlin, T. (2007). *The metaphysics within physics*. Oxford University Press.
Mitchell, S. D. (2009). *Unsimple truths: Science, complexity, and policy*. University Of Chicago Press.
Pietsch, W. (2013). The limits of probabilism. In V. Karakostas & D. Dieks (Eds.), *The European philosophy of science association proceedings 2011*. Springer.
Scott, A. C. (2007). *The nonlinear universe: Chaos, emergence, life* (1st ed.). Springer. sec. 6.8.8 & 8.3.4.
Spekkens, R. (2012). The paradigm of kinematics and dynamics must yield to causal structure. *arXiv*, 1209.0023 [quant.ph].
Wallace, D. (2003). Everett and structure. *Studies in History and Philosophy of Modern Physics, 34*(1), 87–105.
Weatherall, J. O. (2011) The motion of a body in Newtonian theories. *Journal of Mathematical Physics, 52*(3), 032502-032502-16.
Wigner, E. P. (1954). Conservation laws in classical and quantum physics. *Progress of Theoretical Physics, 11*(4), 437–440.
Wigner, E. P. (1979). *Symmetries and reflections*. Ox Bow Press.
Woodward, J. (2005). *Making things happen*. Oxford University Press.

What's so Special About Initial Conditions? Understanding the Past Hypothesis in Directionless Time

Matt Farr

Abstract It is often said that the world is explained by laws of nature together with initial conditions. But does that mean initial conditions don't require further explanation? And does the explanatory role played by initial conditions entail or require that time has a preferred direction? This chapter looks at the use of the 'initialness defence' in physics, the idea that initial conditions are intrinsically special in that they don't require further explanation, unlike the state of the world at other times. Such defences commonly assume a primitive directionality of time to distinguish between initial and final conditions. Using the case study of the time-asymmetry of thermodynamics and the so-called 'past hypothesis'—the hypothesis that the early universe was in a state of very low entropy—, I outline and support a deflationary account of the initialness defence that does not presuppose a basic directionality of time, and argue that there is a relevant explanatory asymmetry between initial conditions and the state of systems at other times only if certain causal conditions are satisfied. Hence, the initialness defence is available to those who reject a fundamental direction of time.

Keywords Time · Past hypothesis · Direction of time · Explanation · Initial conditions · Cosmology · thermodynamics

1 Introduction

Eugene Wigner once remarked that '[t]he sharp distinction between Initial Conditions and Laws of Nature was initiated by Isaac Newton and I consider this to be one of his most important, if not *the* most important, accomplishment', adding that prior to Newton, 'there was no sharp separation between the two concepts' and that after Newton this sharp separation 'was taken for granted and rarely even

M. Farr (✉)
Department of History and Philosophy of Science, University of Cambridge, Cambridge, UK
e-mail: mail@mattfarr.co.uk

Y. Ben-Menahem (ed.), *Rethinking the Concept of Law of Nature*,
Jerusalem Studies in Philosophy and History of Science,
https://doi.org/10.1007/978-3-030-96775-8_8

mentioned' (Wigner 1995, p. 334). The terminology of 'initial' conditions is so widespread within physics as to make its usage seemingly unremarkable. But the very word 'initial' connotes the idea of time as directed, deriving from the Latin for 'the beginning'; something could not be 'initial' without being chronologically first in some sequence of events, and something could not be chronologically first without a direction of time relative to which it is first. However, there are many motivations within fundamental physics for holding time to lack a direction,[1] with the appearance of a direction of time being an emergent feature of higher-level physics. So what is the relationship between initial conditions and the direction of time, and how do philosophical views that reject a directionality of time make sense of the role of initial conditions in physics?

This chapter looks at a particular kind of argument I'll call the 'initialness defence'. The general idea is that initial conditions, as opposed to the state of the system at other times, are explanatorily special in virtue of temporally preceding all other temporal stages of the relevant system, and not being the result of evolution from earlier states. In ordinary experiments, one can consider the initial conditions—the state in which the relevant system is prepared in the experimental set-up—to be a set of information that one does not seek to explain in terms of the theory being tested. Conversely, the final conditions of the experiment, namely the measured state at the end of the experiment, is to be explained in terms of the relevant theory together with the initial conditions. This explanatory asymmetry between initial conditions and final conditions is present in cosmology, such as the cosmological explanation of the second law of thermodynamics, the details of which I'll turn to in a moment. The idea in the cosmological case is that one can design a cosmological model with some assumed initial conditions, such as the extremely low-entropy early universe, and appeal to this to explain some particular observed regularity, in our case the steady increase of thermodynamical entropy towards the future. In such cases, the initial conditions are ordinarily treated as explanatorily special.

Initialness defence. The earliest state of some system is explanatorily special because it is the only state that did not evolve from some earlier state of that system.

Initialness defences commonly assume a primitive directionality of time in order justify distinguishing between initial and final conditions in this way. But what exactly is the relationship between the initialness defence and the direction of time? Do initialness defences presuppose a specific metaphysical view of time direction? And can metaphysical views that hold time to be *adirectional*—to lack a preferred direction—employ initialness defences? In what follows, I assess these issues using the case study of the time asymmetry of thermodynamics.

The standard statistical mechanical account of why macroscopic systems obey the second law of thermodynamics relies on the assumption that in the past, entropy

[1] See Farr (2020).

was far lower than it is now. Contemporary cosmology posits an extremely low entropy and thus low probability initial state of the universe, which following Albert (2000) is commonly termed the *past hypothesis*. There has been much disagreement as to whether the extremely low-probability initial state posited by the past hypothesis stands in need of explanation (e.g. Price 2004 argues it does; Callender 2004b argues it does not), and what would constitute an explanation of such a thing (e.g. Callender 2004a questions whether there could in principle be such an explanation).[2] One thought implicit in a number of considerations of this issue is a version of the initialness defence, that the past hypothesis is explanatorily special due to being a hypothesis about the *earliest* temporal part of the universe, and so not causally dependent upon earlier states. Prima facie, realists about time direction—those who take the directionality of time to be some basic fact about the world, such as Maudlin (2007)—can make use of the initialness defence, but antirealists about time direction—those who hold either that time has no direction, or that the directionality of time is an emergent feature of the macroscopic world, such as Price (1996)—cannot, since on their picture of the world there is nothing intrinsically special about either temporal 'end' of the universe. I argue this dichotomy is false. Instead, antirealists about time direction, or 'C-theorists' as I will call them, can and should adopt a restricted version of the initialness defence.

The chapter is structured as follows. Section 2 introduces the background to the debate, introducing the problem of time asymmetry in statistical mechanics and the role played by the past hypothesis. Section 3 introduces the B- and C-theories of time, and the debate between realist and antirealist accounts of the direction of time. In Sect. 4 I assess Maudlin's B-theoretic version of the initialness defence in the context of the past hypothesis, and Huw Price's C-theoretic rejection of it. Section 5 presents a C-theoretic 'deflationary' version of the initialness defence, motivated by a position outlined by Hans Reichenbach, and argues the deflated initialness defence preserves the ontological simplicity of the C-theory whilst marking out the past hypothesis as explanatorily special. Section 6 is the conclusion.

2 Explaining the Arrow of Time

2.1 The Time Asymmetry of Thermodynamics

Imagine an ordinary irreversible process like the smashing of a wine glass: I take a wine glass out of the cupboard, drop it on the floor, and it shatters into tiny pieces. Such a process is (hopefully) not especially familiar but it's certainly far more familiar than the temporal opposite process; the process run from future to past involves a bunch of shards of glass spontaneously starting to jiggle on the kitchen

[2] See Baras and Shenker (2020) for a recent assessment of the dialectic between Price and Callender, and what it means to *explain* the low-entropy past.

floor due to an injection of impetus from heat on the floor causing the shards to jump together, forming a smooth wine glass, before bouncing up into my hand. Whereas the original process certainly sounds a plausible candidate for what can happen in a kitchen, the second process does not. This asymmetry between two processes here reflects a greater asymmetry of things in time described by the second law of thermodynamics. One of the most familiar statements of the second law is Rudolf Clausius' (1864, p. 44) 'Die Entropie der Welt strebt einem Maximum zu,' or 'the entropy of the world tends towards a maximum.' This introduces the notion of entropy, which is taken to be a measure of the energy of a system from which work cannot be extracted. More familiarly, we can state the second law as follows:

Second Law of Thermodynamics. The entropy of an isolated subsystem of the world does not decrease.[3]

Or, denoting entropy by the letter S,

$$\frac{dS}{dt} \geq 0 \tag{1}$$

2.2 *Boltzmannian Statistical Mechanics*

The latter half of the nineteenth century saw the development of statistical mechanics in the work of (among others) Maxwell, Boltzmann and Gibbs, with the central idea being that thermodynamic regularities can be accounted for in terms of the statistics of the constituents of sufficiently large mechanical systems, or in Gibbs' case the statistics of ensembles of mechanical systems. The discussion of this chapter is based on the Boltzmannian approach to statistical mechanics.[4] To understand the Boltzmannian statistical mechanical account, we need a bit of mathematics. The state of a classical Hamiltonian system of n particles is given by the specification of the positions and momenta of all the particles in the system. To simplify, we can use the concept of a *phase space*, denoted by the symbol Γ.

[3] There exist many different statements of the second law (see Uffink 2001), and many different definitions of entropy (see Maroney 2009). As Price (2002) notes, the precise version of the second law, and the definition of 'entropy' are relatively inessential to the issue of the relation between time direction and thermodynamics. The example of the wine glass is sufficient to illustrate the kind of time asymmetry with which this chapter is concerned.

[4] There are two main approaches to statistical mechanics, owing to the work of Boltzmann and Gibbs respectively. I shall work within the Boltzmannian framework. Gibbsian statistical mechanics (Gibbs 1902) is distinct from Boltzmannian statistical mechanics in a number of ways. The Gibbsian approach is characterised by considering an *ensemble* of systems, each with slightly different microscopic states. As Frigg (2009) notes, both approaches are widely used within the literature but generally for different purposes: the Gibbsian approach is used in practice (i.e. experiments); the Boltzmannian approach is used in consideration of foundational issues (e.g. philosophical issues concerning the reduction of thermodynamics to statistical mechanics). See Wallace (2020) and Robertson (2020, 2021) for recent defences of the Gibbsian approach.

A phase space is an abstract mathematical space in which each point represents a complete kinematically possible microstate of the system in question—in this case the position and momentum values of each particle.[5] For a system of n particles, the phase space has $6n$ dimensions, with a point in the $6n$ phase space picks out a unique set of position and momentum values (in three-dimensional physical space) for each particle in the system. Given this, the evolution of a many-particle classical system can be illustrated by a curve in Γ, and thus the dynamics describes curves in Γ. A point in Γ picks out a *microstate* of the system in question. In thermodynamics, a state of a system is given in terms of macroscopic (observable) parameters, such as the pressure, volume and temperature of a gas. One of the key observations in the development of statistical mechanics is the identification of the thermodynamic *macrostate* of a system with a *region* of Γ. In this sense, the macrostate of a system supervenes on its microstate—there can be no change in the system's macrostate without a change in its microstate—but not *vice versa*.

Central to the Boltzmannian account is the application of the 'standard' (Lebesgue) measure to the 'microcanonical ensemble'—that each microstate on a hypersurface of fixed energy on the phase space is assigned equal probability.[6] Given this, regions of phase space can be assigned measurable *volumes*. Boltzmann famously provided the following statistical mechanical definition of the entropy, S, of a system as proportional to natural logarithm of the volume of its phase space corresponding to its macrostate, where, given the standard measure, W corresponds to the *amount* of microstates corresponding to the system's macrostate, and k is Boltzmann's constant ($1.38062 \times 10^{-23} J K^{-1}$).

$$S = k \ln W \tag{2}$$

This tells us that the larger the volume of phase space corresponding to some macrostate of a system X, the greater the entropy of X. It follows from this that for any gas, given no further constraints on its macrostate than the Boltzmannian

[5] In order for a point in phase space to play this role, the phase space must have the right dimensionality. The state of a one-particle system is provided by two quantities: the position and the momentum of the particle. If the particle is in a three-dimensional *physical* space, then its position and momentum each have coefficients in each spatial dimension: the position is provided by the triple (q_x, q_y, q_z) corresponding to its location in each spatial dimension (x, y and z); and the momentum is provided the triple (p_x, p_y, p_z). Thus, for a one-particle system, the phase space has 6 dimensions—three for position; three for momentum—such that each point in this space denotes a unique set of position and momentum values, with the entire space covering every possible state of the one-particle system.

[6] This assumption is one of the central problems concerning Boltzmannian statistical mechanics—see Sklar (2006) for a survey. The justifiability of Boltzmannian microcanonical probability distribution has historically concerned Liouville's Theorem (which holds that regions of phase space of fixed energy are invariant under Hamiltonian evolution) and the Ergodic Hypothesis. See Frigg (2008) for a discussion of the relation between the ergodic hypothesis and Boltzmannian statistical mechanics.

statistical considerations, at any time the gas is most likely to be in the highest-entropy available macrostate and hence in equilibrium.

To complete the necessary details of the Boltzmannian account, we need to make an assumption about the dynamics of points in phase space—*micro*states—namely that they are quasi-ergodic.[7] The ergodic hypothesis holds that the infinite time averages of systems are such that the proportion of time spent in each macrostate corresponds to the volume of the macrostate given by the standard measure. What this means is that for any point in Γ (except for a set of measure zero), its evolution over infinite time is such that the time spent in each of its available macrostate is directly proportional to its volume.[8] Given the assumption that systems are quasi-ergodic, it follows that systems in small regions of phase space (low entropy) tend to evolve towards larger regions of phase space (higher entropy), and spend the majority of time in the largest region of phase space, which corresponds to the equilibrium state—maximal entropy. It follows that given a non-maximal entropy macrostate, it is overwhelmingly probable that its microstate is such that its evolution leads towards macrostates of higher entropy.

As such, we can see quite simply that the reason non-equilibrium systems tend towards higher entropy is that there are more *ways* for a system to have high entropy. Given this, as Wallace (2011, p. 1) puts it, the overwhelming tendency of systems to approach equilibrium is 'basically just a consequence of the geometry of phase space.' For a system of low entropy left to evolve in phase space, it will invariably tend towards higher entropy simply because high entropy macrostates take up much larger regions of phase space than do low entropy macrostates. This gives us two kinds of 'normal' or 'typical' states for systems; ones that, assuming nothing else, systems can be expected to be in without requiring further explanation:

Normal Macrostate. A normal macrostate for a system is one that corresponds to maximal entropy (its 'equilibrium state').

Normal Microstate. A normal microstate for a system in non-equilibrium is one that evolves towards equilibrium.[9]

2.3 *What Is the Past Hypothesis?*

The problem with this kind of statistical mechanical reasoning is that it works too well. Towards the future, where we expect entropy to increase, the reasoning works

[7] Frigg and Werndl (2011) argue that a weaker property that of ergodicity—'epsilon ergodicity'—is sufficient for Boltzmannian statistical mechanics and avoids the problems of the KAM and Marcus-Meyer theorems.

[8] The status of the ergodic hypothesis has been the subject of a particularly massive literature that does not concern us in the present chapter. See Sklar (1993, ch. 5) for a review.

[9] A normal microstate for a system *in equilibrium* is one that stays in equilibrium (in both temporal directions).

fine. But if we try to use the same reasoning towards the past then we should also think that it's far more likely for any system in a non-maximum-entropy state to have evolved from a higher-entropy state and not from a lower-entropy state, contrary to our beliefs and apparent observations. This is a major problem insofar as the statistical method only gives us a time asymmetry of entropy if we explicitly assume that in the past entropy was far lower than it is now. Without such an assumption, the statistical method implies that both the future and past are higher entropy than the present, so we end up with a sceptical problem about the past (implying that our apparent records and memories of the past are false) and are unable to ground the time asymmetry of the Second Law.

Boltzmann suggested that such a low-entropy assumption about the past is a crucial aspect of the statistical explanation of the second law:

> The second law of thermodynamics can be proved from the mechanical theory if one assumes that the present state of the universe [...] started to evolve from an improbable state and is still in a relatively improbable state. (Boltzmann 1967)

The idea suggested by Boltzmann in this passage is that we need to put in an assumption of a low-entropy past in by hand, and that once we do that then the statistical method gives back the tendancy for entropy to increase monotonically towards the future.[10] The same conclusion is reached by Richard Feynman in his *Lectures on Gravitation*:

> The success of [scientific inferences] indicates that the world did not come from a fluctuation [...] Therefore I think it is necessary to add to the physical laws the hypothesis that in the past the universe was more ordered [...] than it is today—I think this is the additional statement that is needed to make sense, and to make an understanding of [thermodynamic] irreversibility. (Feynman et al. 1971)

This idea that we need to make an assumption of a low-entropy past to complement the Boltzmannian statistical method is elaborated on by Albert (2000), where he terms the idea the 'past hypothesis', which has become the common terminology in the literature. The idea is that the past hypothesis complements the statistical mechanical account of the entropy gradient by stipulating a low entropy constraint on a past macrostate of the world. Thus, for some non-maximal entropy system, such as a half-melted ice cube on a warm plate, the statistical mechanical *pre*diction is unconstrained—we predict that the ice cube will be more melted in 5 min—, but the *retro*diction is constrained insofar as it is conditional upon the stipulated low-entropy past macrostate. Call this macrostate (following Callender 2004a) the *Past State*. We can define the past hypothesis as follows:

Past Hypothesis. The hypothesis that at some point in the distant past, the world occupied an extremely low-entropy macrostate—the Past State.

---

[10] However, this remark is not elaborated on in much detail by Boltzmann, and he elsewhere entertained the hypothesis that the low entropy past is simply the result of a large fluctuation from thermal equilibrium made probable by the universe being significantly older than standard cosmological models assume, and for the most part in thermal equilibrium.

On Albert's (2000) account, the past hypothesis corrects the faulty retrodictions of statistical mechanics, namely the overwhelming likelihood that the past was higher entropy, implying that all apparent records and memories of a lower-entropy past are misleading. The big problem with the past hypothesis is the sheer improbability of the Past State. Penrose (1989) calculates the probability of the universe having been in a state of sufficiently low entropy to be 1 in $10^{10^{123}}$.[11] According to the past-hypothesis-based statistical mechanical explanation of the entropy gradient, the assumption that such a low-probability state occurred is a key part of the explanation of any mundane thermodynamic regularity, such as the inevitably underwhelming final sips of my now-cold flat white.

Boltzmann himself thought the past hypothesis 'is a reasonable assumption to make since it enables us to explain the facts of experience, and one should not expect to be able to deduce it from anything more fundamental' (Boltzmann 1967), emphasising the lack of a need to offer a special explanation of the past hypothesis itself. Various authors have offered a range of explanatory options, from Boltzmann's no-explanation-needed option, to Albert's related suggestion that the past hypothesis possesses a Kantian-type epistemic status in that it is a necessary *precondition* for knowledge of the past, to Callender's (2004a) suggestion that the past hypothesis has the status of a non-dynamical law of nature. But we are here going to assess only the role of the initialness defence in the explanation of the past hypothesis, and to do that we'll focus on Maudlin's initialness defence, Price's rejection of such an approach, and my own preferred 'deflated' version of the initialness defence that is motivated by remarks made by Hans Reichenbach (1956).

So, we are in a position now to present the past-hypothesis-based statistical mechanical explanation of the entropy gradient with which the rest of the chapter is concerned: the universe is, at one temporal end, in a macrostate of very low entropy (the Past State), and there is a monotonically-increasing entropy gradient towards the other temporal end (entailed by applying the Boltzmannian method to the Past State and projecting in one temporal direction) until a maximal-entropy state at the other temporal end, which we may term the *Future State*. On this model, the past hypothesis plays two main problem-solving roles. Firstly, it allows for there to be an entropy gradient at all (without the existence of a low-entropy state, there'd be no room for entropy to increase). Secondly, the past hypothesis corrects the retrodictions of statistical mechanics; we need the past hypothesis to validate our beliefs about a low-entropy past.

Even given the past hypothesis, there is still a major problem. Towards the past we have processes which are anti-thermodynamical in that they are entropy-decreasing and hence highly improbable. Maudlin (2007) argues that the assumption of a fundamental direction of time allows us to make an initialness defence here: the apparently improbable behaviour towards the past is really just probable behaviour towards the future, with the lowest-entropy state (the Past State) being

---

[11] Though it should be used that Penrose's calculation is in the context of black hole thermodynamics, using Bekenstein–Hawking entropy rather than Boltzmann entropy.

the temporally earliest, ensuring all 'real' evolution is entropy-increasing and so thermodynamically normal. On the contrary, Price (2002, 2004), an antirealist about time direction, holds that this improbable behaviour towards the past is reason to think that the past hypothesis requires further explanation. I shall now argue that this sets up a false dichotomy between realists and antirealists about time direction. Instead, I argue that antirealists about time direction, or 'C-theorists' as I will call them, can and should adopt a restricted version of the initialness defence. In order to set this out, I'll first introduce the C-theory of time and what it means to reject that time has a direction.

3 Directionless Time: Understanding the C-Theory

In order to set up the opposing camps regarding the initialness defence, I'll introduce two alternative theories of time: the B-theory; and the C-theory. We can start with simple definitions of the two theories before adding some details to how they differ.

B-theory of time. The world is directed in time from earlier to later.
C-theory of time. The world is not directed in time.

The B- and C-theories derive from McTaggart's (1908) B-series and C-series. McTaggart presented three different understandings of how things could be ordered in time. The A-series orders things in terms of whether they're past present or future, such as the extinction of the dinosaurs being past, the displaying of dinosaur fossils in the Natural History Museum present, and the genetically-modified-dinosaur apocalypse future. The A-series captures how things appear to change in time: for example, your reading of this sentence is, for me (at the time of writing), future, but for you it is present, and as of the time of your reading the next sentence it is past. In this sense events change their A-series properties as time passes. The B-series preserves the directionality of the A-series, ordering things from earlier to later—e.g. my writing of this sentence is earlier than your reading it. But the B-series doesn't change in the way the A-series does; if x is earlier than y, this is a fact that does not change over time in any sense, so there is no need to update or redraw the B-series over time.

McTaggart also introduced the much-less-talked-about C-series, which he took to provide an undirected ordering of things in time, noting that the C-series, 'while it determines the order, does not determine the direction' of a set of events. The implied ordering relation of the C-series in McTaggart (1908) is *temporal betweenness*: a C-series tells us what events are between which other events in time, with this betweenness ordering being insensitive to whether the ordering goes in one time direction rather than the other.

> [T]he *C* series, while it determines the *order*, does not determine the *direction*. If the *C* series runs M, N, O, P, then the *B* series from earlier to later cannot run M, O, N, P, or M, P, O, N, or in any way but two. But it can run either M, N, O, P (so that M is earliest and

> P latest) or else P, O, N, M (so that P is earliest and M latest). And there is nothing [. . .] in the *C* series [. . .] to determine which it will be. (McTaggart 1908, p. 462; my emphasis.)

In the C-series, the event of you reading this sentence is located between your reading of the previous sentence and the next sentence, but this 'temporal betweenness' ordering has no directionality built into it. Building time out of an undirected relation like betweenness gives a picture of time as ordered but undirected, with one direction no more important than the other. A consequence of this is that there is no such thing as the temporal reverse of a C-series of events, since whichever direction one chooses to describe events, the same betweenness relations hold (even 'backwards' in time your reading of this sentence is between your reading of the previous and next sentences).

Unlike the A-theory, the B- and C-theories contain no special 'present moment' of time—the time we consider present can be considered to be indexical, like the term 'here', in that what is 'present' is simply what is simultaneous with us at the particular relevant moment of time that we are thinking or speaking. As such, there is no observer-independent flow or passage of time on the B- and C-theories—they are both versions of the so-called 'block universe', wherein events we think of as being in the past or future are as real as events we think of as present. The difference between the B- and C-theories is simply that the B-theory contains a basic directionality of time—it is a fact that the world runs from earlier to later and not later to earlier, that the Big Bang happened at the beginning and not the end of time. On the C-theory, the universe does not contain sufficient temporal structure to distinguish between an earlier-to-later universe and a later-to-earlier universe, and as such there is no such fact about the Big Bang being at the beginning rather than the end of time (Table 1).

Though the terminology might be new, various people have defended versions of the C-theory. Thomas Gold, when setting out a temporally adirectional position motivated by the time symmetry of classical mechanics, noted:

> [T]he description of our universe in the opposite sense of time [...] sounds very strange but it has no conflict with any laws of physics. Our strange description is not describing another universe, or how it might be but isn't, but it is describing the very same thing. (Gold 1966, p. 327)

A decade earlier, Hans Reichenbach, in a book that almost certainly influenced Gold's position, held:

> Positive and negative time supply equivalent descriptions, and it would be meaningless to ask which of the two descriptions is true. Reichenbach (1956, pp. 31–32)

Table 1 A-theories, B-theories and C-theories understood in terms of a descending hierarchy of structure attributed to time, from Farr (2020)

	Dynamic	Directed	Ordered
A-theory	✓	✓	✓
B-theory	✗	✓	✓
C-theory	✗	✗	✓

The basic idea behind both claims is that whether we describe things from past to future, or from future to past, we are simply giving two alternative descriptions of the same possible world; there's no such thing as *this world but run backwards in time*; there's no basic underlying direction of time. If you try to imagine this universe running backwards in time, the Reichenbach–Gold view holds that you're just imagining the same universe but from a different and unfamiliar temporal perspective. Crucially, all the same temporal facts (i.e. the temporal betweenness facts) are the same across these two alternative time-directed descriptions.

Though both these quotes are from within the discussion of the time symmetry of classical mechanics, Reichenbach (1956, pp. 128–9) also discusses this C-theoretic attitude within the context of time-asymmetric laws like the second law of thermodynamics, noting that 'it has no meaning to say that entropy "really" goes up or that its time direction is "really" positive', but rather that we conventionally prefer to talk about the universe in the direction in which entropy increases; if we were to describe it in the direction which entropy decreases (from future the past) we wouldn't be getting anything wrong about time, but would simply going against the standard convention and adopting an unfamiliar temporal perspective.

4 Initialness and the Past Hypothesis

With the B- and C-theories of time in hand we can think about how they differ when it comes to explaining the low-entropy past.

4.1 Maudlin's Initialness Defence

Maudlin (2007) is a prominent defender of the version of the B-theory.[12] Though an advocate of the block universe, he takes there to be a primitive directionality of time (which he calls 'passage', though devoid of the metaphysics of the A-theory) in virtue of which later states of the world 'are produced by' earlier states of the world, with no such feature being assigned to the opposite direction of time. On this picture, it is a fact that we live in an entropy-increasing world and not an entropy-decreasing world, something Maudlin takes not to be available to the antirealist about time direction (the C-theorist). Crucially, Maudlin takes this B-theoretic fact to play a key role in explaining the thermodynamic asymmetry in a very specific way.

Maudlin distinguishes between two different ways in which entropy-increasing processes (such as the spreading out of gas in a room, the melting of an ice cube,

[12] Indeed, Maudlin (2007, p. 126, fn. 11) describes his own position as a 'B-series theory' as opposed to a 'C-series theory', fitting with the B/C-theory distinction I've set out.

and the smashing of wineglasses) are atypical and hence needy of explanation. Firstly they are *macroscopically atypical* in that their initial states are low-entropy. But there is also a second kind of atypicality we can call *dynamical atypicality*: the microstates of entropy-increasing systems are atypical insofar as backwards evolution (towards the temporal past) leads towards lower entropy. Imagine that midway through the process of an ice cube melting on a warm plate you freeze-frame the system. Given the assumption that the ice cube was less melted in the past, it follows that if you imagine this process running in reverse, the microstate of the system would have to be vastly improbable insofar as the arrangement of positions and momenta of the molecules would lead the system to a state of macroscopically lower entropy, despite there being vastly more microstates compatible with the present macrostate that would lead towards *higher* entropy in the past. In other words, any apparently ordinary thermodynamical process is dynamically atypical if viewed from future to past. Of this kind of dynamical atypicality, Maudlin remarks:

> [The microstates of such systems are] atypical in a way that can only be characterized in terms of how the state will 'evolve' though time [in that...] temporal evolution in one direction from it will lead, over a very long period of time, to monotonically lower entropy. (Maudlin 2007, pp. 132–3)

Maudlin's solution is to hold that past-to-future evolutions of systems are real and future-to-past evolutions are not, and so there is no sense in which systems in the world are actually dynamically atypical. In other words, once the direction of time (from past to future) is stipulated as a fundamental fact about the world, this entails that macroscopic systems only evolve towards more probable and not less probable states over time. As such, the problem of dynamical atypicality, Maudlin argues, is explained away on the B-theory.

> [Dynamical atypicality] is completely accounted for by how it was generated or produced[; it is] the product of an evolution from [an] initial state[; and] [t]his sort of explanation requires that there be a fact about which states produce which[, which] is provided by a direction of time. (Maudlin 2007, pp. 133–4)

This is not to say that the macroscopic atypicality of the past hypothesis is explained away, but rather that the fact that systems behave anti-thermodynamically towards the past is explained away. On Maudlin's B-theoretic account, the Past State is not the 'result' of temporal/causal evolution; it is this that makes Maudlin's argument a version of the initialness defence. It is thus implied by Maudlin that dynamical atypicality is a genuine problem for the C-theorist, since by rejecting a fundamental direction of time the C-theorist is committed to the idea that future-to-past evolutions are as real as past-to-future evolutions, and so in this way the B-theorist has recourse to an initialness defence that the C-theorist lacks.

4.2 Price's Rejection of the Initialness Defence

So is Maudlin right to hold that the C-theorist is troubled by dynamical atypicality in a way the B-theorist is not? I'll argue that he is not, and that the C-theorist does indeed have access to the right kind of initialness defence here. But first it is important to show that at least some defenders of the C-theory appear to agree with Maudlin's contention. Consider, for example, the following passage from Price:

> Imagine that in recent years physics had discovered that the matter in the universe is collapsing towards a big crunch [...] and that as it does so, something very peculiar is happening. The motions of the individual pieces of matter in the universe are somehow conspiring to defeat gravity's overwhelming tendency to pull things together. Somehow, by some extraordinary feat of cooperation, the various forces are balancing out, so that matter will [...] spread itself out with great uniformity. (Price 2004, p. 230)

But, Price continues,

> [T]his state of affairs is exactly what physics has discovered! I've merely taken advantage [...] of the fact that [...] there is no objective sense in which what we call the future is *really* the 'positive' direction of time. [...] Relabelled [from future to past], the familiar expansion from a smooth big bang becomes a contraction to a smooth big crunch, with the extraordinary characteristics just described. And surely if it is a proper matter for explanation described one way, it is a proper matter for explanation described the other way. (Price 2004, pp. 230–1; emphasis in original)

Though Price presents this in the form of an intuition pump, we can go further and construct it in the form of a deductive argument as follows:

P1. Forwards and backwards time are equivalent [central thesis of C-theory]
P2. An evolution from a low-entropy past state is equivalent to an evolution towards a low-entropy 'future'[13] state [implied entailment of P1]
P3. An evolution towards a low-entropy 'future' state would demand a special explanation [intuition]
C. Therefore, an evolution from a low-entropy past state demands a special explanation [from P2 & P3]

It is not clear whether Price explicitly endorses this exact argument, but these passages function as an intuition pump for P3 based on the assumption of P2, the key premises of the argument. It is P2 that I think can and should be rejected.

5 Deflating the Initialness Defence

5.1 Reunderstanding the C-Theory

Rejecting P2 involves putting forward a different understanding of the C-theory. Such a version is motivated by remarks made by Reichenbach (1956), which I elsewhere (Farr 2020, 2022) term *time direction conventionalism*, according to which there are no time-directed facts, but it is more convenient to describe the world from past to future. Central to the Reichenbachian account is the idea that

[13] Here, a 'future' state is simply the Past State described in negative time.

although past-to-future and future-to-past descriptions of the world are equivalent, it does not follow that it is as true to say that the Past State is a product of entropy-decreasing behaviour as it is to say that the present state is the product of entropy-increasing behaviour. As such, this offers a middle-ground option that preserves intuitions behind both Maudlin's and Price's approaches by holding that there is an objective causal direction of processes in time, but that it is as true to describe things from future to past as from past to future.

Whilst discussing past-to-future and future-to-past descriptions of the entropy gradient, Reichenbach (1956, p. 154) remarks:

> The two languages L_1 [a language in which the positive direction of time is that of increasing entropy] and L_2 [a language in which the positive direction of time is that of decreasing entropy] represent *equivalent descriptions*; one is as true as the other.

But he adds:

> If someone argues that it is a matter of convention to select the direction of growing entropy [as opposed to decreasing entropy] as the direction of time, [their] conception cannot be called false. But [they] must not commit the error often connected with other forms of conventionalism: the error of overlooking the empirical content associated with the use of this convention. (Reichenbach 1956, p. 154)

In order to understand what he has in mind here, it is important to first clarify how Reichenbach views the relationship between the direction of cause–effect and direction of increasing entropy. Reichenbach (1956, p. 153–6) makes a subtle point that has rarely if at all been discussed in the secondary literature, namely that there are two separate conventions that guide our talk about the increase of entropy over time. Firstly there is the *time direction convention*—the tendency to describe the world in the direction of increasing entropy (and so past-to-future) and not in the direction of decreasing entropy. Reichenbach says there's nothing factually wrong about adopting the opposite convention, but it simply goes against our standard way of talking about the world. There are pragmatic grounds for adopting the standard convention; for instance Reichenbach (1956, p. 154) suggests that it matches the 'time direction of psychological experience' (though for criticism of this claim, see Farr 2022).

Reichenbach also speaks of a second convention I'll call the *causal convention*, namely the convention that happenings in the world are explained by the temporal past rather than temporal future. The alternative convention would be to prefer a teleological mode of explanation, what Reichenbach refers to as a 'principle of finality', according to which things are explained in terms of the temporal future. What is key is how the time-direction convention and causal convention go together. See Table 2 for the four options these provide. On Reichenbach's conventionalist account, Options 1 and 4 are equivalent descriptions, and Options 2 and 3 are equivalent descriptions. But neither 1 or 4 are equivalent to 2 or 3.

Reichenbach takes the former of these equivalence classes (1 and 4) to adequately describe the world: the world is *either* a causal entropy-increasing world (Option 1), or a teleological entropy-decreasing world (Option 4), noting that when taking the direction of time to be a convention, one must not overlook 'the empirical content

Table 2 Four options for the C-theorist

	Causal	Teleological
Entropy-increasing	1	2
Entropy-decreasing	3	4

associated with the use of this convention' (ibid). The empirical fact Reichenbach points out is what he calls the 'parallelism of entropy increase' for systems: for quasi-isolated subsystems of the universe, as well as the universe itself, the directions of increasing entropy are parallel in that they increase in entropy *in the same direction* and not in different directions.[14] This allows us to talk of a general positive time direction being defined, which Reichenbach takes to be an 'empirical hypothes[is] which [is] convincingly verified' (Reichenbach 1956, p. 154). This does not pick out the direction of entropy-increase or entropy-decrease on its own, since the parallelism itself is not directed. But Reichenbach ties this parallelism to the notion of causality, noting that 'the convention of defining positive time through growing entropy is inseparable from accepting causality as the general method of explanation' (ibid.), and further:

> Once the direction is assumed in the usual sense [i.e. positive time is the direction of increasing entropy], it is not a matter of personal preference, not a mode of consideration, whether we should describe the world in terms of causes or of ends: it is a physical law that causality, and not finality, governs the universe. (Reichenbach 1956, p. 154)

This can be understood in terms of the asymmetric structure of causal networks, in which correlations between variables that are not the direct cause or effect of each other are explainable in terms of their causal past and not their causal future. Reichenbach discusses this in the context of his principle of common cause (Reichenbach 1956, ch. 19), remarking that while some causal forks are closed to the future, such as when two independent events have a common effect, 'a common effect cannot be regarded as an explanation' (Reichenbach 1956, p. 163) of the correlation. As such, the method for establishing causal direction is tied to the entropy gradient (e.g. causes are lower in the entropy gradient than their effects). This asymmetry is taken by Reichenbach to be due to the second law of thermodynamics in that the existence of a sufficiently low-entropy state makes it possible for there to be the kind of past-directed common cause structure for correlated events, meaning that if the Past State is stipulated as the lowest entropy state, it cannot be considered to be causally influenced by any other temporal states of the world.

This understanding of the Past State gives a deflationary account of the direction of causation, entailing that there is a direction of causation *only* in the presence of the right kind of probabilistic asymmetries, such as irreversible processes, time-

[14] Reichenbach's principle of parallelism of entropy increase: 'In the vast majority of branch systems, the directions toward higher entropy are parallel to one another and to that of the main system' (Reichenbach 1956, p. 136). Equivalently, the directions towards lower entropy are also parallel.

asymmetric screening-off conditions, and other such things common to the causal modelling literature. This position is taken by Reichenbach also in the case of time-symmetrical systems; for example in the case of classical particle mechanics Reichenbach (1956, ch. II) holds that the relevant causal notion is one of 'causal betweenness'—the physical laws are sensitive to which events are causally between others, but so long as there is no basic lawlike temporal asymmetry displayed by systems there is no sense in which some events *causally precede* others; for sufficiently time-symmetric systems, causal-direction talk breaks down. For our purposes, an important consequence of this is that entropy gradients, rather than being *explained* by further causal facts, are what allows for the existence of causal explanations of events. This runs contrary to Maudlin's realism about causal direction, which holds that 'even if the world were always in thermal equilibrium [...] later states would arise out of earlier ones' (Maudlin 2007, p. 177), where by 'arise' Maudlin is referring to causal production, meaning that for Maudlin there would be causal direction facts even in the absence of the kinds of probabilistic asymmetries central to standard causal discovery algorithms like Reichenbach's common cause principle and modern descendents of it, such as the causal Markov condition (see Hausman and Woodward 1999).

The convention of taking the direction of entropy-increase to be the direction of positive time is tied to the convention of causality over finality: once one is fixed, the other follows empirically. We can regard these as on a par—neither alone is strictly an empirical matter, but once one is chosen, the other is not open to choice.

> [T]he convention of defining positive time through growing entropy is inseparable from accepting causality as the general method of explanation. Those who prefer to give an explanation in terms of finality would be compelled to use the opposite time direction and to regard time as going from high-entropy to low-entropy. (Reichenbach 1956, p. 154)

What's *not* a convention is that *either* the world is an entropy-increasing causal world *or* an entropy-decreasing teleological world.

Compare this to Price's intuition pump and the argument we constructed from it. Premise 2 holds that an entropy-increasing causal world is equivalent to an entropy-decreasing causal world, aligning with Price's suggestion that, when described from future to past, we see a world full of coincidences in which local causes force systems to decrease in entropy over time. But the C-theorist is not compelled to accept Premise 2 and can instead adopt the Reichenbachian option. The key difference between the Pricean and Reichenbachian approaches here is whether or not the Past State can reasonably be considered the result of a series of coincidences due to an improbably fine-tuned present state of the world. On the picture Price describes, the description of the world evolving thermodynamically from the Past State is equivalent to the description of the world evolving anti-thermodynamically towards the Past State. But on the Reichenbachian account, we can pull apart the time-direction convention and the causal convention and clarify that it is not true to say that Past State is the causal product of anti-thermodynamic evolution from a higher-entropy state, and so we get rid of the problematic unexplained coincidences apparent in Price's future-to-past description.

On the Reichenbachian account, when we switch from a past-to-future to a future-to-past description of the world, we switch a causal explanation of the entropy gradient for a teleological explanation of the entropy gradient, and as such, in both directions we have an explanation of the entropy gradient in terms of the past hypothesis and not vice versa. In this way we find a best-of-both-worlds option between Price and Maudlin's accounts: (1) we avoid Price's conclusion that an entropy-increasing evolution *from* the Past State requires a special explanation; and (2) we capture the idea from Maudlin's account that Past State is not the 'result' of temporal/causal evolution without committing to an underlying directionality of time. This shows that we can have a C-theory of time—one that holds that past-to-future and future-to-past descriptions of the world are equivalent—that nonetheless supports the idea there is preferred direction of causal explanation (lower-entropy states can be used to explain higher-entropy states but not vice versa).

5.2 *Initialness, Deflated*

This takes us to the central conclusion: there is a 'deflated' version of initialness defence available to the C-theorist. On the one hand, B-theorists like Maudlin take the past-to-future direction as metaphysically privileged, and so the low-entropy past not as needy of explanation as a low-entropy future would be. And on the other, C-theorists like Price give up the initialness defence on the grounds that the description of the world as evolving away from a low entropy past is equivalent to the alternative description of the world as evolving towards a low entropy future. But the best-of-both worlds option given by conventionalism holds instead that when described from future to past, the story of the past hypothesis needs to be understood in teleological rather than in causal terms, so that entropy-lowering processes are explained in terms of a fact about their temporal future. On this view, the past hypothesis is still explanatorily special insofar as other temporal states of the world are explained by way of reference to it and there is no equivalent way in which it is explained in terms of other states of the world.

On the Reichenbachian approach, we get back a C-theoretic, deflationary version of Maudlinian Production; if we are to use the language of causal production to explain the entropy gradient, then the Past State is a producer but not a product of other temporal states of the world. This is something briefly suggested Reichenbach (1956):

> Once the time direction is assumed in the usual sense, it is not a matter of personal preference [...] whether we should describe the world in terms of causes or ends: it is a physical law that causality, and not finality, governs the universe. [...] For the time direction of growing entropy [...] the interaction point is the beginning, not the end, of the evolution of the branch system. The statistical relationships [...] account for our conception that the past *produces* the future, and not vice versa. (Reichenbach 1956, p. 155; my emphasis)

Recall that for Maudlin, a 'primitive',[15] 'irreducible' and 'fundamental'[16] directionality of time plays a crucial role in this initialness defence, explaining the dynamical atypicality of microstates at all times apart from the initial time. Maudlin's argument is an inference to the best explanation in favour of the B-theory. It follows from Maudlin's reasoning that only by assuming such a direction of time (from past to future) can we explain away dynamical atypicality. He adds, in defence of the B-theory:

> [W]e cannot run this trick in reverse. [... W]e cannot specify an independent, generic constraint on the *final* state that will yield (granting the final macrostate is typical) ever decreasing entropy in one direction. (Maudlin 2007, p. 133; emphasis added)

But this is simply not the relevant contrast class. The reason we can explain the Future State[17] in terms of the Past State and not vice versa is not because the Past State is earlier than the Future State. It is instead because, given the Boltzmannian statistical mechanical account, we can explain the Future State by conditionalising on the Past State, and we cannot explain the Past State by conditionalising on the Future State. The explanans here is simply that the Future State is a *normal* macrostate, and the Past State is not. Whether the Past State is 'really' earlier or later than the Future State does not enter into this explanation. If the Past State is earlier than the Future State, then statistical mechanics dictates that the 'forwards' direction of time provides the natural (causal) explanation. However, if we flip to using the future-to-past direction, it is now the 'backwards' direction of time that provides the natural (teleological) explanation.

As such, it is not at all clear why a primitive or fundamental directionality of time needs to enter the explanation here. Consider two different models of the entropy gradient using a primitive directionality of time: one in which entropy increases over time; and one in which entropy decreases over time. The two different models correspond to different B-series. The statistical mechanical explanation of the Future State in terms of the Past State, and the lack of an explanation of the Past State in terms of the Future State, is itself independent of the issue as to which state is 'earlier than' the other. The statistical mechanical explanation *can only go one way—from* past *to* future, and this is respected by Reichenbachian time-direction conventionalism, which holds that regardless of whether we choose an entropy-increasing or entropy-decreasing description, the entropy gradient is explained in terms of the existence of the Past State and not vice versa. The important thing is that *regardless* of whether we use a past-to-future or future-to-past description, the explanation goes *from* the Past State (i.e. the *low entropy* state) *to* the Future State (i.e. the *high entropy* state), and as such there is no clear need for a further background directionality of time.

---

[15] Maudlin (2007, 172).

[16] Both Maudlin (2007, p. 3 & p. 107).

[17] Recall that the 'Future State' is the hypothetical maximal entropy state in our distant future.

6 Summing up

We started with a puzzle about the nature of initial conditions: in what sense do laws of nature require there to be genuinely 'initial' conditions, and so for time to have some basic or fundamental preferred direction? By focusing on the problem of the low-entropy early universe, we considered a potential role for a fundamental directionality of time—that the dynamical atypicality of the world when viewed from future to past can be explained away only on the assumption that really time runs from past to future. I have argued that dynamical atypicality is not as serious a problem as it appears, and that the C-theorist of time can perfectly well make sense of the Past State as explanatorily special without there being a fundamental direction of time. On the version of the C-theory defended, the low-entropy early universe is not intrinsically special through being 'temporally first' or anything to that effect. Rather, due to the relationship between entropy and causality, it is the existence of a low-entropy state that allows there to be a causal direction of systems and hence a clear direction of explanation. As such, initialness defences are available to those that reject that time has a direction.

References

Albert, D. Z. (2000). *Time and chance*. Massachusetts: Harvard University Press.

Baras, D., & Shenker, O. (2020). Calling for explanation: The case of the thermodynamic past state. *European Journal for Philosophy of Science, 10*(3), 1–20

Boltzmann, L. (1967). On Zermelo's paper "On the mechanical explanation of irreversible processes". In S. Brush (Ed.), *Kinetic theory*. Irreversible processes (Vol. 2). Pergamon.

Callender, C. (2004a). Measures, explanations and the past: Should 'special' initial conditions be explained? *British Journal for Philosophy of Science, 55*(2), 195–217.

Callender, C. (2004b). There is no puzzle about the low entropy past. In C. Hitchcock (Ed.), *Contemporary debates in philosophy of science* (pp. 240–256). Oxford: Blackwell Publishing.

Clausius, R. (1864). *Abhandlungungen über die mechanische Wärmetheorie* (Vol. 1). Braunschweig: F. Vieweg.

Farr, M. (2020). C-theories of time: On the adirectionality of time. *Philosophy Compass, 15*(12), e12714.

Farr, M. (2022). Conventionalism about time direction. *Synthese, 200*(1), 1–21. https://doi.org/10.1007/s11229-022-03540-1

Farr, M. (2022). Perceiving direction in directionless time. In K. M. Jaszczolt (Ed.), *Understanding human time*. Oxford University Press.

Feynman, R. P., Morinigo, F. B., & Wagner, W. G. (1971). *Feynman Lectures on Gravitation* (Vol. 13). Pasadena, CA: California Institute of Technology.

Frigg, R. (2008). A field guide to recent work on the foundations of statistical mechanics. In D. Rickles (Ed.), *The Ashgate companion to contemporary philosophy of physics*, Chapter 3 (pp. 99–196). London: Ashgate.

Frigg, R. (2009). What is statistical mechanics? In C. Galles, P. Lorenzano, E. Ortiz, & H. J. Rheinberger (Eds.), *History and philosophy of science and technology*. Encyclopedia of Life Support Systems (Vol. 4). Eolss: Isle of Man.

Frigg, R., & Werndl, C. (2011). Explaining thermodynamic-like behavior in terms of epsilon-ergodicity. *Philosophy of Science, 78*(4), 628–652.

Gibbs, J. W. (1902). *Elementary principles in statistical mechanics*. Woodbridge: Ox Bow Press.
Gold, T. (1966). Cosmic processes and the nature of time. In R. G. Colodny (Ed.), *Mind and cosmos* (pp. 311–329). Pittsburgh: University of Pittsburgh Press.
Hausman, D., & Woodward, J. (1999). Independence, invariance and the causal Markov condition. *The British Journal for the Philosophy of Science, 50*(4), 521–583.
Maroney, O. (2009). Information processing and thermodynamic entropy. In E. N. Zalta (Ed.), *The Stanford encyclopedia of philosophy* (Fall 2009 ed.).
Maudlin, T. (2007). *The metaphysics within physics*. Oxford: Oxford University Press.
McTaggart, J. M. E. (1908). The unreality of time. *Mind, 17*(68), 457–474.
Penrose, R. (1989). *The emperor's new mind: Concerning computers, minds, and the laws of physics*. Oxford: Oxford University Press.
Price, H. (1996). *Time's arrow and Archimedes' point: New directions for the physics of time*. Oxford: Oxford University Press.
Price, H. (2002). Boltzmann's time bomb. *The British Journal for the Philosophy of Science, 53*(1), 83–119.
Price, H. (2004). On the origins of the arrow of time: Why there is still a puzzle about the low-entropy past. In C. Hitchcock (Ed.), *Contemporary debates in philosophy of science* (pp. 219–239). Oxford: Blackwell Publishing.
Reichenbach, H. (1956). *The direction of time*. Berkeley: University of California Press.
Robertson, K. (2020). Asymmetry, abstraction, and autonomy: Justifying coarse-graining in statistical mechanics. *The British Journal for the Philosophy of Science, 71*(2), 547–579.
Robertson, K. (2021). In search of the holy grail: How to reduce the second law of thermodynamics. *The British Journal for the Philosophy of Science*, Online first.
Sklar, L. (1993). *Physics and chance: Philosophical issues in the foundations of statistical mechanics*. Cambridge: Cambridge Univ Press.
Sklar, L. (2006). Why does the standard measure work in statistical mechanics? In V. F. Hendricks, K. F. Jörgensen, J. Lötzen, & S. A. Pedersen (Eds.), *Interactions*. Boston studies in the philosophy and history of science (Vol. 251, pp. 307–320). Springer Netherlands. https://doi.org/10.1007/978-1-4020-5195-1-10
Uffink, J. (2001). Bluff your way in the second law of thermodynamics. *Studies in History and Philosophy of Science Part B: Studies in History and Philosophy of Modern Physics, 32*(3), 305–394.
Wallace, D. (2011). The logic of the past hypothesis. *PhilSci Archive*. http://philsci-archive.pitt.edu/8894/
Wallace, D. (2020). The necessity of Gibbsian statistical mechanics. In V. Allori (Ed.), *Statistical mechanics and scientific explanation: Determinism, indeterminism and laws of nature*, Chapter 15 (pp. 583–616). World Scientific.
Wigner, E. P. (1995). Events, laws of nature, and invariance principles. In *Philosophical reflections and syntheses* (pp. 321–333). Springer.

Remarks About the Relationship Between Relational Physics and a Large Kantian Component of the Laws of Nature

Sheldon Goldstein and Nino Zanghì

Abstract Relational mechanics is a reformulation of mechanics (classical or quantum) for which space is relational. This means that the configuration of an N-particle system is a shape, which is what remains when the effects of rotations, translations and dilations are quotiented out. This reformulation of mechanics naturally leads to a relational notion of time as well, in which a history of the universe is just a curve in shape space without any reference to a special parametrization of the curve given by an absolute Newtonian time. When relational mechanics (classical or quantum) is regarded as fundamental, the usual descriptions in terms of absolute space and absolute time emerge merely as corresponding to the choice of a gauge. This gauge freedom forces us to recognize that what we have traditionally regarded as fundamental in physics might in fact be imposed by us through our choice of gauge. It thus imparts a somewhat Kantian aspect to physical theory.

Keywords Relational physics · Shape space · Kantian component of laws · Classical gauges · Quantum gauges · Bohmian mechanics · Non-normalizable · Wave function of the universe · Emergence of interaction

---

S. Goldstein (✉)
Departments of Mathematics, Physics and Philosophy, Hill Center, Rutgers, The State University of New Jersey, Piscataway, NJ, USA
e-mail: oldstein@math.rutgers.edu

N. Zanghì
Dipartimento di Fisica dell'Università di Genova and INFN sezione di Genova, Genova, Italy
e-mail: zanghi@ge.infn.it

Y. Ben-Menahem (ed.), *Rethinking the Concept of Law of Nature*,
Jerusalem Studies in Philosophy and History of Science,
https://doi.org/10.1007/978-3-030-96775-8_9

1 Relational Physics

We shall address here three basic questions concerning relational physics: What is it? What is its point? What can we learn from it? These issues have been addressed in a long paper (Dürr et al. 2020) published in 2020 with our friend and colleague Detlef Dürr, who passed away in January 2021.

Relational physics has been revived in relatively recent times by Julian Barbour and Bruno Bertotti, who, in a very inspiring and influential paper published at the beginning of the eighties (Barbour and Bertotti 1982), transformed a long standing philosophical controversy about the nature of space and time into a well-defined physical problem. Our main focus here will be on gauge freedom and objectivity versus subjectivity, probability, and the non-normalizability of the quantum state.

1.1 Relational Space

The philosophical issue about the nature of space dates back to the dispute between Isaac Newton, who favored and argued for the need of an absolute theory of space and time, and Gottfried Wilhelm Leibniz, who insisted upon a relational approach, also defended by Ernst Mach in the nineteenth century. To put it briefly, the gist of the relational approach is that overall position, orientation, and size are not relevant. This can be clarified by means of a very elementary and simplified model of the universe.

Suppose we are given the configuration of a universe of N particles. And suppose we translate every particle of the configuration in the same direction by the same amount. From a physical point of view it seems rather natural to take the relational perspective that the two configurations of the universe so obtained are physically equivalent or identical. Similarly for any rotation. Going one step further, one regards two configurations of the universe differing only by a dilation, i.e. by a uniform expansion or contraction, as representing in fact the same physical state of the universe. The space of all genuinely physically different possible configurations so obtained—taking into account translations, rotations, and dilations—is usually called *shape space*. The name shape space is indeed natural: only the shape of a configuration of particles is relevant, not its position or orientation or overall size.

Shape space should be contrasted with *absolute configuration space*, the totality of configurations of N points in Euclidean three-dimensional space.

1.2 Relational Time

Once shape space is taken as fundamental, one should address how shapes evolve. While positing an absolute Newtonian time as in classical mechanics is compatable with relational space, it is natural to consider a relational notion of time as well. According to relational time, global changes of speed of the history of the universe give physically equivalent representations, so that a history of the universe is just a curve in shape space without any reference to a special parametrization of the curve given by any sort of absolute Newtonian time.

Relational time should be contrasted with absolute or *metrical time*. By metrical time we refer to any objective physical coordination of the configurations along a geometrical path in a configuration space with the points of a one-dimensional continuum: a (continuous) mapping from the continuum onto the path. The continuum is usually represented by the real numbers, but it need not be. However, it should be physically distinct from the particular continuum that is the path itself.

Understood in this way, metrical time does not exist, from the relational point of view, for the universe as a whole. However, for subsystems of the universe metrical time naturally emerges: the continuum with which the geometrical path corresponding to the evolution of the subsystem is coordinated can be taken to be the path of its environment, with the obvious mapping between the paths.

1.3 Relational versus Relativistic

Relational physics should be contrasted with relativistic physics. A simple point: In relational physics the traditional separation of space and time is retained. While configuration space is replaced by shape space, and time becomes non-metrical, shape space retains an identity separate and distinct from that of (non-metrical) time. This is in obvious contrast with relativistic physics, in which space and time lose their separate identities and are merged into a space-time.

Perhaps the most characteristic feature of relativity is the absence of absolute simultaneity. Not so for relational physics. Since it retains the separation of space and time, an absolute simultaneity is built into the very structure of relational physics as described here. Nonetheless, there is a sense in which simultaneity is lost. With relational time the notion of the shape of the universe at "time t" is not physically meaningful. And with what is meaningful—geometrical paths in the space of possible shapes—one can no longer meaningfully compare or ask about the configurations for two different possible histories at the same time. Given the actual configuration of the universe, it is not meaningful to ask about the configuration of an alternative history at that time without further specification of exactly what that should mean.

One may inquire whether the relational point of view can be merged with or extended to relativity. Can we achieve a relational understanding of space-time? General relativity is certainly a step in that direction, but it does not get us there. Space-time in general relativity is metrical—in a way that neither space nor time are in relational physics. A complete extension, if at all possible, is a real challenge.

Another possibility is that relativity is not fundamental, but is, instead, a consequence of a suitable choice of gauge. This possibility, which is suggested by the work of DeWitt (1970) and Barbour and coworkers (Barbour et al. 2002) would be worth carefully exploring. For a bit of elaboration on this, see Sect. 7.5.3.

1.4 Relational Space, Relational Time, and Gauge Freedom

When physical theories are formulated in shape space, one should consider first the simplest ones, namely the "free" theories based only on the geometrical structures provided by the metric, without invoking any potential. This is in contrast with theories formulated in absolute space, for which free theories can't begin to account for the experimental data. It is then natural to ask: when we represent the theories in absolute space, what form do the laws of motion take? Is the representation unique or are there various representations yielding different-looking laws of motion, some unfamiliar and some more or less familiar? Moreover, do interacting theories emerge with nontrivial interactions, although in shape space the motion is free?

To answer these questions it is helpful to represent absolute configuration space in geometrical terms as a fiber bundle, with shape space as base manifold and the fibers generated by the similarity group, i.e, by translations, rotations and dilations, which acting on configurations yield, from a relational point of view, physically equivalent states. A representation in absolute configuration space of the motion in shape space is then given by a "lift" of the motion from the base into the fibers (more details below).

Such lifts can rightly be called gauges. In the classical case it turns out that in some gauges the law looks unfamiliar but there is (at least) one gauge in which, after performing a time change (representing indeed another gauge freedom when also time is seen as relational), the law of motion is Newtonian with a potential appearing. So, it turns out that relational space and relational time form a harmonious combination. More or less the same is true for the quantum case. And similarly, as hinted above, the same might be true for general relativity.

1.5 Why Go Relational?

One reason for turning to relational physics is this. Only relational distinctions concerning space and time are observable. Thus, so the argument might continue, since the positions of bodies are individuated only by relative distances and angles, from an observational point of view, only shapes or relational times are meaningful, while absolute space, absolute time, and even absolute spacetime, do not have any clear observational meaning. While having some force, this argument is too positivistic to be taken as the main reason to go relational.

Here is another reason. Since the long-standing philosophical controversy about the nature of space and time has now become a well-defined problem in theoretical physics, we have a professional reason to go relational. It seems worth exploring how far one can go in formulating physics in minimal relational terms.

A stronger reason, which is indeed the one we favor, has to do with the status and the nature of the wave function. Since at this stage, this sounds totally incomprehensible, we shift gears and turn to quantum mechanics.

2 Quantum Mechanics

Standard quantum mechanics is problematic. It is often said that the fundamental problem with quantum mechanics is the *measurement problem*, or, more or less equivalently, Schrödinger's cat paradox. However, the measurement problem, as important as it is, is nonetheless but a manifestation of a more basic difficulty with standard quantum mechanics: it is not at all clear what quantum theory is about. Indeed, it is not at all clear what quantum theory actually says. Is quantum mechanics fundamentally about measurement and observation? Or is it about the behavior of suitable fundamental microscopic entities, elementary particles and/or fields? To use the words of Bell, standard formulation of quantum mechanics are "unprofessionally vague and ambiguous." A formulation free of these shortcomings is Bohmian mechanics (Bell 1987; Bohm 1952; Bricmont 2016; Dürr et al. 1992, 2012).

2.1 Bohmian Mechanics

Bohmian mechanics is a theory providing a description of reality, compatible with all of the quantum formalism, but free of any reference to observables or observers in its formulation. In Bohmian mechanics a system of particles is described in part by its wave function, evolving according to Schrödinger's equation, the central equation of quantum theory. However, the wave function provides only a partial description of the system. This description is completed by the specification of the actual positions of the particles. The latter evolve according to the "guiding equation," which expresses the velocities of the particles in terms of the wave function. Thus in Bohmian mechanics the configuration of a system of particles evolves via a deterministic motion choreographed by the wave function.

Bohmian mechanics might be regarded as the minimal completion of Schrödinger's equation, for a non-relativistic system of particles, to a theory describing a genuine motion of particles. For Bohmian mechanics the state of a system of N particles is described by its wave function $\Psi = \Psi(\vec{\mathsf{q}}_1, \ldots, \vec{\mathsf{q}}_N) = \Psi(\boldsymbol{q})$, a complex- (or spinor-) valued function on the space of possible configurations $\boldsymbol{q}$ of the system, together with its actual configuration $\boldsymbol{Q}$ defined by the actual positions $\vec{\mathsf{Q}}_1, \ldots, \vec{\mathsf{Q}}_N$ of its particles. The theory is then defined by two evolution laws. One is *Schrödinger's equation*

$$i\hbar\frac{\partial\Psi}{\partial t} = H\Psi\,, \tag{1}$$

for $\Psi = \Psi_t$, the wave function at time t, where H is the non-relativistic (Schrödinger) Hamiltonian, containing the masses m_α, $\alpha = 1,\ldots,N$, of the particles and a potential energy term V. For spinless particles, it is of the form

$$H = -\sum_{\alpha=1}^{N}\frac{\hbar^2}{2m_\alpha}\vec{\nabla}_\alpha^2 + V\,, \tag{2}$$

where $\vec{\nabla}_\alpha = \frac{\partial}{\partial\vec{\mathrm{q}}_\alpha}$ is the gradient with respect to the position of the α-th particle. The other law is the *the guiding law*, given by the equation

$$\frac{d\vec{\mathrm{Q}}_\alpha}{dt} = \frac{\hbar}{m_\alpha}\mathrm{Im}\frac{\Psi^*\vec{\nabla}_\alpha\Psi}{\Psi^*\Psi}(\vec{\mathrm{Q}}_1,\ldots,\vec{\mathrm{Q}}_N)\,. \tag{3}$$

for $\boldsymbol{Q} = \boldsymbol{Q}(t) = (\vec{\mathrm{Q}}_1(t),\ldots,\vec{\mathrm{Q}}_N(t))$, the configuration at time t (with Im denoting the imaginary part of the complex quantity that follows it). If Ψ is spinor-valued, the products in numerator and denominator in (3) should be understood as scalar products. If external magnetic fields are present, the gradient should be understood as the covariant derivative, involving the vector potential. For spinless particles, setting $\Psi = Re^{iS/\hbar}$ (the polar representation of Ψ), the guiding Eq. (3) becomes

$$\frac{d\vec{\mathrm{Q}}_\alpha}{dt} = \frac{\hbar}{m_\alpha}\,\mathrm{Im}\,\frac{\vec{\nabla}_\alpha\Psi}{\Psi}(\vec{\mathrm{Q}}_1,\ldots,\vec{\mathrm{Q}}_N) = \frac{1}{m_\alpha}\vec{\nabla}_\alpha S(\vec{\mathrm{Q}}_1,\ldots,\vec{\mathrm{Q}}_N) \tag{4}$$

For an N-particle system Schrödinger's equation and the guiding equation, together with the detailed specification of the Hamiltonian H, completely define the Bohmian motion of the system.

2.2 OOEOW

Bohmian mechanics is for us the simplest version of quantum mechanics: Given Schrödinger's equation, it has the Obvious Ontology Evolving in the Obvious Way.

For a theory of *particles*, in addition to the wave function you've got the actual positions of the particles. This is a rather obvious ontology. We usually emphasize this point by saying that the positions of the particles provide the *primitive ontology* of the theory. In so saying we wish to convey that the whole point of the theory—and the whole point of the wave function—is to define a motion for the particles, and it's in terms of this motion that pointers end up pointing and experiments end up having results, the kinds of results that it was the whole point of quantum mechanics to explain. So the connection to physical reality in the theory is via what we're calling the primitive ontology of the theory, in Bohmian mechanics the positions of the particles.

The only thing Bohmian mechanics adds is a first-order equation of motion for how the positions of particles evolve. This equation, the new equation in Bohmian mechanics, is kind of an obvious equation. It is more or less the first thing you would guess if you asked yourself, What is the simplest motion of the particles that could reasonably be defined in terms of the wave function?

There are indeed many (more or less) obvious ways of guessing the velocity field function $\boldsymbol{v} = \boldsymbol{v}^{\Psi}$ in the right hand side of (3). The simplest way, for particles without spin, is the following: Begin with the de Broglie relation $\boldsymbol{p} = \hbar\boldsymbol{k}$, a remarkable and mysterious distillation of the experimental facts associated with the beginnings of quantum theory—and itself a relativistic reflection of the first quantum equation, namely the Planck relation $E = h\nu$. The de Broglie relation connects a particle property, the momentum $\boldsymbol{p} = m\boldsymbol{v}$, with a wave property, the wave vector $\boldsymbol{k}$. Understood most simply, it says that the velocity of a particle should be the ratio of $\hbar\boldsymbol{k}$ to the mass of the particle. But the wave vector $\boldsymbol{k}$ is defined only for a plane wave. For a general wave ψ, the obvious generalization of $\boldsymbol{k}$ is the local wave vector $\nabla S(\boldsymbol{q})/\hbar$, where S is the phase of the wave function (defined by its polar representation, see above). With this choice the de Broglie relation becomes $\boldsymbol{v} = \nabla S/m$, the right hand side of which is our first guess for $\boldsymbol{v}^{\psi}$.

We note that the de Broglie relation also immediately yields Schrödinger's equation, giving the time evolution for ψ, as the simplest wave equation that reflects this relationship. This is completely standard. In this simple way, the defining equations of Bohmian mechanics can be regarded as flowing in a natural manner from the first quantum equation $E = h\nu$.

The quantum continuity equation is the key for a route which is meaningful also for particles with spin. This equation, an immediate consequence of Schrödinger's equation, involves a quantum probability density ρ and a quantum probability current J. Since densities and currents are classically related by $J = \rho\ v$, it requires little imagination to set $\boldsymbol{v}^{\psi} = J/\rho$.

Another way to arrive at a formula for $\boldsymbol{v}^{\psi}$ is to invoke symmetry. Since the space-time symmetry of the non-relativistic Schrödinger equation is that of rotations, translations, time-reversal, and invariance under Galilean boosts, it is natural to demand that this Galilean symmetry be retained when Schrödinger's equation is combined with the guiding equation. As described in Dürr et al. (1992), this leads to a specific formula for $\boldsymbol{v}^{\psi}$ as the simplest possibility.

There are several other natural ways to arrive at $\boldsymbol{v}^{\psi}$, but we shall give no more. All these routes, those we've explicitly mentioned and those to which we've alluded, yield in fact exactly the same formula for $\boldsymbol{v}^{\psi}$, though the explicit form may appear different in some cases.

2.3 The Implications of Bohmian Mechanics

While the *formulation* of Bohmian mechanics does not involve the notion of quantum observables, as given by self-adjoint operators—so that its relationship to the quantum formalism (the familiar axioms of quantum theory) may at first appear somewhat obscure—it can in fact be shown that Bohmian mechanics embodies quantum randomness, as expressed by Born's rule, and the entire quantum formalism based on self-adjoint operators as observables as the very expression of its empirical import (Dürr et al. 2012, Ch.2 and 3). The most notably implication of Bohmian mechanics is the fact that in a universe governed by Bohmian mechanics there are sharp, precise, and irreducible limitations on the possibility of obtaining knowledge, limitations that can in no way be diminished through technological progress leading to better means of measurement. This *absolute uncertainty* is in precise agreement with Heisenberg's uncertainty principle.

We shall review below the key points that lead to these implications.

2.3.1 The Conditional Wave Function

In physics we are usually concerned not with the entire universe but with subsystems of the universe, for example with a hydrogen atom or a pair of entangled photons. The quantum mechanical treatment of such systems involves the quantum state of that system, often given by its wave function—not the wave function of the universe. Yet, the notion of the wave function of a subsystem is rather elusive. In standard quantum mechanics the state of a subsystem is usually described in terms of its reduced density matrix.

However, Bohmian mechanics provides a precise formulation and understanding of this notion in terms of the conditional wave function (Dürr et al. 1992)

$$\psi(x) = \Psi(x, Y)\,, \tag{5}$$

where $\Psi = \Psi(q) = \Psi(x, y)$ is the wave function of the universe, with x and y the generic variables for the configurations of the system and its environment (everything else), respectively, and where Y is the actual configuration of the environment. The conditional wave function of a Bohmian system behaves exactly as one would expect the wave function of a system to behave, with respect to both dynamics and statistics.

2.3.2 The Conditional Probability Formula

In Bohmian mechanics, for a non-relativistic system of particles, the configuration of a system is regarded as random, with randomness corresponding to the quantum equilibrium distribution μ^{ψ} given by $|\psi|^2 dq$. What this actually means, in a deterministic theory such as Bohmian mechanics, is a delicate matter, involving a long story (Dürr et al. 1992) with details and distinctions that we shall ignore here. However, a crucial ingredient for that analysis—for an understanding of the origin of quantum randomness in a universe governed by Bohmian mechanics—is the *fundamental conditional probability formula* for the conditional distribution of the configuration X_t of a system at time t given that of its environment Y_t at that time:

$$P^{\Psi_0}(X_t \in dx\,|Y_t) = |\psi_t(x)|^2 dx, \tag{6}$$

where Ψ_0 is the initial wave function of the universe and P^{Ψ_0} is the probability distribution on trajectories arising from the Bohmian dynamics with an initial quantum equilibrium distribution, and ψ_t is the (normalized) conditional wave function (5) of the system at time t.

A crucial ingredient in proving (6) is *equivariance* (Dürr et al. 1992): if at any time the configuration of a Bohmian system is randomly distributed according to $|\Psi|^2$ at that time then at any other time t it will be distributed according to $|\Psi_t|^2$. Equivariance is an immediate consequence of the continuity equation arising from the Schrödinger equation:

$$\frac{\partial \rho^\Psi}{\partial t} + \operatorname{div} J^\Psi = 0\,, \tag{7}$$

with $\rho^\Psi = |\Psi|^2$, the quantum equilibrium distribution, and

$$J^\Psi = \rho^\Psi \boldsymbol{v}^\Psi\,, \tag{8}$$

the quantum probability current, where $\boldsymbol{v}^\Psi$ is the Bohmian velocity in the right hand side of (4).

2.4 *Bohmian Motion on a Riemannian Manifold*

We conclude this section with a mathematical observation, whose importance will become clear later on when we discuss the quantum mechanics of shapes. Note that, given V, the Bohmian mechanics defined by equations (1), (2), and (4) depends only upon the Riemannian structure $g = g_e$ given by the standard Euclidean metric on configuration space $[g_e]_{ij} = m_{\alpha_i}\delta_{ij}$, where the i-th component refers to the α_i-th particle. In terms of this Riemannian structure, the evolution Eqs. (4) and (2) become

$$\frac{dQ}{dt} = \hbar\ \operatorname{Im} \frac{\nabla_g \Psi}{\Psi} \tag{9}$$

$$i\hbar\frac{\partial\Psi}{\partial t} = -\frac{\hbar^2}{2}\Delta_g\Psi + V\Psi\,, \tag{10}$$

where Δ_g and ∇_g are, respectively, the Laplace-Beltrami operator and the gradient on the configuration space equipped with this Riemannian structure. But there is nothing special about this particular Riemannian structure. Indeed, Eqs. (9) and (10) as such hold very generally on *any* Riemannian manifold. Thus, the formulation of a Bohmian dynamics on a Riemannian manifold requires only as basic ingredients the differentiable and metric structure of the manifold.

3 Quantum Puzzles

The Bohmian perspective allows us to illuminate some quantum puzzles, somewhat hidden in other formulations of quantum mechanics. These puzzles find their natural resolution in a proper relational reformulation of quantum mechanics. We shall elaborate in later sections.

3.1 The Wave Function of the Universe and Entropy

The fundamental equation for the wave function of the universe in canonical quantum cosmology is the Wheeler-DeWitt equation (DeWitt 1967)

$$\mathscr{H}\Psi = 0\,, \tag{11}$$

for a wave function $\Psi(q)$ of the universe, where q refers to 3-geometries and to whatever other stuff is involved, all of which correspond to structures on a 3-dimensional space. In this equation $\mathscr{H}$ is a sort of generalized Laplacian, a cosmological version of a Schrödinger Hamiltonian H. And like a typical H, it involves nothing like an explicit time-dependence. But unlike Schrödinger's equation, the Wheeler-DeWitt equation has on one side, instead of a time derivative of Ψ, simply 0. Its natural solutions are thus time-independent.

That this is so is in fact the *problem of time* in quantum cosmology. We live in a world where things change. But if the basic object in the world is a timeless wave function, how does change come about? Much has been written about this problem of time. A great many answers have been proposed. But what we want to emphasize here is that from a Bohmian perspective the timelessness of Ψ is not a problem. Rather it is just what the doctor ordered.

The fundamental role of the wave function in Bohmian mechanics is to govern the motion of something else. Change fundamentally occurs in Bohmian mechanics not so much because the wave function changes but because the thing Q that it governs changes, according to a law

$$dQ/dt = v^{\Psi}(Q) \tag{12}$$

determined by the wave function. The problem of time vanishes entirely from a Bohmian point of view.

But there is a difficulty that arises in connection with the stationarity of the universal wave function. According to standard formulations of quantum statistical mechanics, thermodynamics seems to depend only upon the wave function. But if the initial wave function is stationary, it can't be responsible for the irreversible behavior of our universe, and in particular of entropy increase. So what is the origin of the arrow of time? And why does entropy increase?

3.2 *Nonnormalizability of the Wave Function of the Universe*

The solutions of Wheeler-DeWitt equation typically fail to be normalizable. So what could the associated non-normalizable "probabilities" given by $|\Psi|^2$ physically mean? How can the quantum equilibrium analysis we hinted at above, that allows us to recover the entire quantum formalism, get off the ground for such a non-normalizable measure?

3.3 *Why vs. What*

From a Bohmian perspective there is another puzzle. One may ask, Why should the motion of configurations be governed by a wave function?

This problem is rather subtle and elusive, hardly understandable at all in the standard formulation of quantum mechanics, as well as in other more precise versions of quantum mechanics, such as Many Worlds or GRW, in which the wave function is all there is. We believe that relational physics sheds light on this problem and shall elaborate on this in Sect. 5. We also wish to note, concerning quantum mechanics, the contrast between asking why and asking what. One often finds derivations of quantum mechanics from some principle or other, which thus seem to answer the why question. But in most, if not all, such cases, the answer leaves us with little or no understanding of what quantum mechanics actually says about physical reality. This is not good. The appropriate time to ask why is after the what question has been satisfactorily addressed. With Bohmian mechanics we are at such a time.

4 Wave Function as Law

The most puzzling issue in the foundations of quantum mechanics is perhaps that of the status of the wave function of a system in a quantum universe. We have suggested elsewhere that the wave function should be regarded as nomological, nomic—that it's really more in the nature of a law than a concrete physical reality.

Thoughts in this direction might arise when you consider the unusual way Bohmian mechanics is formulated, and the unusual way the wave function behaves in Bohmian mechanics. The wave function of course affects the behavior of the configuration, i.e., of the particles. This is expressed by the guiding Eq. (12). But in Bohmian mechanics there's no back action, no effect in the other direction, of the configuration upon the wave function, which evolves autonomously via Schrödinger's equation, in which the actual configuration Q does not appear.

A second point is that for a multi-particle system the wave function $\Psi(\boldsymbol{q}) = \Psi(\vec{\mathrm{q}}_1, \ldots, \vec{\mathrm{q}}_N)$ is not a field on physical space, but on configuration space, the set of all hypothetical configurations of the system. For a system of more than one particle that space is not physical space. What this suggests to us is that you should think of the wave function as describing a law and not as some sort of concrete physical reality. After all (12) is an equation of motion, a law of motion, and the whole point of the wave function here is to provide us with the law, i.e., with the right hand side of this equation.

4.1 ψ vs. Ψ

There are, however, problems with regarding the wave function as nomological. Laws aren't supposed to be dynamical objects, they aren't supposed to change with time, but the wave function of a system typically does. And laws are not supposed to be things that we can control. But the wave function is often an initial condition for a quantum system. We often, in act, prepare a system in a certain quantum state, that is, with a certain wave function. We can in this sense control the wave function of a system. But we don't control a law of nature. This makes it a bit difficult to regard the wave function as nomological.

But with regard to this difficulty it's important to recognize, as already suggested, that there's only one wave function we should be worrying about, the fundamental one, the wave function Ψ of the universe. In Bohmian mechanics, the wave function ψ of a subsystem of the universe is defined in terms of the universal wave function Ψ according to (5). Thus, to the extent that we can grasp the nature of the universal wave function, we should understand as well, by direct analysis, also the nature of the objects that are defined in terms of it, and in particular we should have no further fundamental question about the nature of the wave function of a subsystem of the universe.

Note that from this perspective, the most important equation in quantum mechanics, the time-dependent Schrödinger's equation should be regarded as a phenomenological equation describing how the conditional wave function ψ of a subsystem evolves—under suitable conditions of decoupling from its environment—in a universe having a suitable stationary universal wave function Ψ.

4.2 *A Kantian Component of Laws of Nature*

Laws are supposed to be simple. But here a distinction should be made. Consider the velocity field in the right hand side of (12),

$$v^{\Psi}(Q) = v(\Psi, Q) .$$

If we regard Ψ as a law, v^{Ψ} should be a simple function of Q. On the other hand, if we regard Ψ as *part* of a law, $v(\Psi, Q)$ should be a simple function of Ψ and Q.

So we would like to find a compelling law of motion, or a compelling principle yielding such a law, that can be expressed in terms of a v^{Ψ} for a suitable Ψ. We don't have that. But, as we shall explain, the flexibility and gauge freedom afforded by relational physics might help us reach our goal.

Ideally, the suitable Ψ referred to above should be simple. Failing that, it should at least obey a simple law, such as the Wheeler-DeWitt equation. From a fundamental point of view, it might be a complete accident that Ψ obeys such an equation. It might just happen to do so. The fact that the equation is satisfied might have nothing to do with why the fundamental dynamics is of the form (12). But as long as Ψ does satisfy the equation, by accident or not, all the consequences of satisfying it would follow.

The point here is to distinguish what is really fundamental from what we have traditionally regarded as fundamental. The latter might, in fact, be imposed by us. We express this by saying that the laws of nature might have a Kantian component. And that Kantian component might be much larger than suggested so far, as will become clearer in the next section devoted to relational quantum mechanics.

5 Relational Quantum Mechanics

At the end of Sect. 1 we said that the stronger reason for us to go relational has to do with understanding the status and the nature of the wave function. At the beginning of Sect. 3, we claimed that the quantum puzzles presented there find their natural resolution in a proper relational reformulation of quantum mechanics.

It's time to intertwine all these threads. And the key question to start with is the one we asked in the last paragraph of the previous section, What is really fundamental?

5.1 Shape Space as the Fundamental Configuration Space

Let us go back to the toy model of an N-particle universe introduced in Sect. 1.1. The space of all its genuinely physically different possible configurations is *shape space*. Mathematically, this is described as follows.

5.1.1 Shape Space

The totality of configurations $\boldsymbol{q} = (\vec{\mathsf{q}}_1, \dots, \vec{\mathsf{q}}_N)$ of N points in Euclidean three-dimensional space forms the configuration space $\mathcal{Q} = \{\boldsymbol{q}\} = \mathbb{R}^{3N}$ of an N-particle system. We shall call $\mathcal{Q}$ the *absolute configuration space*. On $\mathcal{Q}$ there is a natural action of the similarity transformations of Euclidean space, namely rotations, translations and dilations, since each of them acts naturally on each component of the configuration vector in physical Euclidean space. The totality of such transformations form the group G of *similarity transformations* of Euclidean space. Since the shape of a configuration is what remains when the effects associated with rotations, translations and dilations are filtered away, the totality of shapes, i.e., the *shape space*, is the quotient space $Q \equiv \mathcal{Q}/G$, the set of equivalence classes with respect to the equivalence relations provided by the similarity transformations of Euclidean space.

As such, shape space is not in general a manifold. To transform it into a manifold some massaging is needed (e.g., by excluding from $\mathcal{Q}$ coincidence points and collinear configurations), but we shall not enter into this.[1] Here we shall assume that the appropriate massaging of $\mathcal{Q}$ has been performed and that Q is a manifold. Since the group of similarity transformations has dimension 7 (3 for rotations + 3 for translations + 1 for dilations), the dimension of $Q \equiv \mathcal{Q}/G$ is $3N-7$ for $N \geq 3$. For $N = 1$ and $N = 2$ shape space is trivial (it contains just a single point). $N = 3$ corresponds to the simplest non-trivial shape space; it has dimension 2.

Accordingly, it is helpful to represent absolute configuration space $\mathcal{Q}$ in geometrical terms as a fiber bundle, with shape space Q as the base manifold and with the fibers generated by the similarity group G group. The points on each fiber represent physically equivalent states. See Fig. 1.

5.1.2 Metrics on Shape Space and Best Matching

The topology of shape space is well defined by its construction as a quotient space, but topology, of course, does not fix a metric. A metric should provide more, namely a natural notion of distance on Q. And since each point in Q represents a class of configurations of N particles related by a similarity transformation, the distance

[1] For more details on this issue, see, e.g., (Le and Kendall 1993) and the references therein.

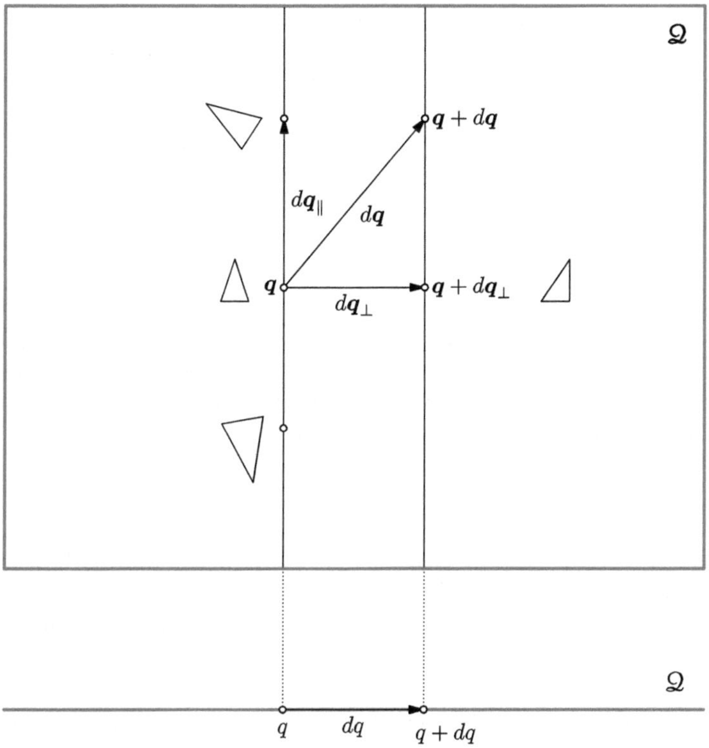

Fig. 1 Absolute configuration space $\mathfrak{Q}$ and shape space $\mathcal{Q}$ (for a system of three particles). The fiber above shape q consists of absolute configurations $\boldsymbol{q}$ differing by a similarity transformation of Euclidean space and thus representing the same shape q. Real change of shape occurs only by a displacement to a neighboring fiber $q+dq$. Only the orthogonal component $d\boldsymbol{q}_{\perp}$ of $d\boldsymbol{q}$ represents real change, while the vertical displacement $d\boldsymbol{q}_{\parallel}$ does not contribute; $\boldsymbol{q}+d\boldsymbol{q}_{\perp}$ is the absolute configuration in the fiber above $q+dq$ closest to q in the sense of the g_B-distance (best matching)

between two elements of $\mathcal{Q}$ induced by the metric should not recognize any absolute configurational difference due to an overall translation, or rotation, or dilation. In other words, it should provide a measure of the *intrinsic* difference between two absolute configurations (that is, not involving any consideration regarding how such configurations are embedded in Euclidean space).

It turns out (Dürr et al. 2020) that *a metric on absolute configuration space* $\mathfrak{Q}$ *that is invariant under the group G of similarity transformations of Euclidean space, given by a suitable "conformal factor" (to be explained below), defines canonically a metric on shape space* $\mathcal{Q}$.

To understand why this is so, consider the representation of absolute configuration space $\mathfrak{Q}$ as a fiber bundle with each fiber being homeomorphic to G and $\mathcal{Q}$ being its base space (see Fig. 1). So, if g is a metric invariant under any element of G, the tangent vectors at each point $\boldsymbol{q} \in \mathfrak{Q}$ are *naturally* split into "vertical" and "horizontal," where by "naturally" we mean that the splitting itself is invariant under the action of G. The vertical ones correspond to (infinitesimal) displacements along the fiber through $\boldsymbol{q}$ and the horizontal ones are those that are orthogonal to the fiber, i.e., to the vertical ones, according to the relation of orthogonality defined by g. More precisely, if $d\boldsymbol{q}$ is an infinitesimal displacement at $\boldsymbol{q}$, we have

$$d\boldsymbol{q} = d\boldsymbol{q}_{\parallel} + d\boldsymbol{q}_{\perp} \quad \text{with} \quad g(d\boldsymbol{q}_{\parallel}, d\boldsymbol{q}_{\perp}) = 0$$

(see Fig. 1), with $d\boldsymbol{q}_{\parallel}$ vertical and $d\boldsymbol{q}_{\perp}$ horizontal.

The corresponding Riemannian metric on $\mathcal{Q}$ is defined as follows. Let q be a shape, $\boldsymbol{q}$ any absolute configuration in the fiber above q, and $d\boldsymbol{q}$ any displacement at $\boldsymbol{q}$. Since g is invariant under the group G, the length of $d\boldsymbol{q}_{\perp}$ has the same value for all absolute configurations $\boldsymbol{q}$ above q. Thus we may set the length of dq equal to that of $d\boldsymbol{q}_{\perp}$ and hence obtain the Riemannian metric g_B on $\mathcal{Q}$

$$g_B(dq, dq) = g(d\boldsymbol{q}_{\perp}, d\boldsymbol{q}_{\perp}) \,. \tag{13}$$

The subscript B stands for Barbour and Bertotti, as well as base and best matching. In fact, The distance on $\mathcal{Q}$ induced by g_B is exactly the one resulting from applying Barbour's best matching procedure. Consider two infinitesimally close shapes, q and $q + dq$, and let $\boldsymbol{q}$ be any absolute representative of q, i.e., any point in the fiber above q. The g_B-distance between these shapes is then given by the g-length of the vector $d\boldsymbol{q}$ such that (i) $d\boldsymbol{q}$ is orthogonal to the fiber above q and (ii) $\boldsymbol{q} + d\boldsymbol{q}$ is an absolute representative of $q + dq$. It follows that $\boldsymbol{q} + d\boldsymbol{q}$ is the absolute configuration closest to $\boldsymbol{q}$ in the fiber above $q + dq$. Thus the g_B-distance is the "best matching" distance.

5.1.3 Invariant Metrics on Absolute Configuration Space

So we need to show how to construct an invariant metric g on $\boldsymbol{\mathcal{Q}}$. Let g_e be the mass-weighted Euclidean metric on $\boldsymbol{\mathcal{Q}}$ with positive weights m_α, $\alpha = 1, \ldots, N$, (the masses of the particles), in particle coordinates $\boldsymbol{q} = (\vec{\mathsf{q}}_1, \ldots, \vec{\mathsf{q}}_\alpha, \ldots, \vec{\mathsf{q}}_N)$ given by

$$ds^2 = \sum_{\alpha=1}^{N} m_\alpha d\vec{\mathsf{q}}_\alpha \cdot d\vec{\mathsf{q}}_\alpha \,, \tag{14}$$

i.e., with $[g_e]_{ij} = m_{\alpha_i}\delta_{ij}$, where the i-th component refers to the α_i-th particle. The corresponding line element is

$$|d\boldsymbol{q}| = \sqrt{\sum_{\alpha=1}^{N} m_\alpha d\vec{\mathsf{q}}_\alpha \cdot d\vec{\mathsf{q}}_\alpha} \,. \tag{15}$$

The metric defined by (14) is invariant under rotations and translations, but not under a dilation $\boldsymbol{q} \to \lambda \boldsymbol{q}$, where λ is a positive constant. Invariance under dilations is achieved by multiplying $|d\boldsymbol{q}|^2$ by a scalar function $f(\boldsymbol{q})$ that is invariant under rotations and translations and is homogeneous of degree -2. We call f a *conformal factor*. So, for any choice of f,

$$g = f g_e\,, \quad \text{i.e,} \quad g(d\boldsymbol{q}, d\boldsymbol{q}) = f(\boldsymbol{q})|d\boldsymbol{q}|^2\,, \tag{16}$$

is an invariant metric on $\mathcal{Q}$, yielding the metric on shape space

$$g_B(dq, dq) = f(\boldsymbol{q})|d\boldsymbol{q}_\perp|^2\,. \tag{17}$$

For the associated line element we shall write

$$ds = |dq| = \sqrt{g_B(dq, dq)} = \sqrt{f(\boldsymbol{q})}\,|d\boldsymbol{q}_\perp|\,. \tag{18}$$

5.1.4 Conformal Factors

Many choices of conformal factors are possible. One that was originally suggested by Barbour and Bertotti is[2]

$$f(\boldsymbol{q}) = f_a(\boldsymbol{q}) \equiv \left(\sum_{\alpha<\beta} \frac{m_\alpha m_\beta}{|\vec{\mathrm{q}}_\alpha - \vec{\mathrm{q}}_\beta|} \right)^2\,. \tag{19}$$

Another example is

$$f(\boldsymbol{q}) = f_b(\boldsymbol{q}) \equiv \mathrm{L}^{-2}\,, \tag{20}$$

where

$$\mathrm{L}^2 = \sum_\alpha m_\alpha \vec{\amalg}_\alpha^2 = \frac{1}{\sum_\alpha m_\alpha} \sum_{\alpha<\beta} m_\alpha m_\beta |\vec{\mathrm{q}}_\alpha - \vec{\mathrm{q}}_\beta|^2 \tag{21}$$

with $\vec{\amalg}_\alpha = \vec{\mathrm{q}}_\alpha - \vec{\mathrm{q}}_{\mathrm{cm}}$, the coordinates relative to the center of mass

$$\vec{\mathrm{q}}_{\mathrm{cm}} = \frac{\sum_\alpha m_\alpha \vec{\mathrm{q}}_\alpha}{\sum_\alpha m_\alpha}\,. \tag{22}$$

[2] Here and in the following examples the conformal factors are modulo dimensional factors.

$I \equiv \mathrm{L}^2$ is sometimes called (but the terminology is not universal) the moment of inertia of the configuration $\boldsymbol{q}$ about its center of mass. This quantity is half the trace of the moment of inertia tensor M,

$$\mathrm{L}^2 = \frac{1}{2}\,\mathrm{Tr}\,\mathsf{M}\,. \tag{23}$$

We recall that $\mathsf{M} = \mathsf{M}(\boldsymbol{q})$, the tensor of inertia of the configuration $\boldsymbol{q}$ about any orthogonal cartesian system x,y,z with origin at the center of mass of the configuration $\boldsymbol{q}$, has matrix elements given by the standard formula

$$M_{ij} = \sum_{\alpha=1}^{N} m_\alpha(\rho_\alpha^2\delta_{ij} - \rho_{\alpha i}\rho_{\alpha j})\,, \tag{24}$$

where $i, j = x, y, z$, $\rho_{\alpha x} \equiv x_\alpha$, $\rho_{\alpha y} \equiv y_\alpha$, $\rho_{\alpha z} \equiv z_\alpha$, and $\rho_\alpha^2 = x_\alpha^2 + y_\alpha^2 + z_\alpha^2$.

A choice of conformal factor that has not been considered in the literature is

$$f(\boldsymbol{q}) = f_c(\boldsymbol{q}) \equiv \mathrm{L}^{-\frac{8}{7}}(\det\mathsf{M})^{-\frac{1}{7}}\,. \tag{25}$$

Since $\det\mathsf{M}$ scales as L^6, $f(\boldsymbol{q})$ given by (25) scales as it should, namely, as L^{-2}. Though at first glance this choice seems unnatural, it is in fact very natural once the motion of shapes is analyzed from a quantum perspective (Dürr et al. 2020).

Finally, we give other two examples:

$$f(\boldsymbol{q}) = f_d(\boldsymbol{q}) \equiv \sum_{\alpha<\beta} \frac{m_\alpha m_\beta}{|\vec{\mathsf{q}}_\alpha - \vec{\mathsf{q}}_\beta|^2} \tag{26}$$

$$f(\boldsymbol{q}) = f_g(\boldsymbol{q}) \equiv \mathrm{L}^{-1}\sum_{\alpha<\beta} \frac{m_\alpha m_\beta}{|\vec{\mathsf{q}}_\alpha - \vec{\mathsf{q}}_\beta|}\,. \tag{27}$$

The first one corresponds to a natural modification of the Newtonian gravitational potential and the second, discussed in Barbour (2003), corresponds to a dynamics very close to that of Newtonian gravity.

5.2 *Free Motion on Shape Space*

All one needs to define the simplest motion on shape space, be it classical or quantum (Bohmian), is the metric g_B on shape space. In the classical case, such a metric g_B directly yields a law of *free motion* on shape space, that is, geodesic motion with constant speed.

In the quantum (Bohmian) case, the situation is similar. Equations (9) and (10) define immediately free quantum motion on shape space with Riemannian metric $g = g_B$ as the motion on shape space given by the evolution equations

$$\frac{dQ}{dt} = \hbar \ \text{Im} \ \frac{\nabla_B \Psi}{\Psi} \tag{28}$$

$$i\hbar \frac{\partial \Psi}{\partial t} = -\frac{\hbar^2}{2} \Delta_B \Psi \tag{29}$$

where Δ_B and ∇_B are, respectively, the Laplace-Beltrami operator and the gradient on the configuration space equipped with the Riemannian metric (17).

We have insisted on free motion, because it is the simplest motion on shape space, which has, surprisingly, a very rich structure. Of course, one may introduce a potential term both in the classical and the quantum case in a straightforward way: (minus) the gradient of the potential in the right hand side of the geodesic equation for the classical case and the usual potential term in the right hand side of Schrödinger's equation.

This is all there is to say about the *formulations* of classical and quantum theories on shape space, modulo a caveat: so defined, classical and quantum motion rely on an external absolute time, which, from a relational perspective, is problematical, see below.

5.3 *Fundamental Level vs. "Human" Level*

Let us draw some morals from the foregoing. First of all, it is rather clear that to define motion on shape space is very easy. The hard work, basically in the quantum case (Dürr et al. 2020), involves the transition from the fundamental level of shape space to what could be called the "human" level of absolute configuration space. We say "human" because it is the description we usually use. Even if shape space is fundamental, we humans don't typically formulate things in terms of shape space. That doesn't come quite so naturally to us, even if shape space is in a deeper sense more fundamental and natural. Traditionally we do things in this way for good human reasons and probably for many others.

The level of absolute configuration space is a sort of emergent level because it is not fundamental. How things are described at that level is imposed by us. It is "human" in the sense that it is based on our choice of description. This is why we consider it a Kantian component of physical theory. As we shall elaborate below, much of what we've taken as fundamental is not fundamental, but imposed by us by a proper choice of gauge. An important example of such a thing is interactions.

5.4 Emergence of Interactions

Given a motion in shape space, there is a huge host of motions in absolute space that are compatible with it, the only constraint being that they should project down to the motion in shape space. In other words, recalling the fiber bundle structure of absolute configuration space described above, we have that a representation in absolute configuration space of the motion in shape space is given by a "lift" of the motion from the base into the fibers.

Such lifts can rightly be called gauges. In the classical case it turns out that in some gauges the law looks unfamiliar but there is (at least) one gauge in which, after performing a time change—representing indeed another gauge freedom when also time is seen as relational—the law of motion is Newtonian with a potential appearing. We call it the *Newton gauge*. The potential depends on the choice of the invariant metric (16) in absolute configuration space, with the various possibilities for the conformal factor, some of them listed at the end of Sect. 5.1.

More or less the same is true for the quantum case, where however the gauge yielding ordinary Bohmian mechanics in absolute configuration space—which we call the *Schrödinger gauge*—emerges only for a stationary, i.e. time-independent, wave function (such as with the Wheeler-DeWitt equation) on shape space. This again involves regarding time as being relational, with an external absolute time playing no physical role.

Also here, while the fundamental physics is given by a free Bohmian dynamics in shape space, in the Schrödinger gauge potential terms appear. One potential term is determined by the scalar curvature induced by the invariant metric on absolute configuration space. Another potential term arises from the gauge freedom we have to lift the Laplace-Beltrami operator from shape space to absolute configuration space, where more gauge freedom arises from allowing transformations of the lifted wave function.

6 Sources of Gauge Freedom

The main sources of gauge freedom are a *lift* from the base into the fibers and a *time change*, both for the classical and quantum case. For the quantum case there is in addition a gauge transformation

$$\Psi \mapsto \Psi' = F\Psi\,, \qquad \text{with } F \text{ real.} \tag{30}$$

The implications of this will be discussed below in detail.

6.1 Classical Gauges

In the classical case, consider the free motion $Q = Q(t)$ on shape space, that is, the geodesic motion with constant speed. A very natural lift of this motion in absolute configuration space is its *horizontal lift*, that is, a motion $\boldsymbol{Q} = \boldsymbol{Q}(t)$ in absolute configuration space that starts at some point $\boldsymbol{q}_1$ on the fiber above q_1 and is *horizontal*, i.e., the infinitesimal displacements $d\boldsymbol{Q}$ are all horizontal. (Note that the final point $\boldsymbol{q}_2$ in the fiber above q_2 is then uniquely determined.) We call this the *invariant (or horizontal) gauge*.

It can be shown (Dürr et al. 2020) that in this gauge, the familiar total momentum $\vec{\mathsf{P}}$, total angular momentum $\vec{\mathsf{J}}$ and (maybe less familiar) dilational momentum D are all zero, i.e.,

$$\vec{\mathsf{P}} = \sum_{\alpha=1}^{N} m_\alpha \frac{d\vec{\mathsf{Q}}_\alpha}{dt} = 0 \tag{31}$$

$$\vec{\mathsf{J}} = \sum_{\alpha=1}^{N} m_\alpha \vec{\mathsf{Q}}_\alpha \times \frac{d\vec{\mathsf{Q}}_\alpha}{dt} = 0 \tag{32}$$

$$D = \sum_{\alpha=1}^{N} m_\alpha \vec{\mathsf{Q}}_\alpha \cdot \frac{d\vec{\mathsf{Q}}_\alpha}{dt} = 0\,. \tag{33}$$

6.1.1 The Newton Gauge

By a suitable change of speed, that is by a suitable time change $t' = t'(t)$, namely,

$$\frac{v}{\sqrt{f}}\frac{dt}{dt'} = \sqrt{2f}\,, \quad \text{i.e.,} \quad \frac{dt'}{dt} = \frac{v}{\sqrt{2}f}\,, \tag{34}$$

we obtain another gauge that we call the *Newton gauge*, a gauge in which the motion is Newtonian, i.e., it satisfies Newton's equation $F = ma$ for suitable F. More precisely, the particle positions $\vec{\mathsf{Q}}_\alpha$, $\alpha = 1, \ldots, N$, forming the configuration $\boldsymbol{Q}$ then satisfy Newton's equations

$$m_\alpha \frac{d^2\vec{\mathsf{Q}}_\alpha}{dt'^2} = -\vec{\nabla}_\alpha V(\vec{\mathsf{Q}}_1, \ldots \vec{\mathsf{Q}}_N)\,. \tag{35}$$

with $V(\boldsymbol{q}) = -f(\boldsymbol{q})$, the conformal factor.

In this regard, one should observe that if time is relational, changes of speed, such as that given by (34), provide equivalent representations of the same motion. Accordingly, the use of one time variable instead of another is a matter of convenience, analogous to the choice of a gauge. The choice of time variable that leads from the invariant gauge to the Newton gauge is the gauge fixing condition

$$E = \frac{1}{2}\left|\frac{d\boldsymbol{Q}}{dt'}\right|^2 + V = 0\,, \tag{36}$$

that is, that the total Newtonian energy be zero. Of course, other gauges could be useful, as discussed next.

6.1.2 The Expansion Gauge

Another gauge that could be useful is what could be called the *expansion gauge*, a gauge in which you do not use a horizontal lift, but instead have a uniform expansion of configurations taking place—in the horizontal gauge the total overall size does not change: the universe would not expand since the total dilational momentum is zero in that gauge (Barbour 2003). In contrast, in the expansion gauge configurations and thus the universe would expand. Why one would consider such a gauge? One reason is that in this way one might simplify the form of the gravitational forces (see below), so that they would become more familiar in this gauge. In this regard, note that since the motion is not horizontal, after suitable time change the Newtonian form of the equation of motion would involve other terms, in addition to the forces generated by the conformal factor. Another reason is that one would obtain in this way a striking explanation of why we think that the universe is expanding.

6.1.3 A Note on Newtonian Gravitation

We have seen that in the Newton gauge, when the physical law on shape space is free motion (or even non-free motion), the potential $V = -f$ appears, where f is the conformal factor. We mentioned some choices for f in Sect. 5.1.4. No such choices, which are necessarily functions homogenous of degree -2, seem to yield exactly the Newtonian gravitational potential U_g. While we believe the detailed exploration of the implications of the models discussed here is worthwhile, we nonetheless regard the models explored in this paper, both classical and quantum, as toy models, so that such an analysis of them, with the expectation of recovering well established physics, might be somewhat inappropriate or premature.

However, it should be observed that some of the conformal factors given in Sect. 5.1.4, e.g., f_a and f_g, indeed give rise to a force law in the Newton gauge that is very close to that of the Newtonian gravitational force. Note for example that for the conformal factor f_g the corresponding potential is of the form $V_g = \mathrm{L}^{-1}U_g$, where U_g is the Newton gravitational potential and L in the Newton gauge is a constant of the motion. The force arising from this potential adds to the Newtonian force a very small centripetal correction that allows $I = \mathrm{L}^2$, the moment of inertia about the center of mass, to remain constant (Barbour 2003). It is worth exploring these possibilities in the expansion gauge.

6.2 Quantum Gauges

Let $Q = Q(t)$ be a Bohmian motion in shape space, that is, a solution of (28) with the wave function Ψ being a solution of Schrödinger's equation (29) on shape space. As in the classical case, we wish to characterize motions in absolute space that are compatible with motions in shape space, that is, motions $\boldsymbol{Q} = \boldsymbol{Q}(t)$ in $\boldsymbol{\mathcal{Q}}$ that project down to $Q = Q(t)$ in $\mathcal{Q}$, i.e., such that

$$\pi(\boldsymbol{Q}(t)) = Q(t)\,, \tag{37}$$

where π is the canonical projection from $\boldsymbol{\mathcal{Q}}$ to $\mathcal{Q}$. Clearly, there are a great many possibilities for compatible motions in absolute configuration space.

As in the classical case, one may restrict the possibilities by considering natural gauges. And as in the classical case, where one looks for gauges such that the absolute motions satisfy Newton's equations, in the quantum case we now look for gauges such that the compatible motions on $\boldsymbol{\mathcal{Q}}$ are themselves Bohmian motions, i.e. motions generated by a wave function in the usual sort of way.

6.2.1 Three Quantum Gauges

Suppose that we proceed as in the classical case and take a horizontal lift of a motion $Q = Q(t)$ in shape space, that is, an absolute motion for which the infinitesimal displacements $d\boldsymbol{Q}$ are all horizontal. Let us now consider the lift to $\boldsymbol{\mathcal{Q}}$ of a wave function Ψ on $\mathcal{Q}$, namely, the wave function $\widehat{\Psi}_1$ on absolute configuration space such that

$$\widehat{\Psi}_1(\boldsymbol{q}) = \Psi(q) \tag{38}$$

for any point $\boldsymbol{q}$ on the fiber above q. Let ∇_g be the gradient with respect to the invariant measure (16). Then the vector $\nabla_g\widehat{\Psi}_1(\boldsymbol{q})$ in $\boldsymbol{\mathcal{Q}}$ is horizontal and the motions on $\boldsymbol{\mathcal{Q}}$ defined by

$$\frac{d\boldsymbol{Q}}{dt} = \hbar\ \mathrm{Im}\ \frac{\nabla_g\widehat{\Psi}_1}{\widehat{\Psi}_1} \tag{39}$$

are horizontal lifts of motions on $\mathcal{Q}$. So, in the quantum case, horizontality is immediate.

Let us now consider the time evolution of the lifted wave function $\widehat{\Psi}_1$ on $\boldsymbol{\mathcal{Q}}$. Let $\widehat{\Delta}_B$ be a lift to absolute configuration space of the Laplace-Beltrami operator Δ_B on shape space, namely an operator on $\boldsymbol{\mathcal{Q}}$ such that

$$\widehat{\Delta}_B\widehat{\Psi}_1 = \Delta_B\Psi\,. \tag{40}$$

Then

$$i\hbar\frac{\partial\widehat{\Psi}_1}{\partial t}=\widehat{\mathsf{H}}_1\widehat{\Psi}_1\,,\quad\text{with}\quad\widehat{\mathsf{H}}_1=-\frac{\hbar^2}{2}\widehat{\Delta}_B\,. \tag{41}$$

It might seem natural to guess that $\widehat{\Delta}_B$ coincides with Δ_g, the Laplace-Beltrami operator with respect to g, but this is wrong; nor is $\widehat{\mathsf{H}}_1$ a familiar sort of Schrödinger Hamiltonian, with or without a potential term.

While $\widehat{\Psi}_1$ need not obey any familiar Schrödinger-type equation, one may ask whether there exists a gauge equivalent wave function that does. By gauge equivalent wave function we mean this: If one writes $\widehat{\Psi}_1$ as $Re^{(i/\hbar)S}$ one sees that the velocity field given by (39) is just $\nabla_g S$, so transformations of the wave function

$$\widehat{\Psi}_1\to\widehat{\Psi}_1'=F\widehat{\Psi}_1\,, \tag{42}$$

where F is a positive function, do not change its phase and thus the velocity.

It turns out that there exists a positive function F such that $\widehat{\Psi}_3=F\widehat{\Psi}_1$ satisfies a Schrödinger-type equation on absolute configuration space for a suitable potential V, namely,

$$i\hbar\frac{\partial\widehat{\Psi}_3}{\partial t}=\widehat{\mathsf{H}}_3\widehat{\Psi}_3 \tag{43}$$

with

$$\widehat{\mathsf{H}}_3=-\frac{\hbar^2}{2}\sum_{\alpha=1}^{N}\vec{\nabla}_\alpha\cdot\frac{1}{fm_\alpha}\vec{\nabla}_\alpha+V \tag{44}$$

$$=-\frac{\hbar^2}{2}\boldsymbol{\nabla}\cdot\frac{1}{f}\boldsymbol{\nabla}+V\,, \tag{45}$$

where $\boldsymbol{\nabla}$ and $\boldsymbol{\nabla}\cdot$ are the gradient and divergence with respect to the mass-weighted Euclidean metric (14), i.e.,

$$\boldsymbol{\nabla}=\left(\frac{1}{m_1}\vec{\nabla}_1,\ldots,\frac{1}{m_N}\vec{\nabla}_N\right) \tag{46}$$

and

$$\boldsymbol{\nabla}\cdot=\left(\vec{\nabla}_1,\ldots,\vec{\nabla}_N\right)\cdot\,. \tag{47}$$

Here f is the conformal factor, and

$$V=V_1+V_2 \tag{48}$$

$$V_1 = -\frac{\hbar^2}{2}\frac{\widehat{\Delta}_B J^{1/2}}{J^{1/2}} \tag{49}$$

$$V_2 = -\frac{\hbar^2}{2} f^{\frac{n}{4}} \Delta_g\left(f^{-\frac{n}{4}}\right), \tag{50}$$

with

$$J = \mathrm{L} f^{7/2} \sqrt{\det \mathsf{M}}\,, \tag{51}$$

where $\mathrm{L} = \mathrm{L}(\boldsymbol{q})$ is given by Eq. (21) and $\mathsf{M} = \mathsf{M}(\boldsymbol{q})$ is the tensor of inertia of the configuration $\boldsymbol{q}$ about any orthogonal cartesian system x,y,z with origin in its center of mass and with matrix elements given by (24), and where $\widehat{\Delta}_B$ is the canonical lift of Δ_B (see Dürr et al. 2020 for details on this and a proof of the aforementioned statements).

There is a 2-gauge, intermediate between the 1-gauge and the 3-gauge, that we shall not describe further here.

6.2.2 The Schrödinger Gauge

We shall now describe what we think is appropriate to be called the *Schrödinger gauge*, the true quantum analogue of the Newton gauge. If we take into account that time is relational, as we should, the fundamental equation for the wave function on shape space is presumably the stationary equation

$$-\frac{\hbar^2}{2}\Delta_B \Psi = \mathscr{E}\Psi\,, \tag{52}$$

where $\mathscr{E}$ is any given fixed constant (for example $\mathscr{E} = 0$).

As before, let $\widehat{\Psi}_1$ be the lift of Ψ to $\mathcal{Q}$, so that $\widehat{\Psi}_1$ satisfies the equation

$$(\widehat{\mathsf{H}}_1 - \mathscr{E})\widehat{\Psi}_1 = 0\,, \tag{53}$$

with $\widehat{\mathsf{H}}_1$ a lift of H as in (41), and the evolution on the absolute configuration space $\mathcal{Q}$ is still given by (39). But now, for relational time, motions following the same path with different speeds are the same motion. So, in the formula for the gradient on the right hand side of (39), $\nabla_g = f^{-1}\nabla$, we may regard f as a change of speed defining a new time variable that for the sake of simplicity we shall still call t (a "time change"). Then in absolute space the guiding Eq. (39) becomes

$$\frac{d\vec{\mathsf{Q}}_\alpha}{dt} = \frac{\hbar}{m_\alpha}\,\mathrm{Im}\,\frac{\vec{\nabla}_\alpha \widehat{\Psi}_1}{\widehat{\Psi}_1}\,. \tag{54}$$

Again, $\widehat{\Psi}_1$ need not obey any familiar stationary Schrödinger-type equation. However, as before, we may exploit gauge freedom to transform (53) into a stationary Schrödinger-type equation. Indeed, we now have an even greater gauge freedom in changing the wave function and the Hamiltonian, see Dürr et al. (2020). In particular, there is a gauge, the Schrödinger gauge, in which (52) becomes

$$\widehat{\mathsf{H}}_S \Phi = 0 \tag{55}$$

with

$$\widehat{\mathsf{H}}_S = -\frac{\hbar^2}{2}\boldsymbol{\nabla}^2 + U\,, \tag{56}$$

where $\boldsymbol{\nabla}^2 = \boldsymbol{\nabla} \cdot \boldsymbol{\nabla}$ is the mass-weighed Euclidean Laplacian, and

$$U = f(V_1 - \mathscr{E}) - \frac{\hbar^2}{8}\frac{n-2}{n-1} f R_g\,, \tag{57}$$

where R_g is the scalar curvature of the invariant metric g and $n = 3N$.

7 Lessons from Relational Physics

In this closing section we shall further consider what we can learn from relational physics.

7.1 Local Beables and Primitive Ontology

We have stressed elsewhere the importance of certain local beables, what we call the primitive ontology of a theory—stuff in space evolving in time—which is what the theory is fundamentally about. Relational physics requires a more flexible understanding of this notion.

To appreciate this point, note that for relational space the state of the universe at a particular location is not, in and of itself, meaningful. In that sense, for relational physics, there are no local beables in the usual sense, and locality itself can't easily be meaningfully formulated.

Similarly one can't meaningfully consider the behavior of individual particles without reference to other particles, since there is no absolute space in which an individual particle could be regarded as moving. And even for a pair of particles, to speak meaningfully of the distance between them, a third particle would be required, to establish a scale of distance. And similarly for galaxies.

Relational physics is, in fact, genuinely holistic, and suggests the holistic character of quantum physics associated with entanglement and quantum nonlocality.

7.2 *Projectivity*

Compare the guiding Eq. (3) with the same equation but with the denominator $\Psi^*\Psi$ omitted. The latter simpler equation might be regarded as in some sense more fundamental, with (3) itself regarded as arising from a convenient choice of gauge corresponding to the freedom of time-change associated with relational time.

The time-parameter corresponding to the use of the denominator in (3) has the nice feature that the dynamics using that time-parameter depends on fewer details of the wave function than would be the case if the denominator were deleted: with the denominator the dynamics depends only on the ray of Ψ, with Ψ and $c\Psi$ yielding the same dynamics for any constant $c \neq 0$. This has a particularly nice implication for the behavior of subsystems.

With this choice of time-parameter the dynamics for a subsystem will often not depend upon the configuration of its environment, with the subsystem evolving according to an autonomous evolution involving only the configuration and the (conditional) wave function of the subsystem itself (Dürr et al. 1992). This would happen when the subsystem is suitably decoupled from its environment, for example for a product wave function when there is no interaction between system and environment. Without the denominator this would not be true, and there would appear to be an additional nonlocal dependence of the behavior of a subsystem on that of its environment that would not be present with a time-parameter associated with the use of the usual denominator.

7.3 *Non-normalizability of the Wave Function*

Note that since it is translation and scaling invariant, the wave function in the Schrödinger gauge or in any of the gauges discussed in Sect. 6.2) are not normalizable. In other words, the corresponding $|\widehat{\Psi}|^2$ measures are non-normalizable. However, since the non-normalizability arises from unobservable (and, from a shape space point of view, unphysical) differences and dimensions it should somehow not be a problem.

Nonetheless, the real question is how the empirical distributions arising from the fundamental shape space level are related to those coming from the physics in a gauge. While the different gauges, such as the Schrödinger gauge, correspond to theories that, we argued, are empirically equivalent to the fundamental shape space theory, that was only in purely dynamical terms. We have not yet addressed the possible differences in empirical distributions that may arise. We would like to see that they don't, i.e., that probabilities naturally associated with relational physics on shape space yield familiar Born-rule probabilities when the physics is lifted to absolute configuration space.

There are several considerations that suggest that the non-normalizability should not be a genuine problem. First of all, as just mentioned, the non-normalizability arises only from non-observable dimensions, suggesting that it should be physically irrelevant. Moreover, it is the universal wave function $\widehat{\Psi}$ (in any of the gauges) that is not normalizable. But the universal wave function is rarely used in practice. In quantum mechanics we usually deal, not with the entire universe, but with small subsystems of the universe. The wave functions with which we usually deal are thus conditional wave functions, and there seems to be no reason why these should fail to be normalizable.

As already said, non-normalizable wave functions tend to occur in quantum cosmology. Such wave functions would normally be regarded as problematical and unphysical (since the formal structures of orthodox quantum mechanics, with their associated probabilities, are crucially based on the notion of a Hilbert space of square-integrable, i.e. normalizable, wave functions). However, for our analysis in Dürr et al. (2020) the nonnormalizability of the measure on configuration space turned out to be crucial for its success (see below). Hence what from an orthodox perspective is a vice is transformed into a virtue in relational Bohmian mechanics.

7.4 *Conditional Wave Function and Path Space*

To substantiate the last statement we mention a few ingredients of our analysis in Dürr et al. (2020), without any pretension of completeness.

7.4.1 The Conditional Wave Function

The very notion of conditional wave function (5) presupposes the possibility of expressing the configuration Q of the universe as a pair (X, Y) consisting of the configuration X of a system, the x-system, and the configuration Y of its environment (the y-system). The holistic character of shape space physics emphazized above does not allow for this, that is, there is no natural product structure

$$\mathcal{Q} = \mathcal{Q}_{sys} \times \mathcal{Q}_{env}$$

for shape space: Here the system is a collection of (labelled) particles with its own shape space $\mathcal{Q}_{sys} = \mathcal{X}$, the set of possible shapes X of the system, and the environment consists of the rest of the particles of the universe, with shape space $\mathcal{Q}_{env} = \mathcal{Y} = \{Q_{env} = Y\}$, with Y the shape associated with the particles (labels) of the environment. The crucial fact is that it is not true that

$$\mathcal{Q} = \mathcal{X} \times \mathcal{Y}.$$

X and Y don't involve sufficient information to determine the complete shape Q. What is missing is the spatial relationship between these shapes.

Nonetheless we have that $\mathcal{Q}$ can be identified with $\mathcal{X}_Y \times_y \mathcal{Y} = \{(X, Y) | Y \in \mathcal{Q}_{env}, X \in \mathcal{Q}_Y\}$, where $\mathcal{Q}_Y = \{Q \in \mathcal{Q} \,|\, Q_{env} = Y\}$. We may then define the conditional wave function for the subsystem, for $Y \in \mathcal{Q}_{env}$ and universal wave function Ψ, by

$$\psi(x) = \Psi(x, Y), \quad x \in \mathcal{Q}_Y. \tag{58}$$

This looks like the usual conditional wave function, but it is important to bear in mind that, unlike with the usual conditional wave function, here x represents the shape of the universe for a fixed Y and there are no obvious natural coordinates to efficiently describe it.

The conditional wave function (58) can be lifted to absolute configuration space and one has the formula

$$\widehat{\psi}(\boldsymbol{x}) = \widehat{\Psi}(\boldsymbol{x}, \boldsymbol{Y}), \tag{59}$$

Here the lifts $\boldsymbol{x}$ of x and $\boldsymbol{Y}$ of Y involve something akin to the choice of a frame of reference in absolute space, and the "hat" on Ψ refers to any of the gauge equivalent representations of the universal wave function that we have described in Sect. 6.2.

7.4.2 The Conditional Probability Formula

One might wonder whether something analogous to (6) holds true. The answer is yes, but with several provisos.

It turn out that in the gauges 1 and 3 described in Sect. 6.2, indeed something similar holds. More precisely, in the 3-gauge the conditional distribution is exactly given by the modulus square of the conditional wave function (59) as in (6). The proof of this is rather intricate and lengthy (Dürr et al. 2020).

In the Schrödinger gauge, however, a conceptual issue arises as to exactly what of physical significance this conditional distribution represents. After all, the transition to the Schrödinger gauge required relational time, but if we take relational time seriously, what is physical is not the configuration Q_t of the universe at some time t, but the geometrical path of the full history of the configuration, with no special association of the configurations along a path with times. In this (more physical) framework, the conditional distribution of the configuration X_t of a subsystem given the configuration Y_t of its environment is not meaningful.

What is meaningful is (i) a probability distribution $\mathbb{P}^{\Psi}$ on the space $\mathscr{P}$ of (geometrical, i.e. unparametrized) paths and (ii) the conditional distribution relative to $\mathbb{P}$ of the configuration X_Y of the subsystem when the path $\gamma \in \mathscr{P}$ has environmental configuration Y, *given* that the path passes through a configuration with environment Y, $Y \in \gamma$. In this more general framework, one arrives at a conditional probability formula analogous to (6), namely,

$$\mathbb{P}^{\Psi}(X_Y \in dx | Y \in \gamma) = |\widehat{\psi}_S(\boldsymbol{x})|^2 d\boldsymbol{x} \,, \tag{60}$$

where the subscript S refers to the Schrödinger gauge. Also the proof of this theorem is rather intricate and lengthy. An assumption of this theorem is what we have called in Dürr et al. (2020) the "existence of clock variables" for the Bohmian dynamics, which is a mild assumption for a sufficiently structured universe.

7.4.3 Non-normalizable Measures

A crucial ingredient of our analysis is the connection between a measure on path space and a (stationary) non-normalizable measure on configuration space. By the very nature of this connection the measure on configuration space and its associated wave function in fact has to be non-normalizable—just to rehearse the point made at the end of Sect. 7.3 that the nonnormalizability of the universal wave function is good!

7.4.4 The Physics of Sub-systems

On the absolute configuration space level the dynamics and the probabilities for subsystems should be of the usual form. While it is true that on the universal level the connection between $|\Psi|^2$ and probability, or, more precisely, typicality, would be broken, this would not be visible in any of the familiar every-day applications of quantum mechanics, which are concerned only with subsystems and not with the entire universe. This is the key to approach the problem of entropy increase raised in Sect. 3.1.

7.4.5 $|\psi|^2$ vs. $|\Psi|^2$

The patterns described by the quantum equilibrium hypothesis will be typical with respect to a measure, not on absolute configuration space, but on shape space, on the fundamental level, which is fine. There is a widespread misconception with respect to Bohmian mechanics that $|\Psi|^2$ for the universe and $|\psi|^2$ for subsystems play, physically and conceptually, similar roles. They do not, since the role of $|\Psi|^2$ is typicality while that of $|\psi|^2$ is probability. If this distinction is too subtle, the fact that, from a relational perspective, these objects live on entirely different levels of description, $|\Psi|^2$ on the fundamental level, i.e., on shape space, and $|\psi|^2$ on absolute configuration space, might make it easier to appreciate how very different they are.

7.5 *More Relational Points*

7.5.1 Identical Particles

There is one rather conspicuous relational aspect that we've ignored. For indistinguishable particles we should have taken one further quotient and enlarged the similarity group G to include the relevant permutations of particle labels. Presumably, this would not have any relevant implications in the classical case, but in the quantum case would lead to the correct quantum description of the dynamics of bosonic and fermionic shapes (Mellini In preparation).

7.5.2 Different Ontologies

We do not regard particles as the inevitable ultimate building blocks of the universe. Far from it. Our insistence on particles is mostly methodological, not metaphysical. The point is that the particle ontology already suffices to highlight the relevant conceptual issues of relational physics. Our analysis—in particular that concerning the wave function of the universe, subsystems, and probability—really depends only on rather general qualitative features of the structure of quantum relational physics, not on the details of any specific ontology.

7.5.3 Geometry

We have assumed that the geometrical relations among the particles are those of Euclidean geometry. But we know that the geometry of space need not be Euclidean and may not be fixed at all, but be dynamically determined by the matter content of the universe.

One is thus naturally lead to geometrodynamics, the dynamics of 3-geometries. According to John Archibald Wheeler, superspace, the totality $\mathcal{Q}$ of 3-geometries, should be regarded as the true arena of general relativity (see, e.g., Misner et al. 1973).

In many respects, superspace is analogous to shape space. The totality $\boldsymbol{\mathcal{Q}}$ of Riemannian metrics on 3-space is analogous to absolute configuration space, and the group G of diffeomorphisms of 3-space is analogous to the similarity group of Euclidean space. The fiber bundle structure $\mathcal{Q} = \boldsymbol{\mathcal{Q}}/G$ (modulo singularities) has been highlighted by DeWitt (1970).

Now DeWitt (1970) and many others after him have regarded four-dimensional spacetime as fundamental and the dynamics of 3-geometries as gauge, arising from the gauge freedom given by the so-called lapse and shift functions. However, one could take the opposite view and consider the 3-geometry as fundamental (analogously to a shape of particles) and four-dimensional spacetime as arising in a suitable gauge from a timeless history of the universe, i.e., a path, in superspace. This view, that has been explored by Barbour and collaborators (2002), leads to the conclusion that relativity (general and special) need not be regarded as fundamental, but can be regarded as arising from a suitable choice of gauge. Again we encounter here a striking manifestation of a Kantian component of physical laws.

7.5.4 Larger Groups

For the case of an N-particle universe, we have already indicated that a natural extension is to include in G the relevant permutations of particle labels. Are there other natural extensions? How far can the gauge group G be extended?

In the case of 3-geometries, Barbour and collaborators have explored the possibility of extending the group of diffeomorphism to include the conformal transformations of the 3-geometries. One arrives in this way at a conformal superspace (Barbour 2012; Gomes et al. 2011; Mercati 2018). They perform some massaging to arrive at general relativity (in a suitable gauge). But maybe there is no need of any massaging, nor any need of arriving exactly at general relativity—to recover it as an approximation of a more fundamental theory is sufficient.

7.6 The Quantum Gauge?

Is there a gauge that could be called the *quantum gauge*? We do not know the answer, but we find that the question itself is very interesting. It is related to the "why"-question we asked at the end of Sect. 3.3.

Let us explain what we mean. We started with the Bohmian motion on shape space governed by the universal wave function Ψ. Everything we discussed in the quantum case had as a basic ingredient the universal wave function Ψ. But one should contemplate the possibility of starting with something that is not really quantum, not a motion guided by a suitable Ψ.

Suppose we have a motion in shape space that is not of the usual Bohmian form, with no wave function Ψ involved in defining the law of motion of the configuration. In other words, suppose we discover a natural motion on shape space that is not quantum at all. But suppose also that when one takes into account the freedom of time change and the freedom of how the motion is lifted to absolute space, that is, that when one exploits all the gauge freedom we have discussed above, it turns out that the law of motion can indeed be cast into the usual Bohmian form, with a wave function guiding it in the usual way. If that were the case, quantum mechanics, while entirely absent from the fundamental (shape space) level, would arise as a choice of gauge in absolute configuration space. If this were the case, we would have another instance, maybe the most surprising, of what we have called a Kantian component of physical laws—that much of what we take as fundamental is not fundamental, but imposed by us by a proper choice of gauge.

In view of this—which at the moment is very speculative—and in view of what has been said before about relativity—much less speculative—it seems to us not unreasonable to entertain the following possibility: that quantum mechanics and relativity, the two great revolutions of twentieth century physics, may not be fundamental at all but imposed on nature by us.

References

Barbour, J. B., & Bertotti, B. (1982). Mach's principle and the structure of dynamical theories. *Proceedings of the Royal Society of London. Series A, 382*, 295–306.

Barbour, J. (2003). Scale-invariant gravity: particle dynamics. *Classical and Quantum Gravity, 20*, 1543.

Barbour, J. (2012). Shape dynamics. An introduction. In *Quantum field theory and gravity* (pp. 257–297). Springer.

Barbour, J., Foster, B. Z., & Murchadha, N. Ó. (2002). Relativity without relativity. *Classical and Quantum Gravity, 19*, 3217.

Bell, J. S. (1987). *Speakable and unspeakable in quantum mechanics*. Cambridge University Press.

Bohm, D. (1952). A suggested interpretation of the quantum theory in terms of "hidden" variables. I. *Physical Review, 85*, 166.

Bricmont, J. (2016). *Making sense of quantum mechanics*. Springer.

DeWitt, B. S. (1967). Quantum theory of gravity. *Physical Review, 160*, 1113.

DeWitt, B. S. (1970). Spacetime as a sheaf of geodesics in superspace. In *Relativity* (pp. 359–374). Springer.

Dürr, D., Goldstein, S., & Zanghì, N. (1992). Quantum equilibrium and the origin of absolute uncertainty. *Journal of Statistical Physics, 67*, 843–907.

Dürr, D., Goldstein, S., & Zanghì, N. (2012). *Quantum physics without quantum philosophy*. Springer.

Dürr D., Goldstein S., & Zanghi N. (2020). Quantum motion on shape space and the gauge dependent emergence of dynamics and probability in absolute space and time. *Journal of Statistical Physics, 180*, 92–134.

Gomes, H., Gryb, S., & Koslowski, T. (2011). Einstein gravity as a 3d conformally invariant theory. *Classical and Quantum Gravity, 28*, 045005.

Le, H., & Kendall, D. G. (1993). The Riemannian structure of Euclidean shape spaces: A novel environment for statistics. *The Annals of Statistics, 21*, 1225–1271.

Mercati, F. (2018). *Shape dynamics: Relativity and relationalism*. Oxford University Press.

Mellini, A. (2022). *Selected topics in shape dynamics and geometrodynamics*. Unpublished doctoral dissertation, University of Rome Tor Vergata.

Misner, C. W., Thorne, K. S., & Wheeler, J.A. (1973). *Gravitation*. W.H. Freeman (1973).

On Two Slights to Noether's First Theorem: Mental Causation and General Relativity

J. Brian Pitts

Abstract It is widely held among philosophers that the conservation of energy is true and important, and widely held among philosophers of science that conservation laws and symmetries are tied together by Noether's first theorem (that rigid symmetries yield conservation). However, beneath the surface of such consensus lie two slights to Noether's first theorem.

First, there is a 325+-year controversy about mind-body interaction in relation to the conservation of energy and momentum, with occasional reversals of opinion. The currently popular Leibnizian view, dominant since the late nineteenth century, claims to find an objection to broadly Cartesian views (and non-epiphenomenalist property dualism) in their implication of energy non-conservation. Here energy conservation is viewed as an oracle, an unchallengeable black box. But Noether's first theorem and its converse show that conservation and symmetry of the laws stand or fall together. Absent some basis for expecting conservation in brains that has a claim on the Cartesian (whose view implies the absence of law symmetries in brains), the objection is circular. An empirically based argument is possible, but is a different argument with little force except insofar as it is rooted in neuroscience.

Second, General Relativity has a 100+-year-long controversy about whether gravitational energy exists and is objectively localized. The usual view is that gravitational energy exists but is not objectively localized, though some deny its existence. Without positive answers to both questions, generally applicable conservation laws do not exist: energy is not conserved. This conclusion is startling

For Springer volume *Rethinking the Concept of Laws of Nature: Natural Order in the Light of Contemporary Science* edited by Yemima Ben-Menahem.

J. B. Pitts (✉)
University of Lincoln, Lincoln, UK

University of Cambridge, Cambridge, UK

University of South Carolina, Columbia, SC, USA
e-mail: bpitts@lincoln.ac.uk.; jbp25@cam.ac.uk

Y. Ben-Menahem (ed.), *Rethinking the Concept of Law of Nature*,
Jerusalem Studies in Philosophy and History of Science,
https://doi.org/10.1007/978-3-030-96775-8_10

in itself and a problem for conserved quantity theories of causation. Yet Noether's first theorem applies to General Relativity, which has uncountably many symmetries of its laws and so has conservation laws, indeed uncountably many of them. Many authors downplay these laws due to their quirky properties; some authors even attempt to explain the laws' supposed nonexistence in terms of an absence of symmetries of the geometry, which is a distraction. Thus Noether's first theorem is widely ignored, left uninterpreted, or distorted in relation to General Relativity. Taking the theorem seriously seems possible, however, restoring the conservation of energy, or rather, energies.

How do these controversies relate? One sometimes finds claims that General Relativity's supposed lack of conservation laws answers Leibniz on behalf of Descartes. Taking seriously the superabundance of formal conservation laws in General Relativity, however, suggests that General Relativity resists (not facilitates) mind-to-body causation. This conclusion can be proven apart from interpretive controversies. The resistance is, however, finite and tends to be swamped by larger world-view considerations.

Keywords Conservation laws · Noether's first theorem · Philosophy of mind · Dualism · Cartesianism · Interactionism · Gravitational energy

1 Introduction

This paper considers two areas where Noether's first theorem tends to be neglected, one in the philosophy of mind, one in physics and the philosophy of physics, and finds the two debates mutually illuminating. In the philosophy of mind, there is a tendency to believe that physics has revealed that energy is conserved, full stop, whereas Noether's first theorem implies (*inter alia*) that energy is conserved if and only if the laws are the same at all times—a claim that sometimes is and ought to be denied by interactionist dualists, against whom energy conservation is often invoked. In General Relativity, on the basis of traditional arguments there is a tendency to downplay the conservation laws that at least formally exist according to Noether's first theorem, but these traditional arguments have become less convincing in the last decade or two. My goal in discussing interactionist dualism is not to make a defense of the position, but simply to critique a very widely received claim that the view has been refuted by physics, in particular by conservation laws.

The energy conservation objection to nonphysical mental causation has been made from the 1690s (Leibniz 1997) to the 2010s (Churchland 2011). According to Leibniz's *Theodicy*,

> ...two important truths on this subject have been discovered since M. Descartes' day. The first is that the quantity of absolute force which is in fact conserved is different from the quantity of movement, as I have demonstrated elsewhere. The second discovery is that the same direction is still conserved in all bodies together that are assumed as interacting, in

> whatever way they come into collision. If this rule had been known to M. Descartes, he would have taken the direction of bodies to be as independent of the soul as their force; and I believe that that would have led direct [*sic*] to the Hypothesis of Pre-established Harmony, whither these same rules have led me. For apart from the fact that the physical influence of one of these substances on the other is inexplicable, I recognized that without a complete derangement of the laws of Nature the soul could not act physically upon the body. (Leibniz 1985, p. 156)

To paraphrase, Descartes's volume $\times$ speed quantity is not conserved, but momentum (mass times velocity, including direction) is conserved, as is 2$\times$kinetic energy (*vis viva*), so the soul's deflecting matter while leaving its speed unchanged violates the newer notion of conservation. Being a rationalist, Leibniz had high aesthetic standards for physics and metaphysics. While co-inventing calculus, Leibniz did not inherit (as we do) an expectation that physical laws would come as differential equations that generically cannot be solved exactly. Hence what he counted as a "complete derangement" might not be nearly so disturbing nowadays.

While philosophy of mind texts often leap from Leibniz to the twentieth century as though nothing happened in between except perhaps more of the same, this objection has waxed, waned and waxed again over time. After much influence for a season in the eighteenth century German world, this objection, resisted explicitly by Crusius and Knutzen (Watkins 1995, 1998) and implicitly by Newton and Euler (the best physicists in the world) (Pitts 2020d), was defeated in the eighteenth century, restoring interactionism as the default position. It probably helped that the conservation of *vis viva* (proto-energy) in that era was a rejected Leibnizian opinion rather than a generally accepted law. In the nineteenth century, as the conservation of *vis viva* was rehabilitated as the conservation of energy, the first law of thermodynamics, the Leibnizian objection reappeared. This reappearance seems to owe as much to cultural opportunism and the *Zeitgeist* (including the carefully cultivated Baconian inductive naivité) as anything else; the question whether one should make exceptions for spiritual influence is clearly untouched by Joule's paddle wheel experiment. Newtonians had always been willing to accept mental forces, which would have falsified momentum conservation, so doing the same for energy conservation would not have been at all difficult. And yet the conservation of energy has been held to have been more damaging to world-views involving spirits (including but not only Christianity) than was Darwin (Turner 1974, p. 27). Some good physicists, including Boussinesq and Maxwell, questioned or rejected the argument (van Strien 2015), but without the intensity of its scientific naturalist proponents. Much of the story has been told for the German-speaking world (Heidelberger 2004; Wegener 2009), though often without the normative coloration that I would give.

While common in the philosophy of mind, the traditional (not general relativistic) argument from conservation against dualism is rarely made by experts in physics even when they discuss the mind. Thus noted philosopher of physics Jeremy Butterfield writes:

> ... [A] traditional argument against interactionism is flawed, because of this false picture of physics.... The idea is that any causal interaction between mind and matter would violate the principle of the conservation of energy.... But, says the argument, physics tells us that

> energy is conserved in the sense that the energy of an isolated system is constant, neither increasing nor decreasing.... And there is no evidence of such energy gains or losses in brains. So much the worse, it seems, for interactionism. (Though traditional, the argument is still current; for example, Dennett endorses it (1991, pp. 34–35).) This argument is flawed, for two reasons. The first reason is obvious: who knows how small, or in some other way hard to measure, these energy gains or losses in brains might be? Agreed, this reason is weak: clearly, the onus is on the interactionist to argue that they could be small, and indeed are likely to be small. But the second reason is more interesting, and returns us to the danger of assuming that physics is cumulative. Namely: the principle of the conservation of energy is not sacrosanct. ... [A]lthough no violations have been established hitherto, it has been seriously questioned on several occasions. It was questioned twice at the inception of quantum theory.... And furthermore, it is not obeyed by a current proposal ... for solving quantum theory's measurement problem. In short: physicalists need to be wary of bad reasons to think physicalism is true, arising from naivety about physics. (Butterfield 1997)

The next section will elaborate on Butterfield's remark.

To be brief, the main problem is circularity or question-begging: when one understands where conservation laws come from in terms of Noether's first theorem, what seemed to be a scientific fact, categorical energy conservation, is replaced by a biconditional relation between symmetries and conservation laws, with no dialectically useful basis for assuming symmetries. If there are souls that act on bodies (and not in the same way at all times and places), then the symmetries of the laws fail: my willing to raise my arm involves my soul's doing something to my brain during my lifetime, but does not affect William the Conqueror or the Moon (at least not directly in the case of the Moon). So one needs to assume the falsehood of interactionism to justify the symmetries of the laws, from which conservation follows by Noether's first theorem. But then one has begged the question. It is of course possible that neuroscience provides good reasons to reject interactionist dualism, but that is a quite different argument, one based on evidence about the brain.

One might wonder why this paper pays no attention to quantum mechanics and quantum field theory. As explained in more detail elsewhere (Pitts 2020a), one reason is that it is not very clear what to say about conservation in quantum mechanics and quantum field theory, in light of the measurement problem, (on some views) the collapse of the wave function, and the like, as well as the uncertainty relations. Some authors have invoked quantum mechanics' arguable non-conservation to make room for mental causation; I do not need to take a stand on the issue. I think that I have something interesting to say about energy conservation and mental causation (namely, that the Leibnizian argument fails, regardless of what one says about quantum mechanics and quantum field theory), something interesting to say about General Relativity and energy conservation (namely, that the formal conservation laws that one gets by a Noether-type recipe can be taken more seriously than most people think), and something interesting to say about General Relativity and mental causation (namely, that General Relativity makes mental causation harder, not easier, a conclusion that coheres with taking general relativistic conservation laws seriously but does not depend on doing so), but I do not have anything interesting to say about quantum mechanics and energy conservation.

While General Relativity, being a classical rather than quantum field theory, is not a fundamental feature of the world, it is good practice to understand the theory and to try to make its quantum successor "quantum gravity" (whatever that turns out to be) as much like General Relativity as possible—unless of course General Relativity is wrong even apart from quantum mechanics, a conclusion far more widely entertained in the physics literature now than in, say, 1997.

One might also recall the Steady State cosmological theory as an example of an energy non-conserving theory more in the vicinity of GR.[1] This feature was not particularly satisfactory even to the theory's proponents, however, for later versions posited a "creation field" C with negative energy (Narlikar 1993, pp. 241–246), so that energy could be conserved, matter could acquire more energy, and the C field's energy could become more negative.

2 Conservation and Symmetry: The Modern View

From the standpoint of the philosophy of physics, a striking weakness in traditional discussions of conservation and the philosophy of mind is the failure to connect with the now-standard (especially 1910s+) relation between symmetries and conservation laws that was put into a recognizably modern form by Herglotz, Mie, Born and Noether (Born 1914; Herglotz 1911; Mie 1913; Noether 1918; Pitts 2016a). (Some sources on the history of the relationship between symmetries and conservation laws exist (Houtappel et al. 1965; Kastrup 1987; Kosmann-Schwarzbach 2011).) In the 1910s the nineteenth century mathematical tradition of analytical mechanics (Lagrange, Hamilton, Jacobi) at last became a standard part of physics, and the ideas were widely understood as applied to continua (whether elastic media or fields such as the electromagnetic and gravitational potentials). I have explained this textbook material (Goldstein 1980, chapter 12; Davis 1970) as clearly as I could (at times working with Alin Cucu) in application to the philosophy of mind (Cucu and Pitts 2019; Pitts 2020a, 2021), so I will be brief here. One infers the laws from a mathematical function of space and time, now called the Lagrangian density $\mathcal{L}$, by taking a curious kind of derivative of the Lagrangian density with respect to the fields/potentials and their temporal and spatial derivatives (rates of change). These are called the Euler-Lagrange equations, which are field equations and/or equations of motion. Given the Lagrangian density, one can discern the conservation (or otherwise) of energy and momentum by inspection. In the early 1910s Max Born expressed how rigid translation symmetries imply conservation laws:

> The assumption of Mie just emphasized, that the function [the Lagrangian density $\mathcal{L}$] is independent of x, y, z, t, is also the real mathematical reason for the validity of the momentum-energy-law. ... We assert that for these differential equations, a law, analogous

[1] Thanks to Patrick Duerr for helpful remarks here and elsewhere. All conclusions are my own.

> to the energy law (3′) of Lagrangian mechanics, is always valid as soon as one of the 4 coordinates x_α does not appear explicitly in [$\mathcal{L}$]. (Born 1914)

In this context the "momentum-energy law" is not a global statement that $E = constant$ at each time or the like, but a local statement. In particular, it is a claim (for each time and place) that the temporal rate of change of the density of the conserved quantity in question (energy or a component of momentum) plus the tendency of that quantity to spew out of a place (its "divergence," the sum of the spatial rates of change of the energy or momentum flux) equals 0. A bit of the mathematics will appear later. To give an analogy, if people are walking into a room faster than people are leaving it, then the number of people in the room increases. This seems obvious; what does it exclude? It excludes instantaneously disappearing in one place and reappearing elsewhere, whether at the same time or later; more generally, it forbids disappearance into nothingness and forbids appearance out of nothingness. Instead energy and momentum flow in accord with the laws of motion/field equations derived from the Lagrangian density. (The same can be said of angular momentum, which follows from a rotational symmetry of the Lagrangian density.) So, at any rate, claims the continuity equation. In some simple cases, with spatially isolated matter, one can add up (spatially integrate) the continuity equation over the whole universe and infer that the total energy is constant, a more familiar claim from secondary school chemistry, but one with less content and less assurance of making sense than the continuity equation (Peebles 1993, p. 139).

The continuity equation, being an equation (or a handful of them for a handful of conserved quantities), says much more than the $E = constant$ darling of many philosophers of mind, so the prospects for a conservation law-based objection to broadly Cartesian mental causation seem brighter. In particular, one strategem that one sometimes sees among dualists, having energy nonconservation due to the mind be compensated by the opposite nonconservation elsewhere, is clearly excluded. This proposal was never worth much anyway. Where does this nonconservation happen—the center of the Earth? How does this other place know to zag when my soul zigs? Matter presumably would not know what my soul is doing; is there a slave-soul that does the opposite of what I do? Is the machinery run by angels? And why find an apparently unobservable compensating violation of conservation laws at all relevant or comforting, given that the conservation of energy is supposed to be empirically motivated by observable events? Surely it was always better to bite the bullet (as a few hardy souls have) even in the late nineteenth, twentieth, and twenty-first centuries and simply deny conservation in the context of mental causation. The preeminent example of doing so intelligently was the Averill and Keating paper (Averill and Keating 1981), which made some contact with the physics literature and the relation between symmetries and conservation laws. While a favorite with some interactionist dualists, a rather beleaguered band until recently, this paper tended to be drowned out by people flaunting their knowledge of secondary-school chemistry. The conservation objection has been very popular, and even interactionists have tended to try to uphold conservation (often unsuccessfully (Cucu and Pitts 2019; Pitts 2020a)).

Hence the question of where and when the continuity equation is true can be postponed no longer. Expressing Born's point a bit more clearly, one can say the following. Wherever and whenever the Lagrangian density does not depend explicitly on time t, energy is conserved in the sense of the continuity equation. Wherever and whenever the Lagrangian density does not depend explicitly on the spatial coordinate x, momentum in the x-direction is conserved; likewise for y and z. On the other hand, wherever and whenever the Lagrangian density depends explicitly on t, x, y, or z, energy or the corresponding momentum component is *not* conserved. The claim that symmetries imply conservation laws was later encapsulated as Emmy Noether's first theorem, though Born's special case is sufficiently general for our purposes. Noether also proved the converse, that conservation laws imply symmetries (Brading 2001; Brown and Holland 2004; Kosmann-Schwarzbach 2011; Noether 1918; Romero-Maltrana 2015). That sounds more surprising, but this converse was already evident for the cases treated by Born.

So what of the dialectic of the energy conservation objection? The objector is making an argument, something that ought to provide some reason for a person initially sympathetic to interactionist dualism to give up that view. Consequently, it would be awkward to use a premise that is exactly as acceptable as the denial of interactionist dualism in order to motivate the denial of interactionist dualism. An argument like

P. Interactionist dualism is false.
C. Interactionist dualism is false.

would clearly not be impressive, despite its validity. But once one understands the biconditional relationship between symmetries and conservation laws due to Lagrangian field theory and Noether's first theorem, one finds the following popular argument similarly unimpressive.

P1. If interactionist dualism is true, then energy and momentum conservation are false.
P2. Energy and momentum conservation are true.
C. Interactionist dualism is false.

This is a valid argument using *modus tollens*. But once one understands Lagrangian field theory and Noether's first theorem, premise 2 ceases to be the deliverance of an oracular black box, Science, and looks instead like an assumption that interactionist dualism is false—which was supposed to be a conclusion rather than a premise. What went wrong? The dualist has no reason to grant that the laws of nature are the same everywhere and always. Your soul, if interactionist dualism is true, causes the Lagrangian density to vary with time and place within your brain so that something physical behaves in a way that it otherwise wouldn't have, in order to implement your decision to raise your arm. Hence the Leibnizian objection is indeed question-begging, a claim typically made only by the heartiest of theists and metaphysicians in recent centuries (Ducasse 1960, p. 89; Larmer 1986; Plantinga 2007), but demonstrably true in light of Lagrangian field theory and Noether's first theorem and its converse. The inverse correlation between immersion in physics and commitment to the Leibnizian argument is not accidental, once one distinguishes between domesticated and genuine physics (Ladyman et al. 2007, p. 24). The foundations of physics show conservation laws to be conditional upon symmetries, symmetries in which there is no reason to believe (in the relevant exceptionless form[2]) if one thinks that souls act upon bodies. Hence when one opens the black box of conservation laws, one finds a mirror reflecting one's own opinion about whether there are souls that act on bodies. Little or no material for an objective argument against interactionism remains. If we take Noether's first theorem seriously, we find that the Leibnizian conservation objection dissipates.

Some readers will be rightly disappointed by so vague an expression as that something physical in your brain behaves in a way that it otherwise wouldn't have. Surely one should say which part of the brain does what, and even provide neuroscientific evidence that this actually happens? Indeed, interactionist dualists would be better situated, the more plausibly they could tell a neuroscientific story that fills in the details, especially if empirical evidence for such a story could be found. My aim is not to defend interactionism, but to remove a popular faulty objection to help the discussion be carried out on the proper terrain, which is *a posteriori* and neuroscientific. But that discussion can hardly occur in a balanced way as long as the authority of physics has been wrongly claimed so as to stack the deck against interactionism. Neuroscience poses a potentially serious problem for interactionism, but one in which conservation laws play no essential role.

The issue of conservation laws in General Relativity also bears upon the philosophy of mind. That connection will be explored towards the end of this essay. But first the issue of conservation in General Relativity itself, a matter of longstanding controversy, merits discussion. There one finds a second slight to Noether's first theorem. As it happens, a new philosophy of mind-type example will help to illuminate the (shortage of) content of a well known general relativistic law sometimes offered as a substitute for the conservation of energy and momentum.

3 Noether's First Theorem and General Relativity: Another Slight

The story about a Lagrangian density $\mathcal{L}$ and conservation's holding insofar as the space-time coordinates t, x, y, z do not appear in $\mathcal{L}$ applies at least formally in Einstein's General Relativity (GR), our usual (though more often challenged

[2] The assumption is that a prudent dualist will hold that nature is uniform except in some small spatio-temporal regions, presumably inside our brains, and that such nonuniformity in our brains is sufficiently large to make a difference but sufficiently small not to be obvious. Nothing in the interactionist dualist position as such prohibits thinking that telekinetic powers are ubiquitous, such that anyone could lift a modest spacecraft out of a swamp like Yoda in *The Empire Strikes Back*, or that our souls render the Lagrangian density so strongly dependent on space and time in our heads that we can radiate as much energy as a nuclear reactor. Such views would have the 'virtue' of falsifiability. But everyone has always known from experience that these claims are false.

since 2000 than previously) theory of gravitation (Papapetrou 1948; Schmutzer 1972). At some level this has been understood since the 1910s, but it is frequently forgotten and its consequences have been kept at arm's length due to the difficulty in interpreting the resulting mathematical expressions. (Some possible leads into the literature are Szabados (2009), Pitts (2010), Chen et al. (2015). The natural initial expectation is that one could use the usual machinery of Lagrangian field theory, whether the entry-level "canonical tensor" (Goldstein 1980, chapter 12) or perhaps an upgraded expression due to Belinfante or Rosenfeld which suffices for other local field theories such as electromagnetism.

It was realized in the 1910s, however, that the canonical tensor, which for GR gives Einstein's "pseudotensor" using Einstein's Lagrangian density,[3] has some rather peculiar properties. In particular, one could apparently make gravitational energy disappear around a heavy round body by choosing coordinates with that aim in mind (Cattani and De Maria 1993; Schrödinger 1918). One could also apparently make gravitational energy appear in empty flat space-time by using spherical coordinates (Bauer 1918; Cattani and De Maria 1993). Indeed one can make the gravitational energy density $\mathfrak{t}^0_0$ vanish at any point in space-time by a choice of coordinates. Presumably nothing real can be created or destroyed through mere descriptive choice. The idea that all coordinate systems are admissible and equally good in GR seemed to imply that anything worth discussing in the theory should be tensorial (broadly construed), that is, should admit changing its components in one coordinate system into another coordinate system in an algorithmic way—which is to say, should be a "geometric object" in the sense of classical differential geometry (Anderson 1967; Duerr 2021; Nijenhuis 1952; Pitts 2006; Read 2022). (A geometric object's transformation law can be affine rather than linear, can be nonlinear, or can involve higher derivatives, for example. The idea arose in the 1930s as a generalization of tensor calculus, which General Relativity made an important branch of mathematics.) A familiar case is the metric tensor, a machine that allows one to infer real spatio-temporal distances (squared) from coordinate displacements, keeping in mind that coordinates might not be distances, but could involve angles, or not be orthogonal, or slosh back and forth in an arbitrary way. The metric tensor components transform like this:

$$g'_{\mu\nu} = g_{\alpha\beta} \frac{\partial x^{\alpha}}{\partial x^{\mu'}} \frac{\partial x^{\beta}}{\partial x^{\nu'}}.$$

One easily sees that

$$\sum_{\mu'=0}^{3} \sum_{\nu'=0}^{3} g'_{\mu\nu} dx^{\mu'} dx^{\nu'} = \sum_{\mu=0}^{3} \sum_{\nu=0}^{3} g_{\mu\nu} dx^{\mu} dx^{\nu} = ds^2,$$

[3] There is a measure of nonuniqueness, which can sometimes be physically interesting, for the Lagrangian density. That seems true for GR (Pitts 2022; Ohanian and Ruffini 1994, p. 647).

the invariant line element. It was concluded that gravitational energy is not "localizable," because the total energy could be viewed as behaving reasonably while the local description from Einstein's pseudotensor t^{μ}_{ν} could not be trusted due to the above-mentioned false negative and false positive and related tendencies. Gravitational energy-momentum pseudotensors not only are not tensors, but also are not even geometric objects (Bergmann 1958; Anderson 1967, p. 428). How, then, one can take their claimed local descriptions of gravitational energy seriously? This nonlocalizability doctrine has remained the majority view to this day.

There later arose a second objection to gravitational energy localization, namely, that the gravitational energy-momentum pseudotensor is not unique (Landau and Lifshitz 1975; Goldberg 1958; Anderson 1967, pp. 423–430). There are various expressions, indeed infinitely many, it turned out eventually, with arguably comparably good credentials. Which, if any, was the real gravitational energy? With two serious problems, localization seemed hopeless to most.

Some philosophers have gone further, proposing the bold but clear nihilist view that there is no such thing as gravitational energy (Duerr 2019a,b, 2021; Hoefer 2000), because nothing real could have such contradictory properties. Hence one should not say, on this view, that binary black holes or other dense bodies radiate away mass-energy during inspiral, and we detect this energy with LIGO and the like. Rather, one should simply say that binary black holes radiate gravitationally and we can detect this radiation. The bookkeeping device of energy is rejected. One can certainly express the content of Einstein's equations without discussing gravitational energy; nothing is lost except some puzzling and anachronistic bookkeeping, one might think.

Still other authors have introduced a red herring in the discussion by hinting of a sort of rival Noether theorem that supposedly licenses non-conservation in GR. These authors bring up the idea of symmetries (or the lack thereof) of the *geometry*, as though a spatio-temporally varying metric were relevant to the existence of conservation laws (Carroll 2010; Hossenfelder 2016; Logunov and Folomeshkin 1977; Maudlin et al. 2020; Motl 2010; Physics Stack Exchange 2017; Siegel 2018). Because the metric of GR lacks symmetries, one should not expect conservation laws, we are told. But Noether's first theorem does not know or care about geometry; it cares only about symmetries of the action (Noether 1918) (equivalently, of the laws) or, if one permits nonvariational fields, whether the nonvariational fields have symmetries (Trautman 1966). If it turns out, e.g., that geometry is conventional—perhaps there is no fact of the matter whether the Einstein 'frame' or the Jordan 'frame' is correct in Brans-Dicke scalar-tensor gravity, partly because there is no fact of the matter whether the variable gravitational 'constant' is geometrical or material—Noether's theorem is perfectly applicable, giving tolerable results either way, because geometry is simply not a relevant category to Noether's theorems. To ask for a field with Euler-Lagrange equations to have symmetries as well in order to have conservation laws, is to require supererogation. At this point one has moved beyond denying that Noether's first theorem has a reasonable interpretation to distorting its content. Hence the nonlocalizability and nihilist views should not be cluttered with such rationalization. There *is* a connection between symmetries of the geometry and conserved quantities *that have uniqueness or coordinate independence properties that most people like*, but that is rather different from having conserved quantities *simpliciter*.

3.1 Giving Up Energy-Momentum Conservation?

One should understand clearly that the cost of either the nonlocalizability view or the nihilist view is simply *renouncing the claim that energy is conserved in general.*[4] Why so? A local conservation law worthy of the name is a law that, when spatially integrated in cases with localized matter, sufficiently rapid falloff of the fields to trivial values, *etc.*, implies secondary school chemistry claims such as $E = constant$, a global conservation law. The kind of law that can yield $E = constant$ is the continuity equation, which implies the absence of sources or sinks. But the only continuity equation that GR implies is one that involves both material stress-energy $T^{\mu\nu}$ (big T), which is usually considered unproblematic, and gravitational (pseudo-?)energy $t^{\mu\nu}$ (little t). Earlier I used the Gothic letter with one index up and one down, $\mathfrak{t}^{\mu}_{\nu}$, which is better for technical reasons (Pitts 2022; Sorkin 1977). Hence the best form of the continuity equation involves the Gothic[5] material energy $\mathfrak{T}^{\mu}_{\nu}$ and the Gothic gravitational energy $\mathfrak{t}^{\mu}_{\nu}$; their sum $\mathfrak{T}^{\mu}_{\nu} + \mathfrak{t}^{\mu}_{\nu}$ satisfies the continuity equation:

$$\sum_{\mu=0}^{3} \frac{\partial}{\partial x^{\mu}} (\mathfrak{T}^{\mu}_{\nu} + \mathfrak{t}^{\mu}_{\nu}) = 0.$$

One can already see that no conservation law is generally forthcoming if one tries to write it in terms of the material stress-energy only; moving the gravitational pseudotensor to the right side gives an equation for the *nonconservation* of *material* energy-momentum:

$$\sum_{\mu=0}^{3} \frac{\partial}{\partial x^{\mu}} \mathfrak{T}^{\mu}_{\nu} = - \sum_{\mu=0}^{3} \frac{\partial}{\partial x^{\mu}} \mathfrak{t}^{\mu}_{\nu} \neq 0.$$

[4] Duerr argues, using a notion of inertial frames, that one can be an antirealist about nonconservation where it occurs formally (Duerr 2019a).

[5] One can (de-)Gothicize and raise and lower indices by matrix manipulations using the metric tensor $g_{\mu\nu}$. An index is moved down or up by multiplication with $g_{\mu\nu}$ or its matrix inverse $g^{\mu\nu}$, respectively. One can make a normal quantity Gothic by multiplication by $\sqrt{-g}$ (g being the determinant of the matrix $g_{\mu\nu}$), thus making it a density of weight 1. Apart from the minus sign due to space-time rather than space, this quantity $\sqrt{-g}$ is familiar in volume integration in spherical coordinates as $r^2 sin\theta$, for example. For the richer world of densities of arbitrary weight, which are often useful and sometimes important, see (Anderson 1967; Schouten 1954).

Thus material energy-momentum fails to be conserved due to the spatio-temporal variation of the gravitational potential/space-time metric, in general. In short, energy and momentum just are not conserved. One can perhaps learn to accept such a conclusion, discarding what little many people remember from secondary school chemistry (I imagine), but is it necessary or advisable?

Some authors would comfort us with a material energy-momentum balance law, which can be written in a tensorial way (no pseudotensor t) in terms of the material stress-energy and the space-time metric: $\nabla_\mu T^{\mu\nu} = 0$ or some equivalent expression (recalling the games playable with Gothic letters and powers of the metric tensor to move indices). Such an equation is indeed true in GR, important, and even entailed by Einstein's field equations. But it should not provide us that much comfort. First, though some people call this a conservation law (more often "covariant conservation," which is in most circumstances a special form of nonconservation), it generically does not imply the conservation of anything like $E = constant$ (Weyl 1922, pp. 236, 269–271; Landau and Lifshitz 1975, p. 280; Misner et al. 1973, p. 465; Lord 1976, p. 139; Stephani 1990, p. 141). Even playing with Gothic letters optimally, one cannot make this equation turn into the continuity equation.[6] Second, it is not very distinctive of Einstein's equations, following simply from the way that matter and gravity couple without need for Einstein's field equations for gravity (Wald 1984, p. 456). GR *is* distinctive in that $\nabla_\mu T^{\mu\nu} = 0$ separately follows from Einstein's equations, unlike most other theories of gravity (such as massive spin 2 gravities (de Rham et al. 2011; Freund et al. 1969; Hassan and Rosen 2012; Ogievetsky and Polubarinov 1965; Pitts 2016b)); that is quite interesting. The third reason requires a more detailed discussion.

Third, the content of the energy-momentum balance equation just isn't enough to rule out the kinds of phenomena that many people have expected conservation laws to rule out. For example, it is reported that disembodied spirits make rooms cold; philosopher Stephen Braude, who accepts such phenomena and reports witnessing a table tilting in graduate school, argues that such phenomena are compatible with energy conservation due to cold breezes and changes in the weight of the medium (Braude 1986, 1987; Braude et al. 2017). Accomplished psychiatrist Richard Gallagher informs us that in an exorcism, the demon(s) can make the room cold or hot (Gallagher 2020, p. 65). I have pointed out that local conservation laws, being more demanding than the global conservation presumably envisaged by Braude and being equivalent to the uniformity of nature as formulated in classical field theory, would exclude such phenomena (Pitts 2020a). (Gallagher, a Catholic, might take the Evil One and his minions to be exempt from conservation laws.) What seems not to have been pointed out, however, is that the material energy-momentum balance

[6] One sees this point argued invalidly with surprising frequency even in the physics literature, but also in the philosophy literature. For electric charge one *can* turn the conservation equation into the continuity equation by playing with Gothic letters and index position (Anderson 1967; Duerr 2019a). One has to use a weight 1 current with index up (contravariant/tangent) in order to remove as many Christoffel symbols as possible, and then check whether any remain. For energy-momentum, one can get rid of one of the two terms but not the other one in this fashion.

law $\nabla_\mu T^{\mu\nu} = 0$ is perfectly compatible with the idea that spirits act on the physical world *at any and all times and places and in arbitrarily strong ways, as long as it is only on space-time/gravity* $g_{\mu\nu}$. This is evident from inspection of an action of the form

$$S_{gravity}[g_{\mu\nu}, \Psi_{spirits}] + S_{matter}[u, g_{\mu\nu}],$$

where u standards for all physical fields other than $g_{\mu\nu}$; the equation $\nabla_\mu T^{\mu\nu} = 0$ follows from the second term in the usual way, notwithstanding all the spiritual influence on $g_{\mu\nu}$ in the first term. One could make a similar point in terms of Humean non-uniformities rather than spirit-to-matter causation: the course of nature might change in ways that involve the dynamics of $g_{\mu\nu}$ but not the coupling of $g_{\mu\nu}$ to matter, and $\nabla_\mu T^{\mu\nu} = 0$ would still hold. The point is that $\nabla_\mu T^{\mu\nu} = 0$, though a lovely and important equation, is surprisingly weak in content in some respects, making it unable to ground some of the claims that the conservation of energy and momentum would ground and that many people have wanted conservation laws to ground.

3.2 Why Gravitational Energy Might Be Localized After All

In my view, the case for taking gravitational energy seriously has gotten much stronger in the last few decades, because new replies to the two traditional objections have become available. I start with the nonuniqueness objection, for which I see three or four possible answers. An initial reply, which appeals to the widespread acceptance of material energy $T^{\mu\nu}$, is a *tu quoque* response (Pitts 2010): gravitational energy is not qualitatively worse off than supposedly unproblematic material energy. Even a scalar field in flat space-time suffers from nonuniqueness due to a multiplicity of comparably plausible candidates, due to the "improved" energy-momentum tensor, which has certain advantages (Callan et al. 1970). While 2 is less than ∞, it is still enough to run a parallel argument against the reality of non-gravitational energy. If that argument is not accepted, then at least the nonuniqueness objection to GR pseudotensors is blunted: nonuniqueness is compatible somehow with reality of the energy distribution in question, even if how has not been specified. A second and more powerful reply comes from the work of Nester et al., according to whom different pseudotensors describe different quasi-localizations with physical meaning tied to boundary conditions (Chang et al. 2000; Nester 2004). Thus different pseudotensors are right in different contexts. Why should the same one be required in every context, given the close relationship between pseudotensors and boundaries? A third possible reply is that there is a best One True pseudotensor. Perhaps it is the Papapetrou-Belinfante pseudotensor (Papapetrou 1948) or a higher-tech relative thereof (Petrov and Katz 2002). Clearly this third reply is incompatible with the second, but one can simply offer their disjunction, or even parts of each: maybe some pseudotensors are always wrong, but others are right in one context or another. A fourth reply is rooted in old work (Bergmann 1958): nonuniqueness is not a distinct objection, but only a repeat of coordinate dependence. Bergmann noted that "the totality of all conservation laws ... in one coordinate system is equivalent to one of them, stated in terms of all conceivable coordinate systems." (Bergmann 1958). The "totality of all conservation laws" refers to different pseudotensors. He shows how to find the Einstein and the Landau-Lifshitz pseudotensors in his expression by choosing $\delta x^\sigma = k^\sigma$ (where k^σ is a set of constants) or $\delta x^\sigma = \mathfrak{g}^{\sigma\alpha} k_\alpha$ (where k_σ is a set of constants). Schutz and Sorkin also find both the Einstein and Landau-Lifshitz pseudotensors as special cases of their Noether operator (Schutz and Sorkin 1977). Especially if one has a reply to the coordinate dependence objection in terms of infinitely many energies, reducing the nonuniqueness objection to the coordinate dependence objection helps.

Turning to the coordinate dependence objection, that pseudotensoriality is bad—one could phrase this in terms of pseudotensors' not being even geometric objects (Anderson 1967, p. 428) and hence not physically real—one might consider what service is performed by a coordinate transformation rule (Pitts 2010). It is clear that a transformation rule does not *create* physical meaning. I could take a number at each point in some coordinate system, $x^1 x^2 x^3 sinh x^0$, postulate a transformation rule (perhaps a scalar density of weight $\frac{43}{89}$, also specifying whether to take absolute values),[7] and have a geometric object: components relative to any coordinate system. (I have assumed that my initial coordinate system covers the whole manifold; otherwise the story is slightly more complicated, but no more illuminating.) A transformation rule, rather, indicates equivalence: a geometric object tells the same story in every coordinate system. A physically meaningful (or meaningless) story in one coordinate system is told equivalently, equally meaningful (or meaningless) in another coordinate system. For a gravitational energy-momentum pseudotensor, the ostensible physical meaning comes not from the transformation rule, but from the algorithmic derivation from the Lagrangian, the realistic interpretation afforded to the results for most or all other field theories, and the *prima facie* expectation of a uniform interpretation for all classical local field theories.[8] But is being a geometric object perhaps a *necessary* condition for physical reality?

It is important to ask how many conserved quantities we should expect. Noether's first theorem tells us to associate with each continuous (i.e., not involving a jump like flipping a right hand into a left hand) rigid (i.e., not depending on the spacetime coordinates) symmetry of the Lagrangian density, a current satisfying the continuity equation. The seven most basic such symmetries are time translation invariance, spatial translation invariance (3), and the isotropy of space (3), implying

[7] While this choice is made whimsically, sometimes peculiar fractional choices, such as $\frac{5}{9}$ or $-\frac{5}{11}$, are physically motivated (DeWitt 1967).

[8] Some readers will notice a slight dose of particle physics egalitaritarianism here: General Relativity is presumed to be like other field theories except insofar as it is shown to be otherwise (Feynman et al. 1995; Kaiser 1998; Pitts 2017b, 2020b). I take this view to be the default view and, in this case, not relevantly defeated. There are respects in which this presumption can be defeated, however, such as regarding observables (Pitts 2017a, 2018, 2019).

the conservation of energy, of 3 momenta, and 3 angular momenta. Less obviously *a priori*, there is also a principle of relativity, whether due to Galileo or Einstein, Lorentz, Poincaré, and so on, which implies another 3 conserved currents. Thus one has 10, matching the 10 components of the Poincaré group of special relativistic fields. For massless relativistic fields, there can be a handful of additional conserved currents associated with the Bateman-Cunningham 15-parameter conformal group. The more symmetries there are, the more conserved currents. Turning to GR, one finds that there are *far more* rigid symmetries, indeed infinitely many, of the Lagrangian density (Bergmann 1958), implying infinitely many conserved energy-momentum currents. For any vector field, one can find a coordinate system such that its components in some neighborhood take the form $(1, 0, 0, 0)$ (time translation), $(0, 1, 0, 0)$ (spatial translation), or the like. Hence there are, at least formally, uncountably infinitely many energy conservations and momentum conservations. This point was eventually taken seriously by pointing out that if there are infinitely many rigid symmetries and infinitely many conserved currents, then pseudotensoriality makes good sense in that different coordinate systems are related to different symmetries (Pitts 2009, 2010). Why can't they all be real and distinct? On the other hand, the usual complaint that gravitational energy forms a pseudotensor amounts to an anti-Noetherian demand that there be only one conserved energy (4 conserved energy-momenta) even though there are infinitely many symmetries of the Lagrangian density. While Noether's first theorem is a standard part of Lagrangian field theory, the postulate $\infty = 1$ or $\infty = 4$ does not seem to me a good axiom for field theory. Nihilists, recognizing the troubles associated with postulating $\infty = 1$ or $\infty = 4$, have in effect postulated $\infty = 0$ instead. My own view is that one should avoid postulating how many conservation laws there ought to be, and simply ask Noether's first theorem how many there are; the answer is ∞, which just is not subject to evaluation based on *ad hoc* criteria. To give an analogy, one might be puzzled by the inequivalence under translation (analogous to lack of a coordinate transformation rule) between "María es alta" (tall) and "Mary is short"—unless María $\neq$ Mary, in which case there is no reason to expect equivalent heights. Hence most of the complaint about pseudotensoriality is rooted in failing to listen to Noether's first theorem how many conserved quantities to expect in GR when the theorem is volunteering the answer. By ceasing to slight Noether's first theorem, one largely ceases to worry about pseudotensoriality in general, though there might be related more specific issues to consider. Hence Schrödinger's false-negative worry is answered by noticing that there are lots of gravitational energies surrounding the round heavy body, many or most of which are not 0, so the fact that one of them is 0 hardly shows the absence of gravitational energy and so is of no concern. Bauer's false positive objection has some other answers, which can be quite technical (Pitts 2010, 2022).

Given that there are infinitely many conserved currents, there is no reason to expect equivalent stories in each coordinate system. Hence the failure of a pseudotensor to be a geometric object is no problem: there is no reason to expect conserved quantities derived from inequivalent symmetries to be equivalent (related as a geometric object by a transformation rule).

With plausible ideas available for responding to both of the traditional objections to gravitational energy localization $\mathfrak{t}^{\mu}_{\nu}$, one is in a better position to take seriously energy conservation

$$\sum_{\mu=0}^{3} \frac{\partial}{\partial x^{\mu}} (\mathfrak{T}^{\mu}_{\nu} + \mathfrak{t}^{\mu}_{\nu}) = 0.$$

One can also think a bit more broadly about some facets of GR, such as the fact that Einstein's field equations alone (with no separate postulation of material equations of motion/field equations) imply these conservation laws. Sometimes that is viewed as some kind of defect, as was claimed by Felix Klein (Brading and Brown 2003). On the other hand, in Maxwell's electromagnetism one has the closely analogous result that Maxwell's equations alone (with no separate postulation of the equations for charged matter) imply the conservation of charge; this is an immediate consequence of the antisymmetry of the electromagnetic field strength tensor $F^{\mu\nu}$, the divergence of which equals the charge-current density. But no one thinks that charge conservation is therefore trivial in Maxwell's theory, to my knowledge, though there is an interesting multiplicity of derivations of charge conservation (Brading 2002; Brading and Brown 2003; Deser 1972). Why think that energy-momentum conservation is trivial in GR because Einstein's equations alone imply it? This implication, rather like the presence of infinitely many conserved energy-momentum currents, suggests that GR is *more* conserving of energy-momentum than other theories, not less as one usually hears. Likewise Maxwell's theory is in some sense more conserving of charge than other electromagnetic theories. The obvious competition is Proca's massive electromagnetism, which cheerfully accepts charge conservation or nonconservation, depending on how matter behaves.

4 GR, Energy Conservation, and the Philosophy of Mind

Given the usual views that gravitational energy either is not localized or does not exist, one cannot make sense of the local conservation equation

$$\sum_{\mu=0}^{3} \frac{\partial}{\partial x^{\mu}} (\mathfrak{T}^{\mu}_{\nu} + \mathfrak{t}^{\mu}_{\nu}) = 0$$

and energy is not conserved in general. Some authors, better informed about physics than usual among philosophers, have found in this conclusion an answer to the Leibnizian conservation objection (Collins 2008, 2011; Mohrhoff 1997, 1999).[9] If

[9] See also (Penrose 1994, pp. 334, 344–346) for a related suggestion that this peculiarity of General Relativity might help to address a conservation-related difficulty of spontaneous collapse theories.

conservation already fails given General Relativity, then there is no conservation remaining for interactionist dualism to spoil, so the usual Leibnizian objection is eliminated. This is an impressive aikido-like move, rhetorically, and shows much better grasp of the physics than usual. Unfortunately the truth is the reverse: General Relativity makes Cartesian mental causation *harder*, not easier. So I claimed some time ago based on the Noether- and pseudotensor-related considerations recalled above (Pitts 2010). Many people are not in the habit of trusting such considerations, however, and with some reason. Can one find a more clearly tensorial argument—one that is framed in tensor calculus or at least geometric objects, not in terms of pseudotensors, and hence that clearly has the same meaning in every coordinate system?

In fact one can: a recent paper used the generalized Bianchi identities to show that General Relativity tends to exclude, not facilitate, Cartesian mental causation (Pitts 2020c). In the simplest case, Cartesian mental influence must be spatio-temporally constant, and hence 0. The difficulty may diminish for more complicated models, however. As it turns out, the more field components worth of mental influence there are, the more readily one can evade the Bianchi identities' strictures by simply sacrificing a handful of fields to satisfy the identities by taking up the slack while the soul's other field components act on the world (such as the brain) in whatever fashion a dualist might expect. Hence GR tends to resist mental causation, but its resistance is finite.

How many field components worth of mental influence should one expect a soul to have? One possible answer is that one hasn't the slightest idea. If that is giving up too easily, then it seems difficult to answer this question without attending to larger world-view considerations. Given naturalism, it would seem that the answer is either 0 because such things could not evolve naturally, or not many because it is difficult (though not impossible) to evolve them, or likely not enough because the number of them is not brought about in a fashion with the foresight to try to circumvent the Bianchi identity restrictions. Hence it seems not very likely (at best) that souls could act on bodies given naturalism. Of course naturalists generally do not believe in souls anyway, so the new argument from GR is perhaps superfluous. On the other hand, given that most dualists are theists, plausibly God has designed (perhaps *via* evolution) souls and bodies to work together in a world that has Einstein's equations for laws (ignoring quantum gravity, which, however, is presumably relevantly similar except where the quantum aspect makes a difference). If so, then presumably souls have sufficiently many field components to act on the world even given GR. So the new argument from GR doesn't work against dualists who are also theists. Thus the new argument from GR against mental causation is unnecessary for naturalists and ineffective for theists. Almost the only targets whom it hits are dualists who are not theists, such as Karl Popper and John Beloff not that

Penrose invokes gravity to induce collapse of the wave function. This project is also linked to the philosophy of mind.

long ago. But the argument nonetheless helps one to understand the content of GR in a way that otherwise has not been so evident.

One notes that gravitational energy realism (taking pseudotensor laws seriously) has the correct heuristic force—it leads one to expect stronger restrictions on mental causation, and the Bianchi identities vindicate that expectation—whereas nonlocalization and nihilism lead one to expect looser restrictions on mental causation. Hence the new general relativistic objection provides some support for realism about gravitational energy-momentum in GR. Besides its intrinsic interest, such a conclusion might help to answer on objection to conserved quantity theories of causation (on which see (Curiel 2000; Dowe 2000; Fair 1979; Rueger 1998)): that energy, the star example of a conserved quantity, is not conserved after all. To the degree that one can take the conservation of energy seriously in GR, such an objection is no longer decisive, though difficulties remain (Pitts 2020c).

5 Conclusion

In different ways, both philosophers of mind and philosophers of space & time have tended to slight Noether's first theorem, the former by supposing conservation to be categorical when it is really conditional upon symmetries (the failure of which is expected given interactionist dualism), the latter by ignoring or underestimating the conservation laws that GR licenses due to symmetries of its laws. The latter slighting of Noether's first theorem suggests that GR is less conserving of energy than earlier theories, which claim some have invoked as a response to the assumed categorical conservation of the Leibnizian objection. But GR is actually more conserving of energy than earlier theories, a conclusion that is motivated by the infinity of symmetries of the GR laws by Noether's first theorem, but also demonstrable in a tensorial way using the generalized Bianchi identities. Thus GR tends to resist mental causation—though the argument does not make much difference given typical naturalist or theist background beliefs. Surprisingly, thinking about the 325+-year mental causation controversy and the 100+-year gravitational energy controversy sheds considerable light in both directions.

References

Anderson, J. L. (1967). *Principles of relativity physics*. New York: Academic.

Averill, E., & Keating, B. F. (1981). Does interactionism violate a law of classical physics? *Mind: A Quarterly Review of Philosophy, 90*, 102–107.

Bauer, H. (1918). Über die Energiekomponenten des Gravitationsfeldes. *Physikalische Zeitschrift, 19*, 163–165.

Bergmann, P. G. (1958). Conservation laws in general relativity as the generators of coordinate transformations. *Physical Review, 112*, 287–289.

Born, M. (1914). Der Impuls-Energie-Satz in der Elektrodynamik von Gustav Mie. *Nachrichten von der Königlichen Gesellschaft der Wissenschaften zu Göttingen, Mathematisch-Physikalische Klasse*, pp. 23–36. Translated as "The momentum-energy law in the electrodynamics of Gustav Mie" (2007). In J. Renn & M. Schemmel (Eds.) *The genesis of general relativity* (Vol. 4, pp. 745–756). Gravitation in the Twilight of Classical Physics: The Promise of Mathematics. Dordrecht: Springer.

Brading, K. (2001). Symmetries, conservation laws, and Noether's variational problem. PhD thesis, University of Oxford. Supervisor Harvey Brown.

Brading, K. A. (2002). Which symmetry? Noether, Weyl, and conservation of electric charge. *Studies in History and Philosophy of Physics, 33*, 3–22.

Brading, K., & Brown, H. (2003). Symmetries and Noether's theorems. In Brading, K., & Castellani, E. (Eds.) *Symmetries in physics: philosophical reflections* (pp. 89–109). Cambridge: Cambridge University Press.

Braude, S. E. (1986). *The limits of influence: psychokinesis and the philosophy of science*. New York: Routledge & Kegan Paul.

Braude, S. E. (1987). How to dismiss evidence without really trying. *Behavioral and Brain Sciences, 10*, 573–574.

Braude, S., Curry, A., & Tsakiris, A. (2017). Parapsychology researcher Dr. Stephen Braude battles against "sleazy arguments". *Skeptiko: Science at the Tipping Point*, 111. https://skeptiko.com/parapsychology-researcher-stephen-braude/. Accessed 3 May 2018.

Brown, H., & Holland, P. (2004). Dynamical versus variational symmetries: Understanding Noether's first theorem. *Molecular Physics, 102*, 1133–1139.

Butterfield, J. (1997). Quantum curiosities of psychophysics. In Cornwell, J. (Ed.) *Consciousness and human identity* (pp. 122–159). Oxford: Oxford University Press.

Callan, Jr., C. G., Coleman, S., & Jackiw, R. (1970). A new improved energy-momentum tensor. *Annals of Physics, 59*, 42–73.

Carroll, S. (2010). Energy is not conserved. *Discover: The magazine of science, technology and the future*. http://blogs.discovermagazine.com/cosmicvariance/2010/02/22/energy-is-not-conserved/#.WaAUO2d3FyA.

Cattani, C., & De Maria, M. (1993). Conservation laws and gravitational waves in General Relativity (1915–1918). In Earman, J., Janssen, M., & Norton, J. D. (Eds.) *The attraction of gravitation: New studies in the history of general relativity* (Vol. 5, pp. 63–87). Einstein studies. Boston: Birkhäuser.

Chang, C.-C., Nester, J. M., & Chen, C.-M. (2000). Energy-momentum (Quasi-)localization for gravitating systems. In Liu, L., Luo, J., Li, X.-Z., & Hsu, J.-P. (Eds.) *The Proceedings of the Fourth International Workshop on Gravitation and Astrophysics: Beijing Normal University*, October 10–15, 1999 (pp. 163–173). Singapore: World Scientific. arXiv:gr-qc/9912058v1.

Chen, C.-M., Nester, J. M., & Tung, R.-S. (2015). Gravitational energy for GR and Poincaré gauge theories: A covariant Hamiltonian approach. *International Journal of Modern Physics D, 2424*(1530026). arXiv:1507.07300v1 [gr-qc]; also published in Ni, W.-T. (Ed.) (2015). *One hundred years of general relativity: From genesis and empirical foundations to gravitational waves, cosmology and quantum gravity* Singapore: World Scientific.

Churchland, P. M. (2011). Consciousness and the introspection of 'qualitative simples'. *Eidos, 15*, 12–47.

Collins, R. (2008). Modern physics and the energy-conservation objection to mind-body dualism. *American Philosophical Quarterly, 45*, 31–42.

Collins, R. (2011). The energy of the soul. In Baker, M. C., & Goetz, S. (Eds.) *The soul hypothesis: investigations into the existence of the soul* (pp. 123–133). New York: Continuum.

Cucu, A. C., & Pitts, J. B. (2019). How dualists should (not) respond to the objection from energy conservation. *Mind and Matter, 17*, 95–121. https://www.mindmatter.de/resources/pdf/pittswww.pdf.

Curiel, E. (2000). The constraints General Relativity places on physicalist accounts of causality. *Theoria, 15*, 33–58.

Davis, W. R. (1970). *Classical fields, particles, and the theory of relativity*. New York: Gordon and Breach.
de Rham, C., Gabadadze, G., & Tolley, A. J. (2011). Resummation of massive gravity. *Physical Review Letters, 106*, 231101. arXiv:1011.1232v2 [hep-th].
Deser, S. (1972). Note on current conservation, charge, and flux integrals. *American Journal of Physics, 40*, 1082–1084.
DeWitt, B. S. (1967). Quantum theory of gravity. II. The manifestly covariant theory. *Physical Review, 162*, 1195–1239.
Dowe, P. (2000). *Physical causation*. Cambridge: Cambridge University Press.
Ducasse, C. (1960). In defense of dualism. In Hook, S. (Ed.) *Dimensions of mind: A symposium* (pp. 85–90). New York: New York University Press.
Duerr, P. M. (2019a). Fantastic beasts and where (not) to find them: Local gravitational energy and energy conservation in general relativity. *Studies in History and Philosophy of Modern Physics, 65*, 1–14.
Duerr, P. M. (2019b). It ain't necessarily so: Gravitational waves and energy transport. *Studies in History and Philosophy of Modern Physics, 65*, 25–40.
Duerr, P. M. (2021). Against 'functional gravitational energy': A critical note on functionalism, selective realism, and geometric objects and gravitational energy. *Synthese, 199*, S299–S333.
Fair, D. (1979). Causation and the flow of energy. *Erkenntnis, 14*, 219–250.
Feynman, R. P., Morinigo, F. B., Wagner, W. G., Hatfield, B., Preskill, J., & Thorne, K. S. (1995). *Feynman Lectures on Gravitation*. Reading: Addison-Wesley. Original by California Institute of Technology, 1963.
Freund, P. G. O., Maheshwari, A., & Schonberg, E. (1969). Finite-range gravitation. *Astrophysical Journal, 157*, 857–867.
Gallagher, R. (2020). *Demonic foes: My twenty-five years as a psychiatrist investigating possesions, diabolic attacks, and the paranormal*. New York: Harper One.
Goldberg, J. N. (1958). Conservation laws in general relativity. *Physical Review, 111*, 315–320.
Goldstein, H. (1980). *Classical mechanics* (2nd ed.). Reading, MA: Addison-Wesley.
Hassan, S. F., & Rosen, R. A. (2012). Confirmation of the secondary constraint and absence of ghost in massive gravity and bimetric gravity. *Journal of High Energy Physics, 1204*(123), 0–16. arXiv:1111.2070 [hep-th].
Heidelberger, M. (2004). *Nature from within: Gustav Theodor Fechner and his psychophysical worldview*. Pittsburgh: University of Pittsburgh Press.
Herglotz, G. (1911). Über die Mechanik des deformierbaren Körpers vom Standpunkte der Relativitätstheorie. *Annalen der Physik, 36*, 493–533. Translated by Delphenich, D. H. as "On the mechanics of deformable bodies from the standpoint of relativity theory". https://neo-classical-physics.info/uploads/3/0/6/5/3065888/herglotz_-_rel._cont._mech..pdf.
Hoefer, C. (2000). Energy conservation in GTR. *Studies in History and Philosophy of Modern Physics, 31*, 187–199.
Hossenfelder, S. (2016). Dear Dr B: Where does dark energy come from and what's it made of? *BackReAction*. http://backreaction.blogspot.com/2016/10/dear-dr-b-where-does-dark-energy-come.html.
Houtappel, R. M. F., Dam, H. V., & Wigner, E. P. (1965). The conceptual basis and use of the geometric invariance principles. *Reviews of Modern Physics, 37*, 595–632.
Kaiser, D. (1998). A ψ is just a ψ? Pedagogy, practice, and the reconstitution of General Relativity, 1942–1975. *Studies in History and Philosophy of Modern Physics, 29*, 321–338.
Kastrup, H. A. (1987). The contributions of Emmy Noether, Felix Klein and Sophus Lie to the modern concept of symmetries in physical systems. In Doncel, M. G., Hermann, A., Michel, L., & Pais, A. (Eds.) *Symmetries in Physics (1600–1980): Proceedings, 1st International Meeting on the History of Scientific Ideas, Sant Feliu de Guíxols*, Spain, September 20–26, 1983 (pp. 113–163). Seminari d'Història de les Ciéncies, Universitat Autónoma de Barcelona. http://www.desy.de/~hkastrup/reprints/.
Kosmann-Schwarzbach, Y. (2011). *The Noether Theorems: Invariance and conservation laws in the twentieth century*. New York: Springer. Translated by Bertram E. Schwarzbach.

Ladyman, J., Ross, D., Spurrett, D., & Collier, J. (2007). *Every thing must go: metaphysics naturalized*. Oxford: Oxford University Press.
Landau, L. D., & Lifshitz, E. M. (1975). *The classical theory of fields*. Oxford: Pergamon. Fourth revised English edition. Translated by Morton Hamermesh.
Larmer, R. (1986). Mind-body interaction and the conservation of energy. *International Philosophical Quarterly, 26*, 277–285.
Leibniz, G. W. (1985). *Theodicy: Essays on the goodness of god and the freedom of man and the origin of evil*. La Salle, Illinois: Open Court. Translator E. M. Huggard.
Leibniz, G. W. (1997). [First] Explanation of the new system of the com-munication between substances, in reply to what was said of it in the *Journal* for 12 September 1695. In Woolhouse, R. S., & Francks, R. (Eds.) *Leibniz's 'new system' and associated contemporary texts* (pp. 47–52). Oxford: Clarendon Press. Original April 1696.
Logunov, A. A., & Folomeshkin, V. N. (1977). The energy-momentum problem and the theory of gravitation. *Theoretical and Mathematical Physics, 32*, 749–771.
Lord, E. A. (1976). *Tensors, relativity and cosmology*. New Delhi: Tata McGraw-Hill Publishing Co.
Maudlin, T., Okon, E., & Sudarsky, D. (2020). On the status of conservation laws in physics: Implications for semiclassical gravity. *Studies in History and Philosophy of Modern Physics, 69*, 67–81. arXiv:1910.06473v1 [gr-qc].
Mie, G. (1913). Grundlagen einer Theorie der Materie, Dritte Mitteilung. *Annalen der Physik, 40*, 1–65.
Misner, C., Thorne, K., & Wheeler, J. A. (1973). *Gravitation*. New York: Freeman.
Mohrhoff, U. (1997). Interactionism, energy conservation, and the violation of physical laws. *Physics Essays, 10*, 651–665.
Mohrhoff, U. (1999). The physics of interactionism. *Journal of Consciousness Studies, 6*, 165–184.
Motl, L. (2010). Why and how energy is not conserved in cosmology. *The Reference Frame*. https://motls.blogspot.com/2010/08/why-and-how-energy-is-not-conserved-in.html?m=1.
Narlikar, J. V. (1993). *Introduction to cosmology* (2nd ed.). Cambridge: Cambridge University Press.
Nester, J. M. (2004). General pseudotensors and quasilocal quantities. *Classical and Quantum Gravity, 21*, S261–S280.
Nijenhuis, A. (1952). Theory of the Geometric Object. PhD thesis, University of Amsterdam. Supervisor Jan A. Schouten.
Noether, E. (1918). Invariante variationsprobleme. *Nachrichten der Königlichen Gesellschaft der Wissenschaften zu Göttingen, Mathematisch-Physikalische Klasse*, 235–257. Translated as "Invariant variation problems" by Tavel, M. A. (1971). *Transport Theory and Statistical Physics, 1*, 183–207.
Ogievetsky, V. I., & Polubarinov, I. V. (1965). Interacting field of spin 2 and the Einstein equations. *Annals of Physics, 35*, 167–208.
Ohanian, H., & Ruffini, R. (1994). *Gravitation and spacetime* (2nd ed.). New York: Norton.
Papapetrou, A. (1948). Einstein's theory of gravitation and flat space. *Proceedings of the Royal Irish Academy A, 52*, 11–23.
Peebles, P. J. E. (1993). *Principles of physical cosmology*. Princeton: Princeton University Press.
Penrose, R. (1994). *Shadows of the mind: A search for the missing science of consciousness*. Oxford: Oxford University Press.
Petrov, A. N., & Katz, J. (2002). Conserved currents, superpotentials and cosmological perturbations. *Proceedings of the Royal Society (London) A, 458*, 319–337.
Physics Stack Exchange. (2017). https://physics.stackexchange.com/questions/333489/conservation-of-energy-in-general-relativity.
Pitts, J. B. (2006). Absolute objects and counterexamples: Jones-Geroch dust, Torretti constant curvature, tetrad-spinor, and scalar density. *Studies in History and Philosophy of Modern Physics, 37*, 347–371. arXiv:gr-qc/0506102v4.
Pitts, J. B. (2009). Gauge-invariant localization of infinitely many gravitational energies from all possible auxiliary structures, or, why pseudotensors are okay. In *Proceedings of the DPF-2009*

Conference, Division of Particles and Fields. Detroit, MI: American Physical Society, Wayne State University. arXiv:0910.3320 [hep-th].

Pitts, J. B. (2010). Gauge-invariant localization of infinitely many gravitational energies from all possible auxiliary structures. *General Relativity and Gravitation, 42*, 601–622. 0902.1288 [gr-qc].

Pitts, J. B. (2016a). Einstein's physical strategy, energy conservation, symmetries, and stability: "but Grossmann & I believed that the conservation laws were not satisfied". *Studies in History and Philosophy of Modern Physics, 54*, 52–72. PhilSci; arxiv.org/1604.03038 [physics.hist-ph].

Pitts, J. B. (2016b). Universally coupled massive gravity, III: dRGT-Maheshwari pure spin-2, Ogievetsky-Polubarinov and arbitrary mass terms. *Annals of Physics, 365*, 73–90. arXiv:1505.03492 [gr-qc].

Pitts, J. B. (2017a). Equivalent theories redefine Hamiltonian observables to exhibit change in General Relativity. *Classical and Quantum Gravity, 34*(055008). arXiv:1609.04812 [gr-qc].

Pitts, J. B. (2017b). Progress and gravity: Overcoming divisions between general relativity and particle physics and between science and HPS. In Chamcham, K., Silk, J., Barrow, J., & Saunders, S. (Eds.) *The Philosophy of Cosmology* (pp. 263–282). Cambridge: Cambridge University Press. https://arxiv.org/abs/1907.11163.

Pitts, J. B. (2018). Equivalent theories and changing Hamiltonian observables in General Relativity. *Foundations of Physics, 48*, 579–590. https://doi.org/10.1007/s10701-018-0148-1. arXiv.org:1803.10059; PhilSci.

Pitts, J. B. (2019). What are observables in Hamiltonian Einstein-Maxwell theory? *Foundations of Physics, 49*, 786–796. arXiv:1907.09473 [gr-qc].

Pitts, J. B. (2020a). Conservation laws and the philosophy of mind: Opening the black box, finding a mirror. *Philosophia, 48*, 673–707.

Pitts, J. B. (2020b). Cosmological constant Λ *vs.* massive gravitons: A case study in General Relativity exceptionalism *vs.* particle physics egalitarianism. In Blum, A., Lalli, R., & Renn, J. (Eds.) *The renaissance of general relativity in context* (Vol. 16, pp. 189–219). *Einstein studies*. Basel: Birkhäuser. arxiv:1906.02115.

Pitts, J. B. (2020c). General relativity, mental causation, and energy conservation. *Erkenntnis*. https://link.springer.com/article/10.1007/s10670-020-00284-7.

Pitts, J. B. (2020d). The mind-body problem and conservation laws: The growth of physical understanding? http://philsci-archive.pitt.edu/16380/.

Pitts, J. B. (2021). Conservation of energy: Missing features in its nature and justification and why they matter. *Foundations of Science, 26*, 559–584.

Pitts, J. B. (2022). What represents space-time? and what follows for substantivalism *vs.* relationalism and gravitational energy? *forthcoming*. In Vassallo, A. (Ed.) *The foundations of spacetime physics: Philosophical perspectives*. Routledge Studies in the Philosophy of Mathematics and Physics, Routledge.

Plantinga, A. (2007). Materialism and Christian belief. In van Inwagen, P., & Zimmerman, D. (Eds.) *Persons: human and divine* (pp. 99–141). New York: Oxford University Press.

Read, J. (2022). Geometric objects and perspectivalism. In Read, J., & Teh, N. (Eds.) *The philosophy and physics of Noether's Theorems*. Cambridge: Cambridge University Press. Forthcoming. http://philsci-archive.pitt.edu/18911/1/GO_NK_v2.pdf.

Romero-Maltrana, D. (2015). Symmetries as by-products of conserved quantities. *Studies in History and Philosophy of Modern Physics, 52*, 358–368.

Rueger, A. (1998). Local theories of causation and the a posteriori identification of the causal relation. *Erkenntnis, 48*, 25–38.

Schmutzer, E. (1972). *Symmetrien und Erhaltungssätze der Physik*. Berlin: Akademie-Verlag.

Schouten, J. A. (1954). *Ricci-calculus: An introduction to tensor analysis and its geometrical applications* (2nd ed.). Berlin: Springer. http://link.springer.com/book/10.1007%2F978-3-662-12927-2.

Schrödinger, E. (1918). Die Energiekomponenten des Gravitationsfeldes. *Physikalische Zeitschrift, 19*, 4–7.

Schutz, B. F., & Sorkin, R. (1977). Variational aspects of relativistic field theories, with applications to perfect fluids. *Annals of Physics, 107*, 1–43.

Siegel, E. (2018). Ask Ethan: Where is the line between mathematics and physics? *Forbes: Starts with a Bang*. https://www.forbes.com/sites/startswithabang/2018/03/10/ask-ethan-where-is-the-line-between-mathematics-and-physics/#73e55d6753be.

Sorkin, R. (1977). On stress-energy tensors. *General Relativity and Gravitation, 8*, 437–449.

Stephani, H. (1990). *General relativity* (2nd ed.). Cambridge: Cambridge University Press.

Szabados, L. B. (2009). Quasi-local energy-momentum and angular momentum in general relativity. *Living Reviews in Relativity, 12*(4). Cited June 22, 2009.

Trautman, A. (1966). The general theory of relativity. *Soviet Physics Uspekhi, 9*, 319–339. Appendix by L. P. Grishchuk.

Turner, F. M. (1974). *Between science and religion: The reaction to scientific naturalism in Late Victorian England*. New Haven: Yale University Press.

van Strien, M. (2015). Vital instability: Life and free will in physics and physiology, 1860–1880. *Annals of Science, 72*, 381–400.

Wald, R. M. (1984). *General relativity*. Chicago: University of Chicago Press.

Watkins, E. (1995). The development of physical influx in early eighteenth-century Germany: Gottsched, Knutzen, and Crusius. *Review of Metaphysics, 49*, 295–339.

Watkins, E. (1998). From pre-established harmony to physical influx: Leibniz's reception in eighteenth century Germany. *Perspectives on Science, 6*, 136–203.

Wegener, F. D. A. (2009). *A true Proteus: A history of energy conservation in German science and culture, 1847–1914*. PhD thesis, Universiteit Utrecht.

Weyl, H. (1922). *Space-time-matter*. London: Methuen & Company. Translated by Henry L. Brose from 4th edition of *Raum-Zeit-Materie*; reprinted by Dover, New York (1952).

How John Wheeler Lost His Faith in the Law

Alexander Blum and Stefano Furlan

Abstract In 1972, at a symposium celebrating the 70th birthday of Paul Dirac, John Wheeler proclaimed that "the framework falls down for everything that one has ever called a law of physics". Responsible for this "breakage [. . .] among the laws of physics" was the general theory of relativity, more specifically its prediction of massive stars gravitationally collapsing to "black holes", a term Wheeler himself had made popular some years earlier. In our paper, we investigate how Wheeler reached the conclusion that gravitational collapse calls into question the lawfulness of physics and how, subsequently, he tried to develop a new worldview, rethinking in his own way the lessons of quantum mechanics as well as drawing inspiration from other disciplines, not least biology.

Keywords John Wheeler · Black hole · Quantum gravity · Teleology · Anthropic principle · Biophysics

The second half of the twentieth century saw the scientific discipline of physics at its prime. It also saw the beginnings of a perceived decline, with the life sciences being perceived more and more as the scientific avantgarde.[1] One aspect of this

[1] For some early remarks in this direction, see a letter from Wolfgang Pauli to Markus Fierz of 14 September 1954 (von Meyenn 1999, p. 756). See also (Potthast 2007).

A. Blum (✉)
Max Planck Institute for the History of Science, Berlin, Germany

Albert Einstein Institute, Potsdam, Germany
e-mail: ablum@mpiwg-berlin.mpg.de

S. Furlan
Max Planck Institute for the History of Science, Berlin, Germany

University of Geneva, Geneva, Switzerland
e-mail: sfurlan@mpiwg-berlin.mpg.de

Y. Ben-Menahem (ed.), *Rethinking the Concept of Law of Nature*,
Jerusalem Studies in Philosophy and History of Science,
https://doi.org/10.1007/978-3-030-96775-8_11

development is the increasing doubt in the validity and relevance of reductionist, microscopic physical laws; another is the increasing importance in physical theorizing of concepts originating in the life sciences, such as emergence or evolution.[2] In this paper, we will study these developments by following the trajectory of a single, albeit prominent, scientist, the American physicist John Wheeler,[3] who in the early 1970s began to promote the idea that the world was lawless in its foundations (culminating in his statement of "law without law" in the late 1970s) and to explore evolutionary ideas with respect to the cosmological genesis of physical laws.

Wheeler was hardly predestined to one day promote the fundamental lawlessness of physics. To the contrary. In the 1950s, Wheeler developed and promoted a methodology he called daring conservatism, which was based on taking the established laws of physics seriously, upholding them for as long as possible, and testing their implications to the utmost (that is the daring part). The hero of this approach was Niels Bohr, whose atomic theory Wheeler took to be the pinnacle of daring conservatism:

> I know no better example of this daring conservatism than Bohr's analysis of the hydrogen atom. Nothing was further from his approach than the free play with ideas that one finds in the Philosophical Magazine of the 1910's. There one author treated atoms as electric fluids, another made a hydrodynamical model, others changed the laws of electrostatics, and still others altered the laws of electromagnetic radiation to account for the absence of radiation from stable atoms. In contrast, Bohr accepted the well founded law of force between point charges, made use of the usual principles of mechanics, and recognized that the laws of electromagnetic radiation are of far reaching application...[4]

Some twenty years later, his views had changed, dramatically:

> [T]he closed universe of Einstein's general relativity [...] undergoes gravitational collapse. In that collapse, classical space and time themselves come to an end. With their end, the framework falls down for everything that one has ever called a law of physics.
> Nothing that relativity has ever predicted is more revolutionary than collapse, and nothing that collapse puts into question is more central than the very possibility of any enduring laws in physics. (Wheeler 1973, p. 203)

The story of this transformation, from arguing that the laws of physics should be extrapolated to the extreme, to arguing that at bottom there are no fundamental laws, at first glance seems straightforward and in the second quote Wheeler is very explicit about the reasons for his newly-held beliefs: it is the realization that the universe will end in gravitational collapse, in what is now customarily referred to as a "big crunch". Indeed, this seems to fit with popular notions, shaped primarily

---

[2] We will be talking about evolution in this chapter. For the story of how the notion of emergence entered the physical sciences, cf. (Muñoz 2022).

[3] For biographical information on Wheeler, see his autobiography (Wheeler and Ford 2000).

[4] The quote is from a revised but unpublished version of Wheeler's 1954 Richtmyer Lecture to be found in the archive of Wheeler's papers at the American Philosophical Society Library (Mss. B.W564) in Philadelphia (which we will be referring to simply as "Wheeler Papers" in the following), Box 182, folder "Fields and Particles". See also (Blum and Brill 2020, footnote 40).

by big crunch's little sibling, the gravitational collapse of a massive star into a black hole; after all, doesn't everyone know that in black holes all bets are off?

But the story is not quite as simple. For starters, the popular imagination of black holes is to a large part shaped by the ideas of Wheeler himself (who famously promoted the use of the term "black hole"). So one needs to be very careful when taking the radical implications of gravitational collapse as self-evident. In this chapter, we aim to outline how it was Wheeler's rather idiosyncratic takes on general relativity, cosmology, and quantum theory that led him to draw such radical conclusions from gravitational collapse. In particular, we will reconstruct how Wheeler reached the conclusion (a) that *all* known laws of physics were violated in gravitational collapse and (b) that *no* newly discovered laws of physics would step in to replace them; both of these conclusions, it needs to be stressed again, are not at all straightforward.

We will be looking at Wheeler's changing views from several angles. In doing so, we will be covering the same historic period, from the 1950s to the 1970s, several times over. We thus offer here a very simple periodization. Wheeler's thinking is more dynamical than the sharp breaks implied by such a periodization might imply, all the more as he frequently pursued different and even *prima facie* contradictory approaches in parallel; but some basic dates will still be helpful for the chronological orientation of the reader. In the period from 1952 to 1960 Wheeler first formulated his daring conservative program for general relativity and began investigating the question of gravitational collapse. He still expected that total collapse into what we would now call a black hole would not actually occur in nature. In the early 1960s, Wheeler came to accept that gravitational collapse might proceed past the so-called Schwarzschild horizon (Furlan 2021) and that a quantum theory of gravity would be necessary to provide an adequate description of gravitational collapse. This mature phase of his research program goes approximately from 1960 to 1969. Around 1970 he began to question the foundational assumptions of his research program and to embark on his exploration of lawlessness, the origins of which it is the purpose of this paper to explain.

We will begin in Sect. 1 by studying how Wheeler first reached the conclusion that in gravitational collapse established conservation laws of physics (most notably baryon number, which roughly corresponds to the amount of matter) would not just be violated, but, as he put it, "transcended". In Sect. 2, we reconstruct why Wheeler rejected lawful alternatives for analyzing the process of gravitational collapse and how this led him—in combination with the observed transcendence of conservation laws—to promote the view that physical law ended in gravitational collapse. We continue in Sect. 3 by showing how Wheeler's emerging views on the specific process of gravitational collapse affected his views on physics more generally by looking at his cosmology. In Sect. 4, we will give an outlook on Wheeler's ultimately unsuccessful attempts of the 1970s to harness concepts from the life sciences in order to flesh out what a lawless physics might look like, before we give some brief conclusions in Sect. 5.

1 Transcending Conservation Laws

John Wheeler's engagement with general relativity, and his identification of that theory as the ideal vehicle for his daring conservatism, begins in 1952 (Blum and Brill 2020). Early on, Wheeler identified the gravitational collapse of massive stars as one of the central physical problems he was interested in. After all, what better way to extrapolate Einstein's theory of gravity to its extremes than to apply it to the most dense (and thus most strongly gravitating) objects expected to exist in the universe. But of course general relativity was not the only physical theory involved in stellar collapse; after all stars were made of matter. And with regards to matter one law in particular appeared to Wheeler to be essential to the question of gravitational collapse: the law of baryon number conservation.

The idea that there was a conservation law for the amount of matter was of course an old one. Over the course of the first half of the twentieth century, it had become clear that the bulk of the mass of (atomic) matter was concentrated in the nucleus, and that this nucleus was composed of protons and neutrons, collectively known as nucleons or heavy particles. In 1949, Eugene Wigner had recast the law of the conservation of matter as a "conservation law for the number of heavy particles (protons and neutrons)" (Wigner 1949, fn. 9). In the following years, other heavy particles were discovered in cosmic ray and accelerator experiments, and in 1953, Abraham Pais introduced the term "baryon" as "a collective name for heavy particles which not only comprises nucleons" (Pais 1954, p. 157). The name caught on, and Wigner's law is now commonly referred to as "baryon number conservation". We shall also be using this term, even though the baryons Wheeler was dealing with were only nucleons and he himself referred to them only as nucleons or heavy particles.

Wigner had introduced the law of baryon number conservation as a conjecture to explain the fact that the proton (and thus atomic matter in general) does not decay radioactively. Within a few years, the law of baryon number conservation had become a mainstay of particle physics (Borrelli 2015). But what, precisely, was Wheeler's interest in this law?

A central issue in the physics of stellar collapse was which state a star would settle into after having burned out its fuel and collapsing under its own weight. The 1930s had thus seen numerous studies of the equilibrium configurations of superdense, cold matter (Bonolis 2017). The densest configuration of matter that had been studied was a star (or a stellar core) consisting only of neutrons. Taking into account general-relativistic effects in the star's self-gravitation, J.R. Oppenheimer and George Volkoff had found a limiting mass of 0.71 solar masses, above which there was no equilibrium state for a collection of neutrons (Oppenheimer and Volkoff 1939).

Oppenheimer and Volkoff had concluded that stars with a greater mass would not be able to reach hydrostatic equilibrium and would collapse indefinitely in a process further studied by Oppenheimer and Hartland Snyder, a paper now considered a pioneering work on the formation of black holes (Oppenheimer and Snyder 1939).

But Wheeler was skeptical of Oppenheimer's conclusion, because it had not taken into account the conservation of baryon number. Oppenheimer and Volkoff had not talked much about baryon number conservation, for the simple reason that it really hadn't been a thing yet in 1939. But Wheeler saw that taking into account the law of baryon number conservation would have striking consequences for neutron stars in particular and stellar collapse in general. That realization was made possible by the conceptual changes that physics had undergone in the intervening 15 years; but it also reflects an ambition that, in different forms, guided Wheeler's research throughout decades: that of tying together the extremely large and the extremely small. This aspiration found highly speculative expressions already in his early notebooks ("Most exciting idea of each particle being a universe, perhaps identical with our own"; "Looks like a proton on the outside, but is a whole universe on the inside").[5] But it could also manifest itself far more concretely (and we shall see more instances of this further on), as here in the attempt of bringing together the microscopic law of baryon number conservation with the macroscopic process of stellar evolution.

Of course, the gap between micro- and macrophysics was not easily bridged: the conservation of baryon number was observed empirically only for the small nuclei available to terrestrial experimentation, not for the gigantic agglomerations of neutrons that might be found in stellar cores. But bridging such gaps was a central part of Wheeler's program of daring conservatism. So, when Wheeler re-analyzed the work of Oppenheimer and Volkoff by "simply recognizing the 'law' of heavy particle conservation" (1 April 1956, NB4, p. 36), he was conceding (with his quotation marks) that the 'law' of baryon number conservation might not be as time-honored as Coulomb's law; but he still considered its extrapolation to the stellar scale an instance of daring conservatism.

What were now the implications of considering the law of baryon number conservation for neutron cores? It offered an alternate reading of the Oppenheimer-Volkoff limit. Oppenheimer had assumed that the neutron core would become unstable if mass was added beyond the limit. This was illustrated by looking at the free energy as a function of density for different values of the mass: equilibrium was attained for the density minimizing the free energy, but for masses greater than the limiting value the free energy curve did not have any minima. Wheeler, in his first reflections on the Oppenheimer-Volkoff paper, objected to this illustration:

> Discusses free energy but incorrectly in my opinion since figure free energy as a function of t_0 [effectively the density at the center of the neutron core] not for fixed no. of particles but for fixed mass. (NB4, p. 35)

Wheeler was thus arguing (to himself in his notebook) that mass was a bad measure of the fixed amount of matter contained in a collapsing star. After all, mass

[5] These quotes are from Wheeler's first Relativity notebook, p. 128. We will be quoting frequently from Wheeler's series of Relativity notebooks, which contains 19 volumes spanning from 1952 to 1973 and is conserved at the Wheeler Papers. We will be referring to the individual notebooks as NBX, where X is the number of the notebook.

was already in special relativity something that could be transformed into energy and thus radiated away. A better gauge of the amount of matter was baryon number. But if one consequently imagined making a star bigger not by increasing its mass, but by increasing the number of baryons there was a whole different way to read the Oppenheimer-Volkoff limit: not as a maximal mass for stability, but as a maximal mass, period. What this implied was that if one had reached the maximal mass and then added another nucleon, the mass would no longer increase; instead it would presumably decrease (the mass of the nucleon plus the mass lost by the star being radiated away), ultimately reaching some constant mass value.

The most striking case, which Wheeler briefly appears to have liked best, would be if that constant mass were zero and if the star would reach that mass already for a finite number of baryons (NB4, p. 39, see Fig. 1). Adding further nucleons to such an object would make no difference to the mass ("all the mass energy will be radiated") and no appreciable difference at all, except to change the baryon number. And while this would not change any of the immediate properties of the object itself, it might eventually make "a difference": Wheeler imagined "a special one of the gods who will keep track of the number of nucleons in the thing—no one else will care." When the "limiting object" then encountered a limiting "anti-object" (with a large, supercritical negative baryon number), the two would annihilate and result in an object with a possibly subcritical and thus observable baryon number: if, e.g., the number of baryons in the first object minus the number of anti-baryons in the second were 1, the result of their annihilation would be a neutron or a hydrogen atom with baryon number 1. Wheeler imagined that "the special god would smile and say 'Look, see!'"—vindicated, one might say, in his daring-conservative belief in the law of baryon number conservation.

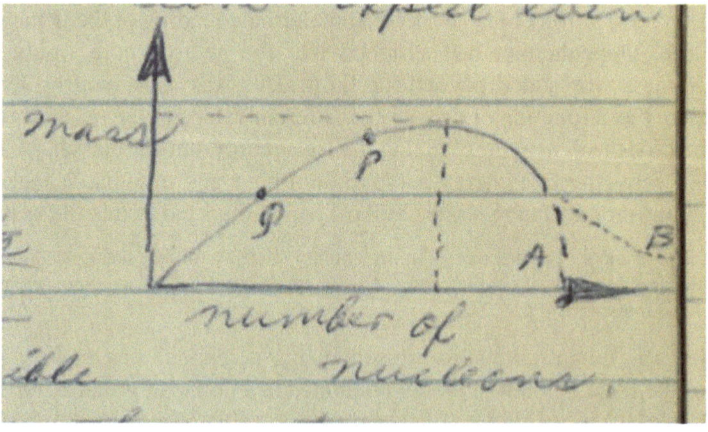

Fig. 1 Mass of star as function of the number of nucleons. The maximal mass is the critical mass of Oppenheimer and Volkoff. Option A represents the possibility that the star reaches a zero mass value already for a finite number of nucleons. In option B zero mass is achieved only asymptotically, for an infinite number of nucleons. NB4, p. 39, 1 April 1956

But could such super-condensed matter really exist? While Wheeler briefly entertained that possibility, he had discovered a paradoxical thought experiment, which ultimately appears to have turned him away from supercondensation:

> When one has nearly the peak mass for stability against collapse, one has a very interesting situation — add one nucleon; mass of star increases by a negligible fraction of mass of one nucleon; hence bulk of mass is given off as heat. When beyond this point, get off more energy than MC^2; hence could have 2 such stars, one the anti of the other and make a profit producing nucleons out of vac[uum]. Expect star will take profit. (NB4, p. 36)

In other words, one could create a nucleon-antinucleon pair from the vacuum, send the nucleon into a critical star and the anti-nucleon into a critical anti-star and get back more energy than was needed to create the pair.[6]

Wheeler had worked out these ideas on a trip from Leiden,[7] where he was spending the year as Lorentz professor, to the US, where he was attending the Rochester conference. On the penultimate leg of his journey, on 2 April 1956, from Washington to his home in Princeton, he reached the conclusion that this was a rather fanciful conjecture and stated (NB4, p. 40):

> "*Either* there exists a super condensed phase *or* law of conservation of nucleons fails — and we don't see former" — This would be my first go around at seeking a conclusion.

Wheeler devoted considerable energy and manpower to stellar collapse over the two following years, but this conclusion remained: no matter the exact dynamical details of gravitational collapse it would involve a violation of the law of baryon number conservation, a point he further emphasized in his talk at the 1958 Solvay conference (Adams et al. 1958). The question of how exactly this violation would occur occupied him for many years. Initially, Wheeler had assumed that the process that violated baryon number would simultaneously prevent the collapsing star from shrinking below its Schwarzschild radius. This meant that the process of baryon number violation would cause the entire star to dissociate, which was the only way to prevent further collapse. Wheeler envisioned that the neutrons in the neutron star might turn into massless neutrinos (NB5, p. 135). This would only happen under the high densities obtained in gravitational collapse and would "uphold [...] most other conservation laws" (Wheeler 1961, p. 127).

In the early 1960s, Wheeler came to accept the possibility that astronomical objects might collapse beyond their Schwarzschild radius. This did not, however, remove the issue of baryon number; indeed, the Schwarzschild radius had played no role in Wheeler's argument for the violation of baryon number conservation. But now Wheeler no longer had to assume that the whole star exploded; rather, he now

---

[6] It should be noted that this is not a perpetual motion machine, as the two stars (star and anti-star) only have finite mass-energy to give away. After they have settled into a constant (possibly zero) mass state, Wheeler's process would only increase the baryon number of the star and the negative baryon number of the anti-star, with no energy being required or gained. Wheeler still considered this process an "idealized machine to manufacture matter at no cost in energy" (Adams et al. 1958, p. 129) and an argument against baryon number conservation.

[7] On Wheeler's stay in Leiden, see (van den Heuvel 2020).

envisioned that in collapse the nucleons were "crushed out of existence", losing their integrity as matter through a new form of radioactive decay and the emission of "δ-radiation" (named in analogy to the three known forms of radioactivity, α, β, and γ radiation).[8] Wheeler appears to have envisioned that the δ radiation actually carried away the baryon number, thereby ultimately implying a conservation of that quantity. It remained unclear, however, what the δ rays would do with that nucleon number, i.e., how they would interact with other matter, so that this detail was ultimately inconsequential. And, in any case, these speculations were short-lived. Kip Thorne recalls that he "discouraged" Wheeler from including them in a 1965 book they co-authored (Thorne 2019, p. 15), which then only contained somewhat vague references to a new form of radioactivity in the final pages (Harrison et al. 1965).

Specific suggestions aside, we see that Wheeler was looking for a *mechanism* with which baryon number conservation was violated in gravitational collapse. He was thus looking for a new law to take the place of baryon number conservation. In the late 1960s, however, he came to realize that the violation of baryon number conservation in gravitational collapse might occur in a very different manner. The mid-1960s were the beginning of what Kip Thorne has called a "golden age of black hole research"[9] and saw a flurry of new theoretical results. We will mention the so-called singularity theorems later on. But more important for Wheeler's attitude toward baryon number conservation was the gradual emergence of the so-called (by Wheeler) no-hair theorem, that a star when collapsing to a black hole will lose almost all of its particularities[10] and be characterized only by its mass (which determines its size), its total charge, and its angular momentum. All other characteristics will be washed out in collapse, such as complicated charge distributions that might lead to dipole moments or magnetic fields.

The first steps toward the theorem had been made by Werner Israel in early 1967, who had shown that a non-rotating, non-charged star would collapse into a perfectly spherical black hole, characterized only by its mass.[11] The completion of the theorem to include charge and rotation (and exclude anything else) was still a few years off, but Wheeler conjectured early on that this would eventually be proven (though, as we shall see, he still thought that magnetic moment might be included in the list of properties of black holes). And this raised an immediate question: what about baryon number? The question was all the more pressing, as particle physicists had, in the interceding years, established further conservation laws of

---

[8] Notes taken by Dieter Brill of a talk given by Wheeler at the Texas Symposium on Relativistic Astrophysics, 16–18 December 1963. Personal possession of Dieter Brill. We would like to thank Dieter Brill for making these notes available to us.

[9] Thorne (1994), p. 258. Thorne credits William H. Press for the expression.

[10] Indeed, before Wheeler settled on the name no-hair, he would speak of no-particularities, see (Wheeler 1971a), where the image on p. 190 already contains the new phrasing "A black hole has no hair", while the text (which was presumably strongly based on the talk given two years earlier) has, on p. 191, "A black hole has no particularities".

[11] Israel (1967). See also (Thorne 1994, p. 279) for more on the history of the no-hair theorem.

similar type, in particular the conservation of lepton number, the number of *light* particles, such as electrons and neutrinos (which Wheeler consequently also referred to as neutrinoicity). Wheeler addressed this question in his notebooks (NB15, p. 60, 7 July 1968):

> The Great Question
> A few days ago trying to formulate my deepest concerns [...], I said, the great question is, do black holes and elementary particles belong to the same breed of cats. A black hole has mass, angular momentum, charge, magnetic moment – but so far as I know, nothing else. Baryon number, neutrinoicity, strangeness – all these as far as I can see, go out in the collapse.

He then re-iterated the thought experiment with the special god, though now using neutrinos instead of baryons (NB 15, p. 61):

> If a black hole "made of" neutrinos collides with one "made of" antineutrinos, is the outcome different from the collision of the black hole "made of" neutrinos and one of the same macroscopic quantities made of [neutrinos], or photons, or ordinary matter, or some mixture thereof?

He was beginning to suspect that even if matter went into the black hole unharmed (i.e., without losing its identity through something like delta radiation, or as he put it in his notebook "all down, none out"), the special god would no longer be vindicated in keeping track of how many baryons or neutrinos went in; baryon number as a property of matter would simply vanish in collapse since it was not a possible property of black holes, no special mechanism needed: "all junk dropped down a black hole leaves it smooth" (NB15, p. 106, 11 August 1968).[12]

The way in which baryon number conservation was violated in black holes thus appeared to be very different from what Wheeler had expected, and indeed from earlier cases of putative laws of nature being violated in new phenomena and by new effects. One year later, in a talk given in Italy on the occasion of the centennial of Mendeleev's periodic table, Wheeler's opinion had solidified (Wheeler 1971a, p. 192–194):

> Gravitational collapse deprives baryon number and lepton number of all significance. "Baryon number is conserved." "Lepton number is conserved." Of all principles of physics these familiar conservation laws belong among the most firmly established. Yet with gravitational collapse the content of these conservation laws also collapses. The established is disestablished.

But he wanted to be entirely sure and devoted detailed studies to both the conservation of lepton number (Wheeler 1971b) and of baryon number, the latter conducted by his PhD student Jacob Bekenstein (1972). Based on these studies, Wheeler arrived at a twofold argument. One, there was no possibility of seeing the particle fall into the black hole (or of preventing it from falling in), because even light was not fast enough to catch up with a particle on its way in. And two, there was no way of determining the lepton or baryon number of a black hole after

[12] A little further on, in December 1968, we also find the first use of the expression "no-hair" (NB15, p. 213), though Wheeler would not start promoting the name until somewhat later.

the fact: while the charge of a black hole could be determined in principle by the electrostatic field at large distances from the black hole, the quanta mediating the shorter-range (weak and strong nuclear) interactions that characterized leptons and baryons beyond their electric charge would not be able to cross the black hole's horizon:

> It is true that no attempt to observe a change in baryon number has ever succeeded. Nor has anyone ever been able to give a convincing reason to expect a direct and spontaneous violation of the principle of conservation of baryon number. In gravitational collapse, however, that principle is not directly violated; it is transcended. It is transcended because in collapse one loses the possibility of measuring baryon number, and therefore this quantity can not be well defined for a collapsed object. (Ruffini and Wheeler 1971, p. 36)

In gravitational collapse, something happened to the conservation laws of particle physics that could no longer be called violation and for which Wheeler coined the term "transcendence". The validity of these laws could no longer be ascertained *in principle*. We can clearly see this putting a dent in the applicability of physical laws in gravitational collapse. It is thus an essential starting point for Wheeler's embrace of lawlessness, and he would later invoke it as such (Wheeler 1973). But of course the question of particle-physics conservation laws is somewhat peripheral to the question of gravitational collapse, which as a physical process is primarily governed by general relativity, the physical theory that Wheeler had placed at the center of his daring conservatism. We will thus discuss in the next section how general relativity itself got caught up in the lawlessness of gravitational collapse.

2 The Rise and Fall of Geometrodynamics Revisited

In the mid-1950s, Wheeler embarked on his daring-conservative, general-relativity-based research program, which he referred to as "geometrodynamics".[13] From the start, this entailed more than just applying general relativity to extreme physical situations, such as gravitational collapse. Wheeler's hope had always been to use general relativity to construct a new foundation for microscopic particle physics. Particulate matter was to be reconceptualized as consisting merely of gravitational and electromagnetic fields concentrated within a small region. These particle-like field disturbances went by the name of "geon" (gravitational-electromagnetic entity) and their localized field energy could be interpreted as the mass of a particle; "mass without mass" in Wheeler's words. Similarly, electric charge was

[13] On Wheeler's program, as described in this section, see Wheeler (1964). Wheeler would sometimes equivocate and use "geometrodynamics" to refer to (a specific formulation of) general relativity, rather than to his research program, which attempted to explain the nature of elementary particles from geometry. Misner (1974) would differentiate between "visionary geometrodynamics" (Wheeler's program) and a more mundane "physical geometrodynamics" (along with a more formal "mathematical geometrodynamics"). We will use the term "geometrodynamics" to refer to "visionary geometrodynamics" throughout.

to be reconceptualized in terms of the topology of space; there would be handles connected to the bulk of space at two points and electric field lines trapped in those handles would emanate from those points as if there were an electric charge present. These handles were called "wormholes" and possessed "charge without charge".

It was clear on many levels that these structures built from classical fields were fully inadequate to represent (let alone explain the existence of) the known elementary particles. One might mention the fact that Wheeler was able to derive an upper limit for the charge-to-mass ratio of a wormhole on the order of $10^{-10}C/kg$ (Misner and Wheeler 1957, p. 593); the electron has a charge-to-mass ratio on the order of $10^{11}C/kg$. Another issue was the instability of these structures. Indeed, numerical calculations performed by Wheelers's former PhD student Richard Lindquist together with Susan Hahn of IBM indicated that the "throat" of a wormhole would inevitably collapse and leave behind a singular configuration with infinite spacetime curvature (Hahn and Lindquist 1964). Geons originally appeared to gradually vanish by leaking their energy (Wheeler 1961, p. 144) until they were believed to collapse as violently as massive stars.[14]

Wheeler was thus convinced early on that a *quantum theory* of geometrodynamics would be needed in order to turn his ideas for "mass without mass" and "charge without charge" into viable (and stable) models of elementary particles. This expectation was further reenforced by the so-called singularity theorem, established by Roger Penrose (1965), which stated that in the gravitational collapse of stars (and of the universe, as proven by Stephen Hawking (1966)) singular points with infinite space-time curvature would occur. Wheeler anticipated that quantum theory would be able to prevent such instances of collapse, just as it had prevented the collapse of the hydrogen atom through Bohr's concept of the stationary state.

This expectation, that a quantum theory of gravity will be required to deal with the apparently singular behavior of gravitational collapse, is still widely held. Under this assumption, gravitational collapse would be very similar to blackbody radiation: a phenomenon where the current theory (general relativity or electrodynamics) gives nonsensical results (singularity or ultraviolet catastrophe) and this situation is alleviated by introducing a new (quantum) theory to replace it.[15] There would be nothing very interesting happening here, as regards the status of laws of nature. At best one could ponder the status of approximate laws and the relation between the old and the new theories.

Why then did Wheeler ultimately invoke gravitational collapse as the reason for giving up the (quantum) geometrodynamical program and embracing lawlessness? Wheeler's abandonment of geometrodynamics was first analyzed, essentially while it was happening, by John Stachel (1974) in a paper fittingly entitled "The Rise and Fall of Geometrodynamics".[16] Here, Stachel listed a number of "conceptual

[14] See NB15, p. 178, 10 November 1968: "Do I not now believe that geons are probably *unstable* with respect to *collective* gravitational collapse..."

[15] As is well known, this is *not* the order of events in the historical discovery of quantum theory.

[16] A story similar to Stachel's, but in more compact form, is given by Aaron Wright (2014, p. 153), who describes Wheeler as being "dissatisfied with progress" in his program.

difficulties" (p. 39) that led Wheeler to abandon his program in the early 1970s. The central difficulty he highlighted was that of "spin without spin". In order to craft particles from fields, Wheeler not only had to reproduce their mass and their charge, but also their spin. Wheeler realized that (a) spin could be constructed from geometry if dynamical changes in the topology of space were possible, but that (b) such topological changes were disallowed by "classical differential geometry", forcing him to abandon all hope that geometrodynamics would be able to provide realistic particles with half-integer spin.

There is certainly some truth in this story: Wheeler's ambitions could not be realized within the restricted framework of general relativity, so he gave up on the program of laying a new geometrical foundation for the constitution of matter. But there are some confusing elements in this story: Wheeler had pointed out the difficulty of "spin without spin" already in the early 1960s (Wheeler 1963b); what had changed 10 years later? Note further that this story then has nothing at all to do with gravitational collapse. And it gives no indication as to why Wheeler would go on to propagate, as quoted in the introduction, the breakdown of "the very possibility of enduring laws in physics." We will thus in the following attempt to show that, to understand Wheeler's turn to lawlessness, one must understand the "fall of geometrodynamics" not as Wheeler abandoning his program because it could not be realized within the restricting framework of quantum gravity, but as Wheeler abandoning quantum gravity because it could not accommodate his program.

This may seem somewhat strange at first. How can one "abandon" quantum gravity, the apparently necessary reconciliation of general relativity and quantum mechanics? To understand this, we need to take a step back again to look more closely at Wheeler's program of quantum geometrodynamics and its wider context. For Wheeler was not the only one pursuing the question of quantum gravity; there was in fact a diversity of approaches to this subject, already in the first decades after World War II. We can ignore for our story the approach then prevalent among particle physicists, of transforming general relativity into a quantum theory of "gravitons" in full analogy to the photons of quantum electrodynamics. Even within the emerging relativity community there were (at least) two starkly different attitudes toward quantum gravity, which we can associate with two physicists, Wheeler and Bryce DeWitt (Blum and Hartz 2017).

For DeWitt, quantum gravity was a problem posed by the simultaneous presence in physical theory of the "two tree-like giants, the quantum theory, and the general theory of relativity" (DeWitt 2011, p. 64). He made "no apology" for pursuing this problem for in his "opinion it is sufficient that the problem is there, like the alpinist's mountain." (DeWitt 1957, p. 377). There were substantial formal and conceptual difficulties involved in combining quantum theory and general relativity, these had to be conquered, and then one would see what the result would be. Wheeler also appreciated the grandeur of reconciling the two great theories of modern physics and would later speak of them as "the two skyscrapers that overpower every other structure in the city."[17] But for Wheeler, this reconciliation was not an end in itself; rather it was an opportunity to further develop the guiding idea of geometrodynamics and thus, as we have outlined, to give a new foundation for all of physics.

More specifically, quantum gravity provided the means to stabilize his geometrodynamically constructed particles and ensure that they were not far too big. It also provided plausibility to his ideas of constructing particles through the topological features of space. He believed that a quantum theory of gravity would involve microscopic fluctuations not just in the metric structure of space-time, but also in its topological structure, resulting in spacetime acquiring a "foam-like structure" at a microscopic level (Wheeler 1957, p. 609).[18][19] One would thus have microscopic virtual wormholes popping in and out of existence, just like the virtual particles of high-energy physics. While this did not prove that charged particles were related to wormholes, it did imply, as he remarked in his notebook, that wormholes were "not an opium dream" (NB4, p. 15, 14 March 1956). Wheeler's speculations also went beyond the realm of particle physics: he believed, for example, that quantum gravity would imply fluctuations in the dimensionality of space and might eventually explain why macroscopic spacetime has four dimensions, no more, no less (Blum and Lessel 2022).

For a while it looked as if these two visions would be referring to the same theory, just to different approaches. DeWitt (and others, such as Peter Bergmann) was trying to reach a quantum theory of gravity by applying the quasi-algorithmic procedure of quantization to Einstein's theory of gravity, ironing out the many formal difficulties encountered on the way. Wheeler was cutting to the chase, assuming central features of quantum gravity from the outset (such as quantum foam), and pondering how they might be used to lay a new foundation for all of physical theory. While convergence of these approaches was not out of the question, the differences in the two approaches were quite manifest. Witness the exchange between Wheeler

[17] The quote is from "The chemistry of geometry", p. 847. This paper is from 1969 or 1970 and was apparently intended for use as a chapter in (Misner et al. 1973). It can be found at https://jawarchive.files.wordpress.com/2012/02/thechemistryofgeometry1969.pdf.

[18] Wheeler did not initially use the term "quantum foam" for this foam-like structure, but ultimately adopted it after it had entered common usage and employed it in the title of his autobiography (Wheeler and Ford 2000).

[19] It should be mentioned that the length scale of quantum gravity, what Wheeler would later christen the Planck scale, was actually too *small* to be associated with the known elementary particles. Wheeler was thus forced to imagine those elementary particles to be "a collective state of excitation of the foam-like medium". He was confident that the large gap between the scale of the known particles and the Planck scale would arise in analogy to superconductors, where the transition to superconductivity is coherent on scales of 10^{-4} cm (Pippard 1953) much larger than the lattice spacing of 10^{-8} cm. We would like to thank Rocco Gaudenzi for helping us to clarify this point.

and DeWitt after the former's 1963 talk on gravitational collapse and Delta Rays, mentioned in the last section:

> Bryce: Why not quantize imploding object instead? I can get Friedman universe to bounce! (after down to 10^{-33} cm)
> Wheeler: [I] couldn't agree more, but don't we expect elementary particle phenomena forced upon us [?]
> Bryce: But it doesn't require anything new!

DeWitt was satisfied with quantum gravity filling the apparent gap in physical theory encountered in the central singularity of gravitational collapse, by predicting a rebound and avoiding the appearance of infinite quantities. For Wheeler, on the other hand, merely preventing the appearance of a singularity was not sufficient, though he was certain that quantum foam would take care of the issue (Fuller and Wheeler 1962, fig. 3). Wheeler expected from the study of gravitational collapse and quantum gravity deep insights into the structure and physics of elementary particles (Wheeler 1968b, p. 271), another instance of Wheeler's belief in a connection between the macroscopic and the microscopic.

In the late 1960s, however, the program of quantizing gravity reached an unexpected degree of concreteness, or, as Wheeler put it, "maturity" (NB14, p. 19, 30 April 1966), as DeWitt constructed a Schrödinger-type equation for quantum gravity (DeWitt 1967), an equation that would become known, quite ironically, as the Wheeler-DeWitt equation. This was of course hardly a complete theory of quantum gravity, as many essential mathematical questions regarding this equation were (and are) still open. Even so, it clearly revealed that Wheeler's expectations would not be met in this approach. The Wheeler-DeWitt equation described only changes in the metric structure of space, not in its topology; and the dimensionality of spacetime, rather than appearing as a dynamical quantity, appeared to be hard-wired into the equation. Of course this was to be expected from a theory that arises from a direct quantization of general relativity; Wheeler liked to emphasize that the Wheeler-DeWitt equation can be derived directly from the Hamilton-Jacobi formulation of general relativity,[20] just like the Schrödinger equation can be derived directly from the Hamilton-Jacobi formulation of classical mechanics.[21] But the absence of a dynamical topology (and dimensionality) flew in the face of Wheeler's expectations for quantum geometrodynamics.

At first Wheeler was undeterred, and we find him describing his "pursuit of gen rel, how I still had great hopes" (NB14, p. 15, 9 March 1966), and more specifically

---

[20] See, e.g., the preparatory notes for a talk in Munich, dated 14 July 1966, in NB14, p. 54: "Not very diff[erent] in gen[era]l character from way do other q[uantum] prob[lem]s. Outline Hamilton Jacobi. Give DeWitt eq[uation]." The Hamilton-Jacobi formulation of general relativity was developed by Asher Peres (1962).

[21] It should be noted that this is not the derivation used by DeWitt in the paper introducing the Wheeler-DeWitt equation.

imagining ways how to get topology change back in through the integration of electrodynamics:

> Could it be that electromagnetism has to do with slipover from one topology to another; that that [sic] – at quantum level – "puts ticks enough into state function"... (NB14, p. 20, 30 April 1966)

But Wheeler struggled to establish a connection between the emerging formalism and his program. He articulated the frustration he felt at this gap when preparing a talk for the end of his presidency of the American Physical Society (NB14, p. 114, 4 January 1967):

> The general story of the dynamics I know insofar as Hamilton Jacobi formalism is involved. And why isn't it enough to tell that. One isn't under obligation to answer all possible questions! [...] Many can ask, do humans have right to wade in on such "high and holy issues". But what alternative do we have?
> Are obliged to discuss the relation to elementary particles? Einstein's dream?

He soldiered on for a while. Topology change was central to his entire research program, providing a potentially unifying picture of macroscopic gravitational collapse and microscopic quantum foam:

> Collapse is going on all the time at the virtual level. Every change of topology has in some sense a connection with collapse. (NB14, p. 120, 8 January 1967)

Wheeler thus tried somehow to fit changes in topology into the new framework:

> Make a comment about topology changes analogous to that which Eddington made about temp[erature] required for thermonuc[lear] reations when he said: if center of sun isn't hot enough, go and find a hotter place. *Must* take place. Kick the math to fit the phys[ics], not conversely. (NB14, p. 141, 27 January 1967)[22]

He also tried to formulate the Wheeler-DeWitt equation without reference to the dimensionality of space (NB15, p. 106ff, 11 August 1968). But ultimately it became clear all this would not work out; the new framework was too close to classical general relativity, which knew nothing of changing topologies and dimensionalities. Wheeler gradually had to give up the hope that quantum gravity could provide the framework for his vision of quantum geometrodynamics. In this sense, Stachel's story of the fall of geometrodynamics is correct. But this does not mean that he accepted the limits of the newly emerging formalism and tried to work within them. What he did instead is best described in Wheeler's own words by an allegorical story he included in the 1973 textbook *Gravitation*, which in many ways represents the sum of his two decades working on general relativity:

> There are reputed to be two kinds of lawyers. One tells the client what not to do. the other listens to what the client has to do and tells him how to do it. From the first lawyer, classical differential geometry, the client goes away disappointed, still searching for a natural way to describe quantum fluctuations in the connectivity of space. Only in this way can he hope to describe electric charge as lines of electric force trapped in the topology of space. Only in this way does he expect to be able to understand and analyze the final stages of gravitational

[22] See also NB15, pp. 33ff, 20 May 1968.

collapse. Pondering his problems, he comes to the office of a second lawyer with the name "Pregeometry" on the door. Full of hope, he knocks and enters. What is pregeometry to be and say? Born of a combination of hope and need, of philosophy and physics, and mathematics and logic, pregeometry will tell a story unfinished at this writing, but full of incidents of evolution so far as it goes. (Misner et al. 1973, p. 1203)

The reference to "classical differential geometry" is somewhat striking. In 1967, Wheeler had already been well aware of these limits (Geroch 1967),[23] but could still anticipate that "[a] new world opens out for analysis in quantum geometrodynamics." (Wheeler 1968b, p. 284); a few years later, it was enough to invoke the limits of classical differential geometry, for the limits of a quantum theory of gravity based on the Wheeler-DeWitt equation were just the same as those of the classical theory.[24] We also see that Wheeler—faced with the limits of classical differential geometry and a quantum theory of gravity that was not able to overcome these limits—did *not* give up on his program, but instead turned to a different "lawyer" who would tell him "how to do it."

We will talk more about the new lawyer's suggestions in the following sections, but we first take a moment to assess how all of this ties in with gravitational collapse and the embrace of lawlessness. In the previous section, we have seen that Wheeler had established a connection between gravitational collapse and the transcendence of microscopic conservation laws. But as regards the fundamental spatio-temporal dynamics of gravitational collapse, there was a clear and traditionally lawful way out of the apparent singularities, namely to embrace the emerging limited formalism of quantum gravity: DeWitt had plausibly derived matter bouncing back after gravitational collapse from the Wheeler-DeWitt equation (DeWitt 1967). But Wheeler now rejected this option, because his *research program* told him otherwise. He held on to the tenets of his program, in particular the concept of microscopic foam-like fluctuations of topology, but conceded that these were not to be had in the framework of quantum gravity. And in thus rejecting quantum gravity, Wheeler was led to instead embrace the catastrophe of total collapse as physical and real:

Black hole squeezes out all but mass, angular momentum and charge. Collapse of universe squeezes out even these. Not 2nd quantization but higher level collapse. (NB18, p. 80, 10 November 1970)

Wheeler decided that gravitational collapse was not something that needed to be fixed, but the beginning of a new era in physics:

Why stand timidly at threshold of this new domain of thought? Why not walk boldly in and exploit its richness? Is it only Columbus who shall see the gold of Mexico? (NB18, p. 124, 10 February 1971)

[23] See also NB14. p. 120–121, 8 January 1967, which contains notes of a phone call with Geroch concerning the impossibility of topology change.

[24] This view of quantum gravity as hopelessly rooted in the classical theory is made most explicit in (Patton and Wheeler 1975, Section 4). Even here we see Wheeler using scare quotes when employing the term "quantum gravity" for a theory that "take[s] as given Einstein's Riemannian space and his standard classical geometrodynamics law".

This brings us back to the talk quoted in the introduction, which Wheeler gave in September 1972 at a symposium in Trieste celebrating Paul Dirac's 70th birthday (Wheeler 1973). Here, total collapse was fully embraced as an unavoidable consequence of relativity: "Today we look at the breakage that relativity has made among the laws of physics." (p 203). And beyond this breakage were not to be new laws, but something else:

> The golden trail of science is surely not to end in nothingness. There may be no such thing as the 'glittering central mechanism of the universe' to be seen behind a glass wall at the end of the trail. Not machinery but magic may be the better description of the treasure that is waiting. (Wheeler 1973, p. 203)

We can understand Wheeler's turn away from a lawful machinery as a result of both the transcendence of the conservation laws and the fact that he was explicitly rejecting the lawful alternative of mainstream quantum gravity.[25] Quantum gravity was not able to provide him with a final theory of physics, a theory that would be able not just to describe gravity up to the smallest scales, but also encompass all of particle physics. So, if a final theory was not to be had within the framework of physical law, one had to leave that framework. The last step on the "staircase of law, and law transcended" (Fig. 2), after "space and time themselves are transcended as categories" was not ultimate law, but something entirely different. But what was the relevance of the anticipated "magic", if it was only to reveal itself in gravitational collapse, hidden behind a horizon? To understand this, the impact of the anticipated lawlessness, we need to backtrack and take a look at Wheeler's cosmology.

3 Wheeler's Cosmology

So far, when speaking of gravitational collapse, we have primarily considered the collapse of very massive stars. But from the very beginning, Wheeler's interest in gravitational collapse was defined also by its cosmological relevance: for gravitational collapse happens not just to stars but can, in general relativity, also happen to the dynamical universe as a whole. This process, which can be considered the inverse of the big bang, is nowadays commonly referred to as the "big crunch".

The first question we must ask is: why did Wheeler anticipate that the universe as a whole would undergo gravitational collapse? The reason for this anticipation—and, as we shall see, for many other elements of Wheeler's cosmology—lies in the importance that Wheeler ascribed to Mach's principle. This principle is notoriously hard to pin down, but its core statement is rather uncontroversial: inertia is to be explained as resulting from interactions with the overall matter distribution of the

[25] In his talk, Wheeler fielded as further reasons for lawlessness the long-recognized problematic status of energy and angular momentum in general relativity. Later he would also field the problematic status of charge in a closed universe, thus implicating all three remaining properties of a black hole in transcendence (Misner et al. 1973, Section 19.4).

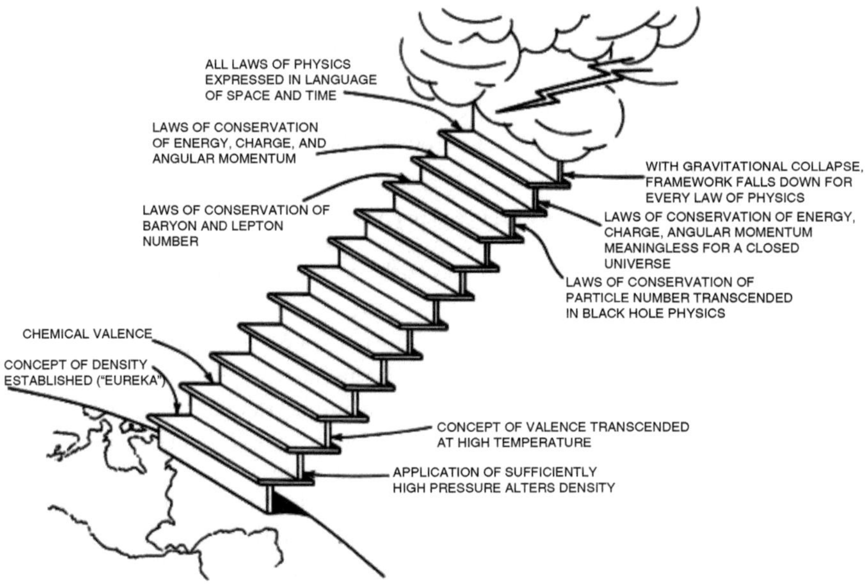

Fig. 2 The staircase of law, and law transcended (Wheeler 1975, p. 282)

universe, and not, as implied by Newtonian physics, as the effect of some absolute structures of space.

Mach's principle[26] had been an inspiration to Einstein in the formulation of general relativity and had indeed been partially integrated in general relativity: first, inertia and gravitation were unified, as both inertial motion and motion under a gravitational force were determined by the metric tensor of spacetime. Second, that metric tensor was to some extent determined by the matter distribution in the universe, making not just gravitational but also inertial motion the result of (field-mediated) interactions with the other masses in the universe.

An important question, already to Einstein, was to what extent the matter content of the universe *fully determines* inertial motion, i.e., to what extent general relativity fully implements Mach's principle. If, for example, the Einstein equations are solved with asymptotically flat boundary conditions, it is really those boundary conditions that determine inertial motion, not the matter content of the universe. Einstein (1917) thus reached the conclusion that in order to be Machian, general relativity needs to assume a spatially closed universe, in which there are no boundary

[26] We will in the following not distinguish between Mach's original formulation (very vague) and Einstein's interpretation (still pretty vague). For a detailed analysis, cf. Norton (2006).

conditions: if one kept on travelling through space in the same direction, one would eventually (after a *very* long journey) end up where one started.

Einstein's views on the connection between Mach's principle and general relativity as well as the cosmological conclusions he drew from it were not part of the mainstream view of general relativity. This was all the more true as Einstein's early cosmological speculations appeared to be invalidated by the discovery of the expansion of the universe. But Einstein's Machian arguments for a closed universe where really quite independent of his arguments for a static universe (to which end he had briefly introduced the cosmological constant). And Machian cosmology continued to exert a great appeal on some, in particular Wheeler; the primary reason being that Wheeler was originally envisioning a reformulation of general relativity as a theory of direct particle interactions, very much in keeping with Mach's relationalist approach (Blum and Brill 2020).

So strong was Wheeler's belief in Mach's principle that he included the assumption of the spatially closed universe among the axioms of general relativity in the 1958 Solvay talk, along with the statement that Einstein's second cosmological assumption of 1917 had been wrong, i.e., there was no cosmological constant, and the universe was changing in time:

> In speaking about Einstein's theory [...] we mean not only the system of differential equations associated with his name, but also two further points [...]:
>
> (1) The universe is closed;
> (2) No "cosmological" term is to be added to the field equations.
>
> It is probably best at this time to accept these points as part of what we mean by the phrase "Einstein's version of general relativity". (Adams et al. 1958, p. 98)

One important implication of the spatial closedness of the universe was that the universe could not go on expanding forever and would ultimately have to start contracting again.[27] We can thus understand how it was Mach's principle that led to Wheeler's expectation that the universe as a whole would undergo gravitational collapse.

But while Mach's principle determined that the universe was closed in space, the question of the boundaries of the universe in time was far less obvious, and in his notebooks of the 1950s we find Wheeler mulling various options in discussion with his PhD student Charles Misner (NB3, p. 238): "3 possibilities? Time closed [...]; time infinite; or time ends sharply". Closed time would of course have been the most immediate complement to the spatial closure of Mach's principle, but Wheeler was doubtful of this option.[28] And a universe with sharp temporal boundaries would have brought the problem of boundary conditions back on the table. A temporally

[27] This statement is predicated also on Wheeler's second axiom, the absence of the cosmological constant, which could, if large enough, prevent recontraction; this was of course the reason for which Einstein originally introduced it.

[28] Closed time is of course a problematic notion in general; it essentially means that if you live long enough you will return to the time where you were born. But Wheeler had more specific qualms about this option, concerning the possibility of the existence of electric charges in a

infinite universe thus seemed the most attractive possibility. But there is a difficulty here: if the universe is closed in space, it will at some point undergo gravitational collapse.[29] Wheeler thus had to assume that this collapse was not the end of the universe, but that the collapsing universe would rebound and start expanding again; an infinite sequence of expansion and collapse; a cyclic cosmology. This was the view of the universe that Wheeler would adopt in the late 1950s and pursue all through the 1960s.

Wheeler was not the first to consider a cyclic cosmology. The general idea of course goes way back. And also in the context of general relativity other physicists had anticipated that a possible gravitational collapse of the universe as a whole would not be the end of all things, but that the universe would somehow rebound and start expanding again (Kragh 2018). But as with the baryon non-conservation discussed in Sect. 1 it was one thing to state that a "big bounce" happened and quite another to find a mechanism for it actually happening. Wheeler initially turned to an analogy with hydrodynamics, hints at which he discovered in a book on "Underwater Explosions".[30] While this book is primarily about the shock waves produced by such an explosion, it also contains a substantial section on the motion of the gas sphere left over by the explosion (the "bubble"). This bubble starts contracting under the pressure of the surrounding water, but then, after being sufficiently compressed, reverses its motion again and then continues to oscillate, emitting secondary pressure waves ("bubble pulses"). The book also presented some very suggestive photographs of the bubble, showing how small departures from

temporally closed universe. These qualms are not sufficiently worked out in the notebooks to be fully reconstructed, but they appear to have carried quite some weight for Wheeler.

[29] It should be noted that this conclusion is nowadays considered not as inevitable as it would have seemed to Wheeler (Barrow et al. 1986).

[30] It is not entirely clear what prompted Wheeler to read this book (Cole 1948) in the first place. The book is a scientific one, but is primarily the outcome of (and relevant for) military research on submarine warfare. Misner recalls that Wheeler had learned about underwater explosions "during his wartime service or in connection with it" (Oral History interview conducted by Chris Smeenk, https://www.aip.org/history-programs/niels-bohr-library/oral-histories/33697). But Wheeler never really worked on submarine warfare; the closest thing was counseling work he did for the aircraft company Convair around 1955, which resulted in the conclusion that nuclear-powered airplanes were not feasible, as opposed to nuclear-powered submarines (Wheeler and Ford 2000, p. 272). But given that the first nuclear-powered submarines had already been commissioned at the time, this work would hardly have needed original research on submarines. It is thus conceivable that Wheeler read the book either for inspiration on the hydrodynamics-GR analogy (which he explored in great detail in Power and Wheeler (1957)) or for pleasure; Wheeler was, after all, a big fan of explosions in general, as recalled by Tullio Regge and Sandrelli (2012, p. 37): "[Wheeler] always used to go around equipped with firecrackers and he had a true, sincere and deep passion for explosions in general. That probably explains his enthusiastic participation both in the Manhattan project and in the later project for the development of the H bomb. Just to give you an idea, from time to time he would explode devices under the university toilets. I will never forget when, one Sunday, we managed to make a lot of stuff explode in the garden of the institute... You would have to be there! To be honest, we were not expecting to create such a mess: the police quickly arrived and we had to convince the officers that we were not terrorists, as they seemed inclined to believe, but mere theoretical physicists who were studying black stars." (our translation)

sphericity were amplified into horns and spikes at the point of maximal contraction. Wheeler considered this a promising "mechanical analog for the universe":

> A look at the bubble at this time [of maximal contraction] suggests, not an ideal mathematical object, but a glove which is turning itself inside out one finger at a time. Does the universe likewise undergo repeated oscillations? show instabilities? have short-lived localized regions of high curvature and high concentrations of matter energy? have no single well-defined time of maximum contraction? (Adams et al. 1958, p. 115f)

But these hydrodynamical analogies were short-lived, as Wheeler soon started emphasizing disanalogies, most importantly that the "universe has no internal pressure to reverse its oscillations" (NB5, p. 252, 16 August 1958). More importantly, results started to pile up that no non-singular bounce would be possible in general relativity, culminating in Hawking's cosmological version of the singularity theorem, the analog of Penrose's result for stellar collapse (Hawking 1966). As Wheeler came to accept the possibility that collapse beyond the Schwarzschild horizon was possible in the early 1960s, he also accepted that quantum theory would be needed to provide a picture of how an apparent collapse to a singularity was avoided and, in the cosmological case, was replaced by a rebound (Wheeler 1962, p. 70).

We thus see that Wheeler's assessment of cosmological gravitational collapse developed quite in parallel to his views on stellar collapse: he honed in on the problem early on, began looking for mechanisms to deal with critical aspects (baryon number non-conservation, need for a rebound), and by the late 1960s accepted that a quantum theory of gravity was necessary for a fuller description, while holding on to his basic intuitions. We will turn soon to the question of how his cosmological considerations changed when he began to turn away from his program of geometrodynamics around 1970. But first, we need to understand how cosmological gravitational collapse was connected with the question of physical law, above and beyond the issues arising in stellar collapse that we discussed in the preceding sections.

The central issue here was that the destruction of matter through the non-conservation of baryon number had, for Wheeler, farther-reaching implications when it occurred on a cosmological scale. The oscillations performed by the universe as a whole implied that the universe would periodically enter periods of high density, akin to those obtained in stellar collapse. In those periods, baryon number would no longer be conserved. Consequently, the universe would enter each cycle with a (possibly vastly) different matter content (Adams et al. 1958, p. 140). This was of course an interesting thought in and of itself.[31] But the total matter content of the universe was tied up with the laws of physics in a way that was again defined by Wheeler's reading of Mach's principle.

As we have already mentioned, Mach's principle postulated that the inertial forces were the result of interactions with distant masses. This very vague idea

[31] And one that would ultimately come to be seen as making the entire setup untenable, cf. Halpern (1995), p. 257.

found some concretization in general relativity, where these interactions were to be related to gravitation; but also here, the way in which *inertia* was to be understood in terms of interaction was not spelled out until 1953 when Wheeler and, in parallel or perhaps with priority, Dennis Sciama (Blum and Brill 2020) used a schematic version of general relativity to derive Newton's law of inertia from a test body's gravitational interaction with the rest of the universe under the condition that

$$\frac{G}{c^2}\frac{M}{R} \approx 1 \tag{1}$$

where G is Newton's gravitational constant, c is the speed of light, M is the total mass of the universe and R its radius.[32] Wheeler and Sciama initially just took this as a condition that was needed to fulfil Mach's principle, but it already clearly hinted at a close relation between the total mass of the universe and the laws obtaining within it.

Another connection between the total mass of the universe and its laws had been proposed by Dirac in 1937.[33] Dirac had pointed at the connection between a number of very large dimensionless numbers. These large numbers were (i) the age of the universe given in atomic units,[34] (ii) the total mass of the universe given in proton masses, and (iii) the ratio between the electric and the gravitational forces in a hydrogen atom. The first and the third were given by Dirac to be on the order of 10^{39}, while the second was on the order of 10^{78}, i.e., $(10^{39})^2$. Dirac provided a tentative explanation for these correlations: both the total mass of the universe and the gravitational constant were functions of cosmological time t, the mass increasing as t^2 (presumably through some process of spontaneous creation of matter), while the gravitational constant was decreasing as t^{-1}, so that the gravitational force grew progressively weaker as compared to the electromagnetic one. For Dirac the temporal evolution of the universe was thus driving a change both in its total mass and in its laws, thereby leading to a correlation between the two.

These ideas were taken up in the mid-1950s by Wheeler's Princeton colleague Robert Dicke, who also brought them to Wheeler's attention. Wheeler's notebooks contain notes taken at a talk given by Dicke at Princeton on 8 November 1956 (NB5, p. 42). They clearly show Wheeler's interest, but also his concern that a varying gravitational constant would be in conflict with the geological history of

[32] The equation in this exact form does not show up in Wheeler or Sciama, but first appears in Dicke's work (Dicke 1958, p. 36), which we shall discuss in a moment. It can be obtained from equations in Wheeler and Sciama under some simplifying assumptions, such as uniform expansion of the universe and a homogeneous distribution of the distant matter on a spherical shell of radius R. The same equation had already been obtained by Einstein (1917)—with the numerical ratio specified to be $\pi/2$—but with a totally different derivation involving the cosmological constant and the assumption of a static universe.

[33] Dirac (1937). See also Kragh (2016).

[34] The unit of atomic time that Dirac used was the time it takes light to traverse the classical radius of the electron.

earth. Wheeler remained skeptical when Dirac's hypothesis was again discussed by others at the already mentioned Solvay conference in June 1958, remarking in his conference notes: "Conclude phys[ical] const[ant]s aren't changing." (NB5, p. 228).

At around the same time, at a talk given to the Philosophical Society of Washington in April 1958, Dicke proposed an inversion of the causal relations underlying the correlations between the large numbers—an inversion based on Mach's principle: Dicke proposed that the mass of the universe should be taken as the primary large number. It determined the smallness of the gravitational constant through something like the Mach-inspired equation 1. The problem was then of course to explain how the age of the universe could be determined by the other large numbers; after all, the age of the universe seemed to be a brute fact if there ever was one. Here Dicke introduced what would later be called "anthropic" arguments:

> The present epoch is conditioned by the fact that the biological conditions for the existence of man must be satisfied. This requires the existence of a planetary system and a hot star. (Dicke 1958, p. 33)

Somehow, Dicke argued, the time needed to achieve the right "biological conditions" would depend on the size of the gravitational constant, thereby establishing the second causal link needed to explain the large number correlations. But it was not these anthropic considerations that initially appealed to Wheeler, although they would become, as we shall see, important later on. Rather, it was the idea of using Mach's principle not just to explain inertia and to deduce the closedness of the universe, but also to interpret natural constants as functions of the mass of the universe. In discussions with Dicke (NB5, p. 284, 24 September 1958), Wheeler remarked that Mach's principle might "change masses of elem[entary] particles or fine structure constant."

This idea came to new life after Wheeler in the early 1960s embraced the conclusion that the big bounce would be a genuinely quantum-mechanical process. Combined with the fact that baryon number (and thus the total mass of the universe) was not conserved in gravitational collapse, this implied that the initial matter content of a given cycle of the universe was merely determined probabilistically, as were then the natural constants determined through Mach's Principle. Wheeler began exploring this idea, evoking geological metaphors that recalled the original difficulties involved in having constants varying with time:

> We are accustomed to the idea of fossil molecules and fossil atomic nuclei. Should we also accustom ourselves to the idea of fossil elementary particles? The piece of wood on which I lay my hand was made in a tree a few years ago from carbon dioxide and water. Its molecules are "chemical fossils" of recent origin. The atomic nuclei of these carbon and oxygen atoms date back to an older epoch some billions of years ago, the time of thermonuclear combustion in the stars. They are "nuclear fossils". In another chemical environment the carbon dioxide and the water would have joined into a molecule quite different from cellulose. Nuclear matter cooked in a different star for a different length of time would have left us more iron and less oxygen as its fossil testament. Are we likewise to expect that the properties of the elementary particles and even their numbers differ between one cycle of the universe and another? (Wheeler 1968a, p. 268–269)

But he was still looking for a mechanism to explain how exactly the essentially random initial conditions determined the observed properties of the elementary particles, a question not answered by the invocation of Mach's principle. He recalled Laplace's Demon, remarking that "[t]he reader of 1814 [...] was wrong if he thought one would uncover the rationale of the initial conditions" (p. 271). The subsequent notebooks thus contain numerous attempts at trying to determine "how can size of universe affect properties of elementary particles" (NB 15, p. 87, 8 July 1968). He considered an analogy with the geodesic dome of Buckminster Fuller, where the size and shape of the dome at large determines the number of pentagons it was composed of, "a line of comparison that might almost be called Keplerian!" He explored the possibility that the identical particles observed today would only gradually emerge in the evolution of the universe:

> I have argued that [particle] properties are not superuniversal but differ from one history of the universe to another. Therefore a particle building phase must occur. Up to now I have always assumed that this phase is to be identified with the moment when the universe has the dimensions $L^* = (\hbar G/c^3)^{1/2}$. However, I see [...] that there is another possibility: that *the particles are not formed until sometime after the universe has commenced its expansion. Time* is needed to ensure that the various "bubbles in the dough" all have the same size. (NB 15, p. 174, 10 November 1968, emphasis in the original)

But everything changed with the fall of geometrodynamics discussed in the previous section. We have now finally reached the point where we can analyze the impact of Wheeler's embrace of lawlessness on his cosmology and then in turn analyze the implications of cosmology for lawlessness. We have identified and traced the origins of several central elements in Wheeler's cosmological thought in the late 1960s, just before the fall of geometrodynamics: the idea of a cyclic cosmology; the impact of cosmology, and in particular of the initial conditions of a given cycle, on elements of physical theory that would normally be considered immutable, in particular the mass spectrum of the elementary particles; and the search for a mechanism that would ensure this connection between the cosmological and the microphysical. With the abandonment of geometrodynamics it was now no longer just the matter content of the universe (and the natural constants determined by it) that was wiped out in the gravitational collapse of the universe; it was joined by "everything that one has ever called a law of physics." In Wheeler's cosmological thinking, this of course implied that all laws of physics were now to be considered cosmologically determined. But it also meant that there was no point in searching for a mechanism for how this determination occurred, because there were no more "superuniversal" laws that such a mechanism could obey.

Of course, the "golden trail of science" did not end there. In the absence of laws to argue from, Wheeler instead began to invoke "the quantum principle", a term he would continue emphasizing (sometimes simply as "the quantum") until the end of his days, in the new millennium (Tegmark and Wheeler 2001). What did it stand for? One might at first be tempted to believe that Wheeler was not actually embracing lawlessness, that he had merely rejected relativity or quantum gravity, and was now instead pursuing a new daring conservatism based on the established framework of quantum mechanics or quantum field theory. This is indeed how he was interpreted, e.g., by Steven Weinberg:

> My friend and teacher John Wheeler has occasionally suggested that there is no fundamental law and that all the laws we study today are imposed on nature by the way that we make observations. (...) Wheeler (...) simply seem(s) to me to be merely pushing back the problem of the final laws. Wheeler's world without law still needs metalaws to tell us how our observations impose regularities on nature, among which metalaws is quantum mechanics itself. [...] I expect that all attempts to do without fundamental laws of nature, if successful at all, simply result in the introduction of metalaws that describe how what we now call laws came about. (Weinberg 1993, p. 186)

This does not do justice to Wheeler's invocation of the "quantum principle".[35] The expression was Wheeler's and it did not refer to any sort of metalaw, like the least action principle, say, from which any part of the formal apparatus of quantum mechanics could be derived directly. Wheeler was referring to a far more abstract, philosophical core idea of quantum physics: the relation between observer and observed system, which Wheeler had long viewed as one of the central lessons of his mentor, Niels Bohr (Wheeler 1963a, p. 30). While the quantum principle was thus grounded in history, it was also transformed by Wheeler:[36] unlike Bohr's observer, Wheeler's observer was to be of cosmological and cosmogonic significance. Wheeler explored this significance in a series of talks he gave in the mid-1970s, tellingly entitled "The universe as home for man" (Wheeler 1975);[37] "Is physics legislated by cosmogony?" (Patton and Wheeler 1975); and "Genesis and observership" (Wheeler 1977).

The question of the observer had played a role in Wheeler's cosmological thought early on, as it raised the question of how to interpret the wave function of the universe (Blum and Lessel 2022). An extrapolation of the Copenhagen interpretation, with its dichotomy between the classical observer and the observed quantum system, appeared to require an observer with a god-like point of view, looking at spacetime from the outside, the system under observation being the universe itself. This was a problematic conclusion, to say the least. The view proposed by Wheeler's 1957 PhD student Hugh Everett III offered an attractive alternative: observers are just subsystems *within* the world. Everett described these

[35] That is not to say that Wheeler was not impressed by the contemporary developments in quantum field theory. The early 1970s was also the time of the establishment of the Standard Model of Particle Physics (Hoddeson et al. 1997), and this development was clearly a further influence in Wheeler's realization that geometrodynamics was an insufficient foundation for all physical theory: "general relativity [...] created a new standard for the surprise of a prediction and for the scope of an explanation. The standard has meantime risen, not least because of the beautiful regularities uncovered in particle physics. General relativity has not kept up with the rise." (Wheeler 1973, p. 236).

[36] As is typical of Wheeler's use of history, see Furlan (2020).

[37] Wheeler presented this talk at a symposium celebrating the 500th birthday of Copernicus. The irony of presenting an anthropocentric cosmology on this occasion was not lost on the participants, with the bacterial physiologist Bernard Davis remarking that it was "a little bit contrary to the Copernican tradition". (Gingerich 1975, p. 582).

subsystems in his thesis draft as "automatically functioning machines, possessing sensory apparatus and coupled to recording devices capable of registering past sensory data" (reprinted in (Everett 1973, p. 64)). Wheeler preferred to think of them as "sufficiently complex to simulate all that human beings do, including acting, observing, and recording" and as having "arisen within the overall system by organic evolution" (Wheeler to Alexander Stern, 25 May 1956, reprinted in (Barrett and Byrne 2012, p. 221)).

After the rejection of Everett's ideas by Bohr and other Copenhagen physicists (Osnaghi et al. 2009), Wheeler left the problem of the embedded observer more or less alone for a decade,[38] before returning to it in the early 1970s in the wake of the fall of geometrodynamics. His take was rather unorthodox, but this should not be taken to imply that he was positioning himself against the Copenhagen interpretation, or in any way positioning himself within the emerging "interpretation debate".[39] "Interpretations" such as Everett's were just one of many ways of approaching the quantum principle (all potentially fruitful), on par with Feynman's path integral or the quantum logic of Birkhoff and von Neumann (Wheeler 1975, p. 288),[40] a principle so many-faceted it was admittedly hard to grasp:

> Quantum principle? What an inadequate name for an overarching feature, or *the* overarching feature, of nature. We understand any other principle of physics in enough completeness to summarize it, beginning with a good name, in a dozen words. But not this. It continually unfolds with fresh meaning. It might almost better be called the "Merlin principle". Merlin the magician, pursued, changed first to a fox, then to a rabbit, and finally to a bird fluttering on one's shoulder. (Wheeler 1975, p. 287)

Indeed, Wheeler's take on the quantum principle is best understood as a combination of several elements: Bohr's complementarity, Everett's ideas on the embedding of the observer, and also the views of his Princeton colleague Eugene Wigner on the role of human consciousness in measurement (on which more in the next section). The resulting statement of the quantum principle was ultimately very simple:

> [B]eyond the rules of quantum mechanics for calculating answers from a Hamiltonian stands the quantum principle. [...] It promotes observer to participator. It joins participator with system in a "wholeness" (Niels Bohr) [...] quite foreign to classical physics. It demolishes the view we once had that the universe sits safely "out there", that we can observe what goes on in it from behind a foot-thick slab of plate glass without ourselves being involved in what goes on. We have learned that to observe even so miniscule an object as an electron we have to shatter that slab of glass. We have to reach out and insert a measuring device. We can put in a device to measure position or we can insert a device

[38] Though it did not disappear entirely from his thinking, cf. Wright (2014), p. 187–188.

[39] On which, see Freire (2015).

[40] In view of the spectrum of alternatives that has crystallized into today's interpretation debate, there is a quite striking absence in the many facets of the quantum principle invoked by Wheeler: Bohmian mechanics. While a comprehensive story of the relationship between Wheeler and David Bohm, also on a personal level, has still to be written, it seems safe to state here that such approaches were excluded by Wheeler quite simply because they pointed in a direction exactly opposite to his own, by trying to restore a classical level beneath the quantum.

> to measure momentum. But the installation of one prevents the insertion of the other. We ourselves have to decide which it is that we will do. Whichever it is, it has an unpredictable effect on the future of that electron. To that degree the future of the universe is changed. We changed it. We have to cross out that old word "observer" and replace it by the new word "participator".[41]

And the influence of this "participator" could also stretch back in time, e.g., when observing "radiative evidence of what went on in the first few seconds after the big bang" (Wheeler 1977, p. 25).[42] Without the observer, possibilities do not become actual existences. It was this ability of observers to actualize possibilities[43] on cosmic scales that allowed them to replace the mechanism for determining the laws of the universe that Wheeler had been looking for. Going even further, it allowed the observer to be responsible for the genesis of the universe itself, to be a "participator in genesis" (Wheeler 1977, p. 25). No endless cycles of the universe with probabilistic initial conditions were needed anymore; the preference for cyclic cosmology had anyway, as we have seen, been rooted in a very lawful reading of general relativity and Mach's principle.[44] In Wheeler's new "self-reference cosmology" observership was sufficient for explaining cosmogony:

> Self-reference cosmology has these features: (1) one cycle only; (2) the laws and constants and initial conditions of physics frozen in at the big bang that brings the cycle and dissolved away in the final extremity of collapse; and (3) a guiding principle of "wiring together" past, present and future that does not even let the universe come into being unless and until the blind accidents of evolution are guaranteed to produce, for some non-zero stretch of time in its history-to-be, the consciousness, and communicating community, that will give *meaning*[45] to the universe from start to finish. (Patton and Wheeler 1975, p. 566–567)

This then was Wheeler's re-imagination of Dicke's anthropic arguments. We might end our story of how John Wheeler lost his faith in the law here. We have seen how the transcendence of law in gravitational collapse and the apparent inability of physics to provide a satisfactory final theory led Wheeler to abandon a traditional view of reductionist physical law and instead embrace the view that on a cosmological level "the observer himself closes up full circle the links of interdependence between the successive levels of structure" (Wheeler 1977, p. 3).

---

[41] Patton and Wheeler (1975), pp. 560–562.

[42] This observation can be viewed as an anticipation of Wheeler's later work on cosmological delayed-choice experiments (Wheeler 1983b, p. 190).

[43] An element of quantum theory for which Wheeler liked to invoke a quote by William James, "Actualities seem to float in a wider sea of possibilities from out of which they are chosen", e.g., in Wheeler (1971a), p. 288.

[44] It should be noted, however, that Wheeler did not abandon his Mach-inspired views on the *spatial* boundary conditions of the universe and continued to assume that the universe was closed. He thus did not question the possibility that the universe would end in a big crunch, and his cosmology was thus still heavily shaped by Mach's Principle.

[45] Here Wheeler included a footnote, quoting the Norwegian philosopher Dagfin Føllesdal: "Meaning is the joint product of all the evidence that is available to those who communicate."

We have also seen that this invocation of the observer can hardly be considered, as Weinberg did, as falling back on (the metalaws) of quantum mechanics.

That is not to say that there is no element of "pushing back of the problem" in Wheeler's post-geometrodynamical thinking. But it was pushed back out of physics and into different fields of thought. When exploring the structure of what he referred to as "pregeometry", Wheeler started drawing on mathematics and propositional logic, culminating in his slogan "it from bit" (Wheeler 1990).[46] This would bring us to the relation between laws of nature and the laws of mathematics and logic, which is beyond the scope of this paper. But there is another aspect of Wheeler's pushing back that relates more directly to laws of nature, in particular to laws from other disciplines of the natural sciences. For while Wheeler's notion that the observer is responsible, as it were, for the creation of universe might at first glance sound like naive idealism, he was not claiming that the universe is a projection or creation of the mind. Mind could not be primary, because it, in turn, had to be brought into being by the universe. The question of how mind (and with it life) were constituted and how they came into being was thus an important element of Wheeler's self-reference cosmology. In the last section we will thus offer a brief gloss of Wheeler's engagement with the laws of (evolutionary) biology.

4 Creating the Observer

Wheeler was not the only one in the early 1970s thinking about what constituted the observer. In NB18, p. 219, 10 June 1971, in a small section quite significantly entitled "Program", we find a reference to a recent work by Wigner (1970), "Physics and the explanation of life". In the 1960s, Wigner had gotten interested in foundational questions of quantum mechanics and had put forth his ideas that human consciousness might be responsible for the collapse of the wave function (Freire 2015, Chapter 4). In the 1970 paper, Wigner observed:

> [T]here is a tendency in both physics, which we consider as the most basic science dealing with inanimate objects, and in the life sciences to expand toward each other. Furthermore, the tendency is strongest in the modern parts of the two disciplines: in quantum mechanics on the one hand, in microbiology on the other. Each feels it cannot get along by relying solely on its own concepts. (p. 39)

Wigner did not venture to propose some specific idea (or mechanism) to bridge the gap. But he did present the two main alternatives: either the "laws of nature, for the formulation of which observations on inanimate matter suffice, are valid also for

[46] See Wright (2014), Section 4.5. It should also be noted that Wheeler's turn to propositional logic was also related to the Merlin principle, another facet of which was the quantum logic of Birkhoff and von Neumann.

living beings" (p. 39)[47] or "the laws of physics will have to be modified drastically if they are to account for the phenomena of life" (p. 41). In the paper, Wigner clearly expressed his preference for the second option.

Which brings us back to Wheeler, whose program was rather ambiguous regarding this point. On the one hand, the physical laws, once constituted in the big bang, were supposed to give rise by themselves to the conscious observer. This was in keeping with his earlier daring-conservative methodology, which might be called reductionist, in the sense of starting from fundamental microscopic laws, while also being exploratory, an open-ended pursuit of the manifold possibilities those fundamental laws provided. Viewed from this angle, invoking a modification of these fundamental laws for the purpose of obtaining consciousness would appear as a "free play with ideas" and thus as the opposite of daring conservatism.

On the other hand, however, in the cosmological reflections Wheeler was developing, physical law and the existence of the universe were based on the existence of sentient observers, a view that seems far closer to the second option. Wheeler thus concurred with Wigner that "[i]f the concept of observation is to be further analyzed, it cannot play the primitive role it now plays in the theory" (p. 43). In the early 1970s, consequently, Wheeler started reaching out to scientists from beyond physics, who might be able to proved insights for fleshing out the concept of observer in his new anthropic cosmology. For instance, he sent "The Universe as Home for Man" to the neurophysiologist John Eccles. Eccles seemed pleased with the paper (Letter to Wheeler of 30 December 1975, Wheeler Papers, Box 8), which appeared to resonate with his own, more philosophical, work in which he argued against the "materialist monists" (Eccles and Popper 1977). Not much of a dialogue, however, seems to have been born from this brief exchange. Richer and more fruitful were the interactions that Wheeler had with two eminent scientists who, like Wheeler, had a background in the physical sciences, but had become interested in questions of life and even consciousness, Manfred Eigen and Ilya Prigogine.

Eigen's talk at the 1972 Trieste Symposium (where Wheeler had first enunciated the end of all physical law) made a great impression on Wheeler. A couple of years later Wheeler wrote to Eigen that "[i]n the Trieste seminar of two years ago [...] no talk more captured my imagination, no talk seemed more to break new ground, and none opened wider vistas than your own" (Wheeler Papers, Box 8). Eigen was then invited to Princeton—to give a colloquium on, Wheeler suggested, "Physical and Chemical Background of the Origin of Life" and for "a Trieste reunion" with "[a]nother one of our Trieste symposium colleagues, Eugene Wigner". Let us try, then, to get an idea of what Wheeler appreciated in Eigen's talk and why he was so keen on exchanging ideas with someone who "belong[ed] just as much to the world of physics as to the world of biology".

[47] Though it is important to specify here that "this assumption need not imply, as is often postulated, that the mind and the consciousness are only unimportant derived concepts which need not enter the theory at all" (p. 41).

Eigen had begun his Trieste talk by stating: "In our days the physicist's conception of nature certainly has to include the phenomenon of life". Eigen went on to illustrate his hypercycle model for the origin of life,[48] not as an ultimate answer, but as a possible physical mechanism worthy of exploration:

> Do the known laws of physics provide a complete basis for an understanding of the phenomenon life? All we can do at the moment is, to disprove the negation of such a claim by quoting counter-examples. In other words we can use physical models to disprove any claim that the known laws of physics are not sufficient to describe the phenomena which specify a living object. The hypercycle is one example. (Eigen 1973, p. 631)

In tentatively exploring what we have referred to as Wigner's first option, Eigen placed an emphasis on known laws—a heuristic strategy that is clearly reminiscent of daring conservatism. Before making definitive proclamations or invoking the need for radical changes, Eigen adopted and insisted on a constructive attitude that tried to squeeze out the potential inherent in established laws. Eigen was offering a picture where some underlying randomness became encapsulated in more and more complex levels of organization, through some sort of selection process. Even if the actual dialogue between Wheeler and Eigen appears, at least in the light of currently available archival material, not to have been that extended, Wheeler would go on to prominently cite Eigen's work at the beginning of his 1977 Varenna lectures, which represented the first sum of Wheeler's post-geometrodynamical thinking and introduced the phrase "law without law" (Wheeler 1979c). Here Eigen was (the only one!) credited with having shown that "[t]he hierarchical speciation of plant and animal life [...] arises out of the blind accidents of genetic mutation and natural selection", i.e., that evolution proceeding in accord with known physical laws could indeed bring forth an observer.[49]

Prigogine's ideas were more radical, advocating the need for a fresh conceptual reformation at the basis of physics, closer to Wigner's second option. These ideas resonated with Wheeler's post-geometrodynamical thinking. Like Eigen, Prigogine had attended the Trieste symposium, where he had given two talks, the main one being on "Time, Irreversibility and Structure", where he, too, spoke on how to get structure from randomness, in this case (quantum)[50] fluctuations:

---

[48] By using systems of non-linear differential equations, Eigen considered a class of self-replicating macromolecules (possibly nucleic acids—the topic of the speech, after all, concerned life and its origins), linked in such a way that each one of them acts as a catalyst for the formation of the next one, with the last one catalyzing the first. "Such a cyclic hierarchy of catalytic processes may be called a 'hypercycle'" (Eigen 1973, p. 623).

[49] Eigen, in turn, would quote the grand finale of Wheeler's Trieste talk, "We can believe that we will first understand how simple the universe is when we recognize how strange it is" in his book "Das Spiel. Naturgesetze steuern den Zufall" (Eigen 1978, p. 244).

[50] Prigogine had worked (George et al. 1972) on the foundations of quantum mechanics, co-authoring a paper with Léon Rosenfeld, in which they aimed to refine Bohr's ideas on measurement—since, as Prigogine later stated, "Bohr's ideas were largely intuitive. Bohr has said deep things, but in a rather obscure fashion" (Buckley and Peat 1979, p. 75). Indeed, his second, shorter talk at Trieste was on the subject of "Measurement Process and the Macroscopic Level of Quantum Mechanics" (Prigogine 1973a).

> Now fluctuations - the spontaneous deviations from some average regime - are a universal phenomenon of molecular origin and are always present in a system with many degrees of freedom. [...]
> A new order principle appears that corresponds essentially to an amplification of fluctuations and to their ultimate stabilization by the flow of matter and energy from the surroundings. We may call this principle 'order through fluctuations'. (Prigogine 1973b, p. 587)

But their main interactions occurred after Wheeler retired from Princeton and moved to the University of Texas in 1976, where Prigogine was a professor.[51] Prigogine's work on non-equilibrium thermodynamics had offered him the hints for his new vision: complex "dissipative" structures, self-organized in (local) defiance of the second law of thermodynamics, had a characteristic lifetime within which their complexity was built and ultimately dissolved; no view of time as a fourth dimension or as a mere parameter that "flows" could do justice to this *creative* role of time, an explicitly Bergsonian notion. Mankind's creativity is thus not an accident that developed meaninglessly in the midst of barren immensities. This was a fundamental insight underlying Prigogine's proclamation of a "Nouvelle Alliance" (Prigogine and Stengers 1979) between scientific and humanistic knowledge, working together to provide a picture of the universe where we have a place and can find a meaning.

Even from this cursory outlook, the affinities with the vision Wheeler began to develop in the early 1970s are evident, as are the common topics. One commonality lies in their view of statistical quantum fluctuations not as undesirable deviations from a given order, but as the basis out of which order and complexity can emerge (quantum foam, order through fluctuations).[52] Wheeler himself would highlight the analogies between Prigogine's dissipative structures and his universe, which was similarly self-organized (lawless) and built and ultimately dissolved complexity (the observer) within its characteristic lifetime from big bang to big crunch (Wheeler 1979c, p. 154).[53] But this juxtaposition also revealed a fundamental disanalogy, which Wheeler also highlighted: for the universe the end in a big crunch is preordained, and we also have *final* boundary conditions. For Prigogine, and even more strongly for Eigen, the resolution of the dichotomy between physical matter on the one hand and life (and, for Prigogine, even meaning) on the other is achieved (or achievable) on a local level, where the laws of physics could be complemented with novel mechanisms (hypercycles, dissipative structures) and (in Prigogine's case)

---

[51] Prigogine highlighted his interactions with Wheeler several years later in Géhéniau and Prigogine (1988).

[52] Wheeler would use similar wording when speaking about Darwin: "It's inspiring to read the life of Charles Darwin and think how the division of plant and animal kingdoms, all this myriad of order, came about through the miracles of evolution, natural selection and chance mutation. To me this is a marvelous indication that you can get order by starting with disorder." (Wheeler 1979b).

[53] Prigogine, in turn, would approvingly refer to Wheeler's observer-participator, highlighting it, e.g., in later and updated translations of *La Nouvelle Alliance* (Prigogine and Stengers 1981, p. 267). He would also frequently refer to Wheelerian themes, including a chapter on irreversibility and spacetime structure in Prigogine (1985) or an appendix on black holes in Nicolis and Prigogine (1989).

new conceptualizations of time to produce life and all that comes with it. Wheeler's vision, instead, was global and made reference to the whole universe. It was in the creation and in the collapse of the universe as a whole that the foundation in physical law vanished, a foundation that remained central to Eigen's and Prigogine's thinking, serving them (as it had Wheeler before the fall of geometrodynamics) as a resource for mechanism. Wheeler, on the other hand, had to do without law and thus had to rely on a form of teleology.

This is precisely where Max Delbrück put his finger. Delbrück, who had turned from physics to biology many years earlier, influenced by Bohr's speculations on the connection between life and quantum mechanics,[54] had received from Wheeler a copy of "Is Physics Legislated by Cosmogony?". In his reply thanking Wheeler, Delbrück raised some objections to the way evolution was being invoked:

> What bothers me with respect to your universe which self-excites into the development (evolution) of mind is this: it ignores some facts we know about the evolution of mind, viz., that the brain (and mind) evolved in *adaptation* to very specific requirements for getting along and permitting its possessor to survive, in the water, on the ground, in the trees, in the cave. [. . .] It did <u>*not*</u> evolve in adaptation to a requirement to invent relativity, Qu[antum] M[echanics], cosmogony, etc.[55]

In other terms, the cosmogonic role of the observer introduces a teleology which is in contrast with the observer as product of Darwinian evolution, with all its accidents and contingencies; yet both aspects were supposed to be present in Wheeler's cosmology. Wheeler's answer is not extant, but an answer of sorts, albeit a paradoxical one, can be found in the very paper he had sent to Delbrück. Wheeler was clearly genuine in wanting to accept the role of the environment and of chance in the genesis and evolution of more and more complex structures. He was, as we have seen, also receptive to attempts at providing, on the basis of modern physics, an intrinsically random mechanism for Darwinian evolution to proceed on; and, perhaps, at providing a physical description of life and its origins, as suggested—at least in terms of viability—by Eigen and Prigogine. But on the global and cosmogonic level that Wheeler was trying to face, that alone could not be enough. That was why in Patton and Wheeler (1975) he had appealed to a deep "wiring together" of past, present and future and invoked the following colorful metaphor for combining teleology and random evolution:

> Chance mutation, yes; Darwinian evolution yes; yes, the general is free to move his troops by throwing dice if he chooses but he is shot if he loses the battle. Deprived of all meaning, stripped of any possibility to exist is any would-be universe where Darwinian evolution brings forth no community of evidence-sharing participators. . . (p. 567)

This does not exactly sound like a scientific justification; it sounds more like a half-ironical play with ideas. Wheeler was clearly struggling to bring across the radical conclusions he had drawn from the long and open-ended intellectual

[54] See, e.g., Delbrück (1949).

[55] Letter from Delbrück to Wheeler, 14 April 1975; Max Delbrück Papers, California Institute of Technology Archives and Special Collections, Box 24, Folder 10.

development we have been reconstructing in this paper.[56] Nevertheless, the idea remained problematic, to use a slight euphemism, and the tension pointed out by Delbrück was there to stay: Wheeler's invocation of a "Darwinian" mechanism, regardless of its convoluted paths, *had to* give rise to the observer—which appears incompatible with a Darwinian world view. What we see emerging here is—it should be noted—a *new* kind of global tension between biology and physics that went beyond any local reconciliation.

This issue was not resolved. Instead, Wheeler's observer would be gradually de-anthropomorphized. Originally his observers had been very human. But the specifics of this observer moved to the background. Once in Texas, Wheeler's focus shifted to the process (rather than the agent) of quantum measurement (Wheeler and Zurek 1983). Already in the Varenna lectures we can notice such a tendency. After remarking that the "strange necessity of the quantum as we see it everywhere in the scheme of physics comes from the requirement that—via observer-participancy—the universe should have a way to come into being", Wheeler highlights the prerequisite for the cosmic game (a metaphor, not unlike Eigen's, that runs through all the Varenna lectures) to start:

> I know that in that empty courtyard many a game cannot be a game until a line has been drawn—it does not matter where—to separate one side from the other. [...] But how much arbitrariness is there in this more ethereal kind of demarcation, the line between "system" and "observing device". (Wheeler 1979c, p. 18).

Instead of stressing the human features of the observer (or consciousness), Wheeler adopted a much more sober, in a sense even minimalistic, view:

> The same stick, when grasped firmly and used to explore something else, becomes an extension of the observer or—when we depersonalize—a part of the measuring equipment. As we withdraw the stick from the one role, and recast it in the other role, we transpose the line of demarcation from one end of it to the other. The distinction between the probed and the probe, so evident at this scale of the everyday, is the without-which-nothing of every elementary phenomenon, of every "closed" quantum process.
> Do we possess today any mathematical or legalistic formula for what the line is or where it is to be drawn?
> No.
> Then what is important about this demarcation?
> Existence, yes; position, no. (Wheeler 1979c, pp. 18–19)

The distinction between the probed and the probe, together with an explicit rebuttal of the centrality of consciousness, reappears in a paper very explicitly entitled "Not Consciousness but the Distinction Between the Probe and the Probed as Central to the Elemental Quantum Act of Observation" (Wheeler 1981). Wheeler was now exploring a different facet of the quantum principle:

> What one word does most to capture the central new lesson of the quantum? 'Uncertainty,' so it seemed at one time; then 'indeterminism'; then 'complementarity'; but Bohr's final

[56] Already in 1967 we find him lamenting in his notebooks: "Why is everyone so silent? Why does it have to be me who says this? Story *crying* to be told. (NB 14, p. 140, 27 January 1967).

> word 'phenomenon'—or, more specifically, 'elementary quantum phenomenon'—comes still closer to hitting the point. [...] In today's words, no elementary quantum phenomenon is a phenomenon until it is a registered ('observed' or 'indelibly recorded') phenomenon, 'brought to a close' by an 'irreversible act of amplification'. (Miller and Wheeler 1984)[57]

There was clearly a strong shift of emphasis, even if Wheeler never rejected his previous anthropic considerations.[58] Even if humankind was no longer the protagonist of his worldview, humankind was still very much his intended audience. Not unlike Prigogine, Wheeler kept trying to evoke a worldview where humans could feel "at home in the universe" and be reassured of its harmony (Wheeler 1994).[59] Indeed, despite affirming the lawlessness of the universe, the resulting picture is not at all one of chaos and meaninglessness, but that of a cosmos where a sort of pre-established harmony (a concept for which Wheeler drew on a tradition that he traced back via Einstein to Leibniz (Wheeler 1983a))[60] arises as the background for the cosmic game.

5 Conclusions

We have followed John Wheeler's intellectual trajectory on the long (and sometimes winding) path from a resolute belief in the laws of physics to an embrace of fundamental lawlessness and the search for new ways of constituting the physical world using concepts from the life sciences. Let us briefly recapitulate: In the late 1960s, Wheeler realized that the status of physical laws seemed to be modified in gravitational collapse. Laws were being "transcended" (Sect. 1). Around the same time, his hopes for constructing a theory of everything from the foundations of current physical theory (general relativity and quantum mechanics) faltered (Sect. 2). These two factors led him to embrace ideas of fundamental lawlessness. Wheeler attempted to fill the resulting void through a novel combination of cosmology and consciousness, his participatory universe. This brought in questions and concepts from biology on two levels, through the constitution of the live and sentient observer and through the need of invoking a deep-time evolution that combined the evolution of the cosmos as a whole with the biological evolution of the observer.

---

[57] On Bohr's use of the term phenomenon, cf. Wheeler (1979a), fn. 91.

[58] He would write, e.g., the foreword to John Barrow and Frank Tipler's "The Anthropic Cosmological Principle" (Barrow and Tipler 1986).

[59] Wheeler's hopes for the effect this reassurance would have on humanity could be, to put it mildly, rather idiosyncratic. As he remarked in some of his first notes on anthropic arguments: "What strange effect would it have on everyone's morality, our resolution in the war in Viet-Nam and everywhere else, our image of ourselves and our children, to have such views accepted." (NB14, p. 173f, 3 June 1967).

[60] See also, SF, "Merging Labyrinths: Leibniz in J.A. Wheeler's Quest", to be published in *Studia Leibnitiana* (2022).

These ideas remained vague and problematic. Ultimately, the fundamental lawlessness that Wheeler embraced when conferring a cosmogonic (and nomogonic) role to his observer-participator (Sect. 3) provided no clear basis for further scientific investigation. His move away from an anthropic observer in the second half of the 1970s (see Sect. 4) also implied a mitigation of the radical lawlessness that we highlighted in this paper. In this later phase (bearing in mind the difficulties of periodizing Wheeler mentioned in the introduction), Wheeler entered for the first time the grey zone between the strict lawfulness of daring conservatism and the total lawlessness of the cosmogonic observer-participator, the grey zone where underlying basic principles are intertwined with the emergence of statistical regularities from a large number of individual events. In this sense, Wheeler over time traversed the entire spectrum of alternatives that defines the debate on laws of nature to this day. But what our historical reconstruction hopefully makes clear is that this was not merely the result of changes in philosophical attitude; it was rather Wheeler's methodology of pushing ideas to their limits—and the novel physical insights that it brought—which led Wheeler, almost dialectically, from a position to another. Similarly, the radical lawlessness he explored might seem whimsical, a "postmodern" play with ideas, when looking at some of Wheeler's writings in isolation. But Wheeler's cross-disciplinary reconfiguration of ideas was not something that he imported from wider cultural contexts; it arose from a crisis within modern fundamental physics itself, where laws were transcended and appeared insufficient to ground all of physics by themselves. It should also be remarked that this crisis was one that did not result directly from the great conceptual upheavals of early twentieth century physics; in spite of their epistemological novelty, relativity and quantum mechanics had been well integrated into and ultimately even strengthened, also in Wheeler's thinking, a lawful (though non-deterministic) view of the physical universe by vastly extending the scope of scientific explanation. It was rather later developments, related to the attempts at combining relativity and quantum theory, that, for Wheeler, set in motion the crisis of law.

Wheeler's turn to radical lawlessness was not just unsustainable (at least if one wanted to continue doing science, as Wheeler evidently did), it may well also have seemed premature at the time. After all, the 1980s saw the rise of string theory, which promised to be both a quantum theory of gravity and a theory of everything, thereby fulfilling Wheeler's ambitions for geometrodynamics.[61] At the same time, advances were made in "quantum gravity proper", which appeared to imply that such a theory might have more to say about the structure of spacetime than Wheeler's first lawyer had let on. But these major breakthroughs are already several decades in the past, and one increasingly gets the impression that new impulses are needed. It may thus well be time to take another look at Wheeler's exploration of various degrees of lawfulness. It is hoped that the insights provided in this paper into the historical origins of Wheeler's fascinating reconfiguration of

[61] Wheeler's only assessment (that we are aware of) of string theory can be found in his autobiography: "intriguing" (Wheeler and Ford 2000, p. 320).

ideas can contribute to a better understanding of their position within the framework of science and their possible role in its future, not least by highlighting the crucial questions that Wheeler ultimately struggled with: if no final theory of physics is to be had, on what ground can the lawfulness of the physical world stand? And how can physical law be reconciled with life, consciousness and even meaning on a *global* scale?

References

Adams, J., Harrison, B., Klauder, J., Mjolsness, R., Wakano, M., Wheeler, J., & Willey, R. (1958). Some implications of general relativity for the structure and evolution of the universe. In *Onzième Conseil de Physique, Brussels 1958, La structure et l'évolution de l'universe* (pp. 97–148). Institut Solvay: Stoops.

Barrett, J. A., & Byrne, P. (Eds.) (2012). *The Everett interpretation of quantum mechanics: collected works 1955–1980 with commentary*. Princeton: Princeton University Press.

Barrow, J. D., Galloway, G. J., & Tipler, F. J. (1986). The closed-universe recollapse conjecture. *Monthly Notices of the Royal Astronomical Society, 223*(4), 835–844.

Barrow, J. D., & Tipler, F. J. (1986). *The anthropic cosmological principle*. Oxford: Oxford University Press.

Bekenstein, J. D. (1972). *Baryon number, entropy, and black hole physics*. Ph. D. thesis, Princeton University.

Blum, A. S., & Brill, D. (2020). Tokyo Wheeler or the epistemic preconditions of the renaissance of relativity. In Blum, A. S., Lalli, R., & Renn, J. (Eds.) *The renaissance of general relativity in context* (Vol. 16, pp. 141–187). Einstein studies. Basel: Birkhäuser.

Blum, A. S., & Hartz, T. (2017). The 1957 quantum gravity meeting in Copenhagen: An analysis of Bryce S. DeWitt's report. *European Physical Journal H, 42*, 107–157.

Blum, A. S., & Lessel, B. (2022). The interpretation debate and quantum gravity. In Freire, O. (Ed.) *Oxford handbook of the history of interpretations of quantum physics*. Oxford: Oxford University Press.

Bonolis, L. (2017). Stellar structure and compact objects before 1940: Towards relativistic astrophysics. *European Physical Journal H, 42*(2), 311–393.

Borrelli, A. (2015). The making of an intrinsic property: "symmetry heuristics" in early particle physics. *Studies in History and Philosophy of Science, 50*, 59–70.

Buckley, P., & Peat, F. D. (1979). *A question of physics: Conversations in physics and biology*. Toronto: University of Toronto Press.

Cole, R. H. (1948). *Underwater explosions*. Princeton: Princeton University Press.

Delbrück, M. (1949). A physicist looks at biology. *Transactions of The Connecticut Academy of Arts and Sciences, 38*, 173–190.

DeWitt, B. S. (1957). Dynamical theory in curved spaces. I. A review of the classical and quantum action principles. *Reviews of Modern Physics, 29*(3), 377–397.

DeWitt, B. S. (1967). Quantum theory of gravity. I. The canonical theory. *Physical Review, 160*, 1113–1148.

DeWitt, C. M. (2011). *The pursuit of quantum gravity: Memoirs of Bryce DeWitt from 1946 to 2004*. Berlin: Springer.

Dicke, R. H. (1958). Gravitation an enigma. *Journal of the Washington Academy of Sciences, 48*(7), 213–223.

Dirac, P. A. M. (1937). The cosmological constants. *Nature, 138*(3512), 323.

Eccles, J., & Popper, K. (1977). *The self and its brain*. Berlin: Springer.

Eigen, M. (1973). The origin of biological information. In *The physicist's conception of nature* (pp. 594–632). Dordrecht: Kluwer Academic Publishers.

Eigen, M. (1978). *Das Spiel. Naturgesetze steuern den Zufall*. München: Piper.

Einstein, A. (1917). Kosmologische Betrachtungen zur allgemeinen Relativitätstheorie. *Sitzungsberichte der Königlich Preussischen Akademie der Wissenschaften zu Berlin*, 142–152.

Everett, H. (1973). The theory of the universal wave function. In DeWitt, B. S., & Graham, N. (Eds.) *The many-worlds interpretation of quantum mechanics* (pp. 1–140). Princeton: Princeton University Press.

Freire, O. (2015). *The quantum dissidents*. Berlin: Springer.

Fuller, R. W., & Wheeler, J. A. (1962). Causality and multiply connected space-time. *Physical Review, 128*, 919–929.

Furlan, S. (2020). Einstein's Mantle, Bohr's Shadow: Glimpses from wheeler's relativity notebook III. In *Proceedings of the 39th Annual Conference; Pisa, 9–12 September 2019*, Pisa (pp. 221–227). Società Italiana degli Storici della Fisica e dell'Astronomia (SISFA): Pisa University Press.

Furlan, S. (2021). John Wheeler between cold matter and frozen stars: The road towards black holes. MPIWG Preprint No. 504.

Géhéniau, J., & Prigogine, I. (1988). The birth of time. In Zurek, W. H., van der Merwe, A., & Miller, W. A. (Eds.) *Between quantum and cosmos: Studies and essays in honor of John Archibald Wheeler*. Princeton Legacy Library (pp. 159–165). Princeton: Princeton University Press.

George, C., Prigogine, I., & Rosenfeld, L. (1972). The macroscopic level of quantum mechanics. *Det Kgl. Danske Videnskabernes Selskab Mathematisk-fysiske Meddelelser, 38*(12), 1–44.

Geroch, R. P. (1967). Topology in general relativity. *Journal of Mathematical Physics, 8*(4), 782–786.

Gingerich, O. (Ed.) (1975). *The nature of scientific discovery*. Washington, DC: Smithsonian Institution Press.

Hahn, S. G., & Lindquist, R. W. (1964). The two-body problem in geometrodynamics. *Annals of Physics, 29*, 304–331.

Halpern, P. (1995). *The cyclical serpent*. Berlin: Springer Science+Business Media.

Harrison, B. K., Thorne, K. S., Wakano, M., & Wheeler, J. A. (1965). *Gravitation theory and gravitational collapse*. Chicago: University of Chicago Press.

Hawking, S. W. (1966). The occurrence of singularities in cosmology. *Proceedings of the Royal Society A: Mathematical and Physical Sciences, 294*(1439), 511–521.

Hoddeson, L., Brown, L., Riordan, M., & Dresden, M. (Eds.) (1997). *The rise of the standard model*. Cambridge: Cambridge University Press.

Israel, W. (1967). Event horizons in static vacuum space-times. *Physical Review, 164*(5), 1776–1779.

Kragh, H. (2016). *Varying gravity: Dirac's legacy in cosmology and geophysics*. Switzerland: Springer.

Kragh, H. (2018). Cyclic models of the relativistic universe: The early history. In Rowe, D. E., Sauer, T., & Walter, S. A. (Eds.) *Beyond Einstein* (pp. 183–204). Basel: Birkhäuser.

Miller, W. A., & Wheeler, J. A. (1984). Delayed-choice experiments and Bohr's elementary quantum phenomenon. In Kamefuchi, S., Ezawa, H., Murayama, Y., Namiki, M., Nomura, S., Ohnuki, Y., & Yajima, T. (Eds.) *Proceedings of the International Symposium "Foundations of Quantum Mechanics in the Light of New Technology"*, Tokyo (pp. 140–152).

Misner, C. W. (1974). Some topics for philosophical inquiry concerning the theories of mathematical geometrodynamics and of physical geometrodynamics. In Schaffner, K. F., & Cohen, R. S. (Eds.) *Proceedings of the Biennial Meeting of the Philosophy of Science Association*, Boston (Vol. 1972, pp. 7–29). Dordrecht: D. Reidel Publishing Company.

Misner, C. W., & Wheeler, J. A. (1957). Classical physics as geometry. *Annals of Physics, 2*, 525–603.

Misner, C. W., Wheeler, J. A., & Thorne, K. S. (1973). *Gravitation*. London: Macmillan.

Muñoz, N. (2022). The development of emergence in modern physics. Ph. D. thesis, Technical University Berlin.

Nicolis, G., & Prigogine, I. (1989). *Exploring complexity: An introduction*. New York: W. H. Freeman.

Norton, J. D. (2006). Mach's principle. In Blackmore, J. T., Itagaki, R., & Tanaka, S. (Eds.) *Ernst Mach's science: Its character and influence on Einstein and others* (pp. 193–243). Tokyo: Tokai University Press.

Oppenheimer, J. R., & Snyder, H. (1939). On continued gravitational contraction. *Physical Review, 56*(5), 455–459.

Oppenheimer, J. R., & Volkoff, G. M. (1939). On massive neutron cores. *Physical Review, 55*(4), 374–381.

Osnaghi, S., Freitas, F., & Olival, F., Jr. (2009). The origin of the everettian heresy. *Studies in History and Philosophy of Modern Physics, 40*, 97–123.

Pais, A. (1954). ω-Space theory. In *Proceedings of the International Conference of Theoretical Physics, Kyoto and Tokyo, September 1953*, Tokyo (pp. 156–163). Nihon-Gakujutsu-Kaigi: The Science Council of Japan.

Patton, C. M., & Wheeler, J. A. (1975). Is physics legislated by cosmogony? In Isham, C., Penrose, R., & Sciama, D. (Eds.) *Quantum gravity: an Oxford symposium* (pp. 538–591). Oxford: Clarendon Press.

Penrose, R. (1965). Gravitational collapse and space-time singularities. *Physical Review Letters, 14*, 57–59.

Peres, A. (1962). On Cauchy's problem in general relativity II. *Il Nuovo Cimento, 26*(1), 53–62.

Pippard, A. (1953). The coherence concept in superconductivity. *Physica, 19*, 765–774.

Potthast, T. (2007). Was bedeutet "Leitwissenschaft" – und übernehmen Biologie oder die "Lebenswissenschaften" diese Funktion für das 21. Jahrhundert? In *Autonomie durch Verantwortung* (pp. 285–318). Leiden: Brill.

Power, E. A., & Wheeler, J. A. (1957). Thermal geons. *Reviews of Modern Physics, 29*(3), 480–495.

Prigogine, I. (1973a). Measurement process and the macroscopic level of quantum mechanics. In Mehra, J. (Ed.) *The physicist's conception of nature* (pp. 697–701). Boston: D. Reidel Publishing Company.

Prigogine, I. (1973b). Time, irreversibility and structure. In Mehra, J. (Ed.) *The physicist's conception of nature* (pp. 561–593). Dordrecht: Kluwer Academic Publishers.

Prigogine, I. (1985). *Vom Sein zum Werden: Zeit und Komplexität in den Naturwissenschaften* (2nd ed.). München: Piper.

Prigogine, I., & Stengers, I. (1979). *La Nouvelle Alliance. Métamorphose de la science*. Paris: Gallimard.

Prigogine, I., & Stengers, I. (1981). *Dialog mit der Natur: Neue Wege naturwissenschaftlichen Denkens*. München: Piper.

Regge, T., & Sandrelli, S. (2012). *L'infinito cercare*. Torino: Einaudi.

Ruffini, R., & Wheeler, J. A. (1971). Introducing the black hole. *Physics Today, 24*(1), 30–41.

Stachel, J. (1974). The rise and fall of geometrodynamics. In Schaffner, K. F., & Cohen, R. S. (Eds.) *Proceedings of the 1972 Biennial Meeting of the Philosophy of Science Association* (Vol. 1972, pp. 31–54). Dordrecht: D. Reidel Publishing Company.

Tegmark, M., & Wheeler, J. A. (2001). 100 years of quantum mysteries. *Scientific American, 284*(2), 68–75.

Thorne, K. S. (1994). *Black holes & time warps: Einstein's Outrageous Legacy*. New York: W. W. Norton.

Thorne, K. S. (2019). John Archibald Wheeler (1911–2008): A biographical memoir. arXiv:1901.06623v1

van den Heuvel, E. V. (2020). Together we rise: The Renaissance of General Relativity, rebuilding Dutch physics, and co-producing American hegemony in the post-war period. Master's thesis, Universiteit van Amsterdam and Vrije Universiteit Amsterdam.

von Meyenn, K. (Ed.) (1999). *Wolfgang Pauli: Wissenschaftlicher Briefwechsel mit Bohr, Einstein, Heisenberg u.a.* (Vol. IV, Part II, pp. 1953–1954). Berlin: Springer.

Weinberg, S. (1993). *Dreams of a final theory: the search for the fundamental laws of nature*. London: Vintage.
Wheeler, J. A. (1957). On the nature of quantum geometrodynamics. *Annals of Physics, 2*, 604–614.
Wheeler, J. A. (1961). Geometrodynamics and the problem of motion. *Reviews of Modern Physics, 33*(1), 63–78.
Wheeler, J. A. (1962). The universe in the light of general relativity. *The Monist, 47*(1), 40–76.
Wheeler, J. A. (1963a). 'No fugitive and cloistered virtue'' – A tribute to Niels Bohr. *Physics Today, 16*(1), 30–32.
Wheeler, J. A. (1963b). Quantum geometrodynamics. *Progress of Theoretical Physics Supplement, 25*, 106–107.
Wheeler, J. A. (1964). Geometrodynamics and the issue of the final state. In DeWitt, C. M., & DeWitt, B. (Eds.), *Relativity, groups, and topology*. New York: Gordon and Breach.
Wheeler, J. A. (1968a). Our universe: The known and the unknown. *American Scholar, 37*, 248–274.
Wheeler, J. A. (1968b). Superspace and the nature of quantum geometrodynamics. In DeWitt, C. M., & Wheeler, J. A. (Eds.) *Battelle rencontres: 1967 lectures in mathematics and physics*, New York (pp. 242–307). Seattle: W.A. Benjamin.
Wheeler, J. A. (1971a). From Mendeleev's atom to the collapsing star. In Verde, M. (Ed.) *Atti del Convegno Mendeleeviano 1969*, Torino (pp. 189–233). Turin: Accademia delle Scienze di Torino.
Wheeler, J. A. (1971b). Transcending the law of conservation of leptons. In *The astrophysical aspects of the weak interaction (Cortona "Il Palazzone", 10–12 June 1970)*, Rome (pp. 133–164). Rome: Accademia Nazionale dei Lincei.
Wheeler, J. A. (1973). From relativity to mutability. In Mehra, J. (Ed.) *The physicist's conception of nature* (pp. 202–247), Dordrecht.
Wheeler, J. A. (1975). The universe as home for man. In Gingerich, O. (Ed.) *The nature of scientific discovery*. Smithsonian International Symposia Series, Washington, DC (pp. 261–296). Washington: Smithsonian Institution Press.
Wheeler, J. A. (1977). Genesis and observership. In Butts, R. E., & Hintikka, J. (Eds.) *Foundational problems in the special sciences: Part two of the proceedings of the fifth international congress of logic, methodology and philosophy of science, London, Ontario, Canada 1975* (pp. 3–33). Dordrecht: D. Reidel Publishing Company.
Wheeler, J. A. (1979a). Beyond the black hole. In Woolf, H. (Ed.) *Some strangeness in the proportion: A centennial symposium to celebrate the achievements of Albert Einstein* (pp. 341–375). Boston: Addison-Wesley.
Wheeler, J. A. (1979b). From the big bang to the big crunch. *Cosmic Search, 1*(4), 2ff.
Wheeler, J. A. (1979c). *Frontiers of time*. Amsterdam: North Holland.
Wheeler, J. A. (1981). Not consciousness but the distinction between the probe and the probed as central to the elemental quantum act of observation. In Jahn, R. G. (Ed.) *The role of consciousness in the physical world* (pp. 87–111). Boulder: Westview Press.
Wheeler, J. A. (1983a). Einstein's second century. In Schmutzer, E. (Ed.) *Proceedings of the Ninth International Conference on General Relativity and Gravitation, Jena, 14–16 July 1980* (pp. 23–38). Cambridge: Cambridge University Press.
Wheeler, J. A. (1983b). Law without law. In Wheeler, J. A., & Zurek, W. H. (Eds.) *Quantum theory and measurement* (pp. 182–213). Princeton: Princeton University Press.
Wheeler, J. A. (1990). Information, physics quantum: The search for links. In Zurek, W. H. (Ed.) *Complexity, entropy, and the physics of information* (pp. 3–28). Redwood City, CA: Addison-Wesley.
Wheeler, J. A. (1994). *At home in the universe*. College Park: American Institute of Physics.
Wheeler, J. A., & Ford, K. (2000). *Geons, black holes & quantum foam*. New York: W. W. Norton.
Wheeler, J. A., & Zurek, W. H. (Eds.) (1983). *Quantum theory and measurement*. Princeton: Princeton University Press.

Wigner, E. P. (1949). Invariance in physical theory. *Proceedings of the American Philosophical Society, 93*(7), 521–526.
Wigner, E. P. (1970). Physics and the explanation of life. *Foundations of Physics, 1*(1), 35–45.
Wright, A. S. (2014). More than nothing: Histories of the vacuum in theoretical physics, 1927-1981. Ph. D. thesis, University of Toronto.

Lawlessness

Yemima Ben-Menahem

Abstract One route through which the possibility of lawlessness has entered scientific and philosophical discourse about laws of nature is Wheeler's celebrated vision of "law without law." I leave Wheeler's question about the origin of laws in general unanswered, but examine three more limited kinds of lawlessness. The first is Feynman's derivation of the principle of least action from an underlying quantum level which, on his account, is (nearly) lawless. The second stems from Davidson's anomalous monism—the thesis that the mental is lawless and irreducible to fundamental physics—and pursues applications of this thesis both within and outside of physics. The third kind of lawlessness derives from the algorithmic notion of randomness and its compatibility with determinism.

Keywords Origin of laws · Richard Feynman · Reduction · Non-reductive physicalism · Randomness · Complexity theory · Determinism · John Archibald Wheeler · Gregory J. Chaitin · Andrey N. Kolmogorov

1 Introduction

I seek to examine several types of modification to the traditional conception of laws of nature, all of which could, in one sense or another, come under the rubric of lawlessness. Although the questions that concern me in what follows (certainly not an exhaustive list of challenges to the traditional concept) have been studied (in this volume or elsewhere), they have not usually been looked at from the more unifying perspective taken here. By using the term 'lawlessness' I do not mean to suggest that there are no laws at all, or that science does entirely without them; I mean that there are (or might be) exceptions to the regime of law—islands of lawlessness—and that these exceptions merit serious consideration. I should note that I do not join the

Y. Ben-Menahem (✉)
Department of Philosophy, The Hebrew University of Jerusalem, Jerusalem, Israel
e-mail: yemima.ben-menahem@mail.huji.ac.il

Y. Ben-Menahem (ed.), *Rethinking the Concept of Law of Nature*,
Jerusalem Studies in Philosophy and History of Science,
https://doi.org/10.1007/978-3-030-96775-8_12

debate on Humean supervenience (on which several contributions to this volume shed new light), namely, I do not address the question of whether laws of nature transcend empirical facts or merely summarize them. Rather, I draw attention to a number of other questions that are less discussed by philosophers in the context of understanding natural laws, but are nonetheless pertinent to our understanding of the concept of law and its place in science. The most general (and daunting) question with respect to laws of nature is whether there is any scientific account of their origin. The boldest attempt to answer this question was made by John Archibald Wheeler and the best account (that I am aware of) of how his position developed can be found in Blum and Furlan's paper in this volume. I will only comment very briefly on Wheeler's question and move to the more limited questions that I would like to address. (1) Are there any examples of the emergence of specific laws from an underlying level that is lawless (or nearly so)? (2) Could lawful behavior under one *description* of phenomena turn out to be lawless under another description of the same phenomena? (3) Dose determinism guarantee lawfulness? My discussion may be too preliminary to do justice to the problems, but might still be useful in enhancing the reexamination that this volume is aiming at.[1]

The explanatory ambition of science is known to extend beyond the explanation of facts to the explanation of laws. As Richard Feynman put it, "In the further development of science we want more than just a formula. First we have an observation, then we have numbers that we measure, then we have a law which summarizes all the numbers. But the real *glory* of science is that *we can find a way of thinking* such that the law is *evident*" (1963, p. 26–23, emphasis in original). The sentiment has been shared by others, from Leibniz who maintained that all of natural science could be ultimately derived from the (self-evident) principle of sufficient reason, to Hermann Weyl who took the fundamental symmetries of physics to be a priori (1952, p. 126). There may of course be disagreement around the concept of self-evidence and its applications, but clearly, whenever a claim (law, principle, theorem, axiom,) is considered to be self-evident (a priori, necessary) the question about its origin or explanation no longer arises. Has science as we know it, however, justified the expectation of rendering its most fundamental laws self-evident? If it has failed to do so, as most present day physicists would probably concede, then the explanatory challenge—either if still pursued in the hope of reaching the goal of self-evidence, or if a different kind of explanation is being sought—should continue to engage us. Seeking an explanation of a phenomenon and looking for its origin may be altogether different issues (as Darwinian evolution makes clear), but in the case of laws of nature, the quest for explanation is not limited to the derivation of laws from more fundamental ones and spills over to Wheeler's mind boggling question of how it came about that the world is (or appears to us to be) subject to laws in the first place.

[1] Parts of what follows (section "Feynman's derivation of the principle of least action" in particular) draw on my *Causation in Science* (Ben-Menahem, 2018).

It has often be noted that in the religious context of the scientific revolution, when the concept of law became central to science, laws could be taken to be legislated by the divine creator or used by him as a blueprint for the creation of the world. If this option is declined, the question regarding the origin of laws immediately arises. The question is particularly pressing for those who see laws as more fundamental than individual facts, but it seems to me that it should also be a matter of concern to the proponents of Humean supervenience. For even if laws are just the best summary of individual facts that we can think of, the question of why facts (appear to us to) exhibit such orderly behavior and whether they have always done so and will continue to do so under all conditions is intriguing. Consider the rules of grammar. One can think of them as legislated by the elders of the tribe way back in the nascent days of the development of language and that languages have then evolved according to these rules. We would then want to know how this group of legislators worked, how they reached their decisions. But we could also think of the rules of grammar as a summary of so many oral and written utterances. The question would then be how sporadic utterances, unguided by grammar, could turn out to obey a concise set of rules. And while in the case of grammar, it is reasonable to settle for a combination of sporadic linguistic behavior and conscious legislation, this option does not seem to make much sense in the context of laws of nature.

Two polar options suggest themselves by way of answering the question regarding the origin of laws. (Wheeler in fact entertained both of these options at different times). One points in the direction of finding a fundamental law or principle from which all other laws follow. Another points to a primordial lawless state of affairs from which laws have emerged. In principle, either one of these options could obtain the ideal of self-evidence, albeit in different ways. The former (if successful) could produce a self-evident fundamental law; the latter (if successful) could demonstrate that the lawless situation is *bound* (as a matter of mathematical necessity, say) to lead to the emergence of laws. Naturally, each of these options also faces difficulties. On pain of regression, the fundamental principle sought by the former option cannot be just another law of physics. What could it then be? Here again there are different possibilities. If the fundamental principle turns out to be a purely mathematical one, the goal of self-evidence would be achieved, but the difference between physics and mathematics would be erased. This way leads to Pythagoreanism, a position that a number of scholars have found attractive on account of the extraordinary effectivity of mathematics in contemporary physics.[2] Those who find Pythagoreanism unsatisfactory, maintaining that physical reality is more concrete than abstract mathematics and not entirely reducible to it, could opt for a third kind of principle, neither an ordinary physical law, nor a purely mathematical one; perhaps a philosophical principle of the kind that Kant deemed transcendental, that is, a principle that every research program and every method of investigation must

[2] Pythagoreanism is advocated, for instance in Steiner (1998), though he acknowledges that mass still resists a mathematical explanation. Roger's Penrose also comes to mind as placing great confidence in the mathematical underpinnings of physics.

presuppose. Although it is not too difficult to come up with conditions necessarily presupposed by scientific practice (reliability and intersubjectivity of observation, for instance) it *is* rather difficult to see how specific laws could be derived from such general principles. Realizing these difficulties of the first option (finding a self-evident fundamental law), Wheeler turned to the second—understanding laws as emerging from lawless origins.

> Are we then to believe that all of physics will one day be expressed in one or more beautiful equations, chiseled as it were on a tablet of granite and standing there from everlasting to everlasting?... Much thought, many discussions and long study lead me to the directly opposite vision. All of physics, in my view, will be seen someday to follow the pattern of ... regularity based on chaos, of "law without law" (1983, p. 398).

As Deutsch (1986) noted, apart from dissatisfaction with the fundamental law option *per se*, Wheeler had another reason for favoring the "law without law" option. The most general laws of physics, though economic, efficient, and beautiful, tend to gloss over the existence and structure of numerous individual processes that give rise to them. Classical thermodynamics, for example, ignores the fundamental level of atoms and molecules obeying Newtonian mechanics. Rather than searching for the most general laws, we should aim at a theory that explains them via their subtle connections with the underlying level of multifarious individual events and processes. Wheeler's confidence in the feasibility of such explanations is based on specific examples that physics and mathematics have already provided.[3] Although each one of these examples applies to a specific domain and thus falls short of answering the general question, they are sufficient to make him confident that this is the way to go. Quantum mechanics (QM), of course, is even more promising than classical examples in manifesting lawlessness. When QM is taken into account, even spacetime, Wheeler was hoping, would eventually be understood as an emergent structure.

> Quantum Phenomena... may seem to obey law when registered in large numbers, but the individual event, the individual yes or no decision, is as lawless as anything on the face of the Earth. Moreover, unlocalized and unlocalizable as the phenomenon is in space and time, it is the only thing we know in all of physics which has the smell of an element logically prior to spacetime on which spacetime—and all the rest of the structure of physics—might be considered to be built (1983, p. 399).

In the vein of Wheeler's examples, I examine (in the next section) a quantum mechanical derivation of a general principle from a lawless multitude of possibilities—Feynman's derivation of the principle of least action.[4] Despite the

[3] In his (1983) Wheeler describes four examples of this kind: Boltzmann's statistical mechanics, the phenomenon of universality and phase transition, Wigner's calculation of nuclear energy levels, and the travelling salesman problem.

[4] For Wheeler, QM provides yet another deep insight into the question of what is real—only registered phenomena are real. Consequently, Wheeler urges the "participatory"conception of reality, which could be said to assign human observers the role of constructing reality. Limiting myself to the problem of laws (and although Wheeler saw them as interrelated) I will not comment of this aspect of Wheeler's program.

ingenuity of Wheeler's approach (and Feynman's derivation) there is a caveat to the "law without law" program: The derivation, of laws generally requires nontrivial assumptions about the lawless state of affairs. These may pertain, for example, to the individual states as we decide which ones are equally probable, or to the probability measure we use.[5] Even when probability theory is considered to be part of logic, and as such self-evident, prior assumptions of this kind are necessary for application of the theory and are decided on a case to case basis. They are therefore contingent. Aware of the problem, Wheeler speaks of "regulating principles" that enable the derivation of laws, principles of which we have as yet only a number of specific examples, but no general formula common to all of them.[6]

In their contribution to this volume, Blum and Furlan trace the development of Wheeler's program from his earlier avowal of fixed universal laws to the polar position of "law without law." A crucial step on this journey, they show, was his recognition of the failure of the conservation of baryon number in black holes. The breakdown of this fundamental law eventually led him (through a series of steps described in their paper) to skepticism about the above quoted vision of laws "chiseled as it were on a tablet of granite and standing there from everlasting to everlasting." Under extreme conditions, the regime of physical law collapses. The mathematical indicators of this collapse, such as singularities in equations, are not seen by Wheeler as curiosities of the formalism that could be sidestepped, but as signifying that the fundamental concepts of physics, including space and time, lose their meaning. Understanding how meaning and law are regained is the tremendous challenge that Wheeler was coping with.

What about 'ordinary' physics in its less extreme settings? Does it fully support the universal law image? Consider the principle of conservation of energy. While often seen as one of the most general and stable principles of physics, the concept to which the conservation principle applies has undergone such significant changes (and is indeed still so widely open) that stability is only apparent. Moreover, the validity of the principle in the general theory of relativity (GTR) has, from early on, been a matter of concern to physicists and has (more recently) also drawn the attention of historians and philosophers of science.[7] As Carl Hoefer (2000) has argued, and as has been shown in great detail by Patrick Duerr in his DPhil dissertation (Duerr, 2020), in GTR there is no well-defined quantity that can be associated with the gravitational field and identified as its energy. (Hoefer actually points to the roots of the problem in Newtonian mechanics). Even though formally, an analogue of the energy conservation law could be formulated in the framework of GTR, the conserved quantity it involves is not analogous to classical energy; in

[5] See, for example, Hemmo and Shenker (2012), Pitowsky (2012), Farr (this volume).

[6] The position Wheeler expounds here, which acknowledges the need for regulating principles, is a late position, constituting some withdrawal from absolute lawlessness as an option. See Blum and Furlan's concluding section.

[7] It turns out that Wheeler saw the problem of energy-momentum conservation in GTR as an additional reason for skepticism about universal laws. See f.n. no 25 in Blum and Furlan's paper.

particular it is not *localized.* Since J. Brian Pitts paper in this volume also discusses the limitations of energy conservation (from a slightly different perspective) I will not go into details. The upshot of these works calls to mind Poincaré's cautionary remarks about the principle of conservation of energy (1952 [1902], chapter VIII), namely, that the principle has a clear meaning in particular cases, but is in danger of losing this clarity under extension and extrapolation. It is as if by generalizing the formula, we diminish its content.[8] Let me now move to the three issues I have raised at the beginning of this introduction: laws emerging from (nearly) lawless origins; lawlessness associated with the description sensitivity of laws; the concept of lawlessness developed by algorithmic complexity theory.

2 Feynman's Derivation of the Principle of Least Action

While Wheeler's four examples of laws emerging from lawless origins were taken from classical physics (Wheeler, 1983), I would like to describe an example taken from QM—Richard Feynman's derivation of the principle of least action, based on his path integral approach to QM. There are at least two reasons to view Feynman's account of the principle as particularly ingenious. First, on the deductive nomological model of explanation, lower-level laws can be explained by deriving them from higher-level laws, but the 'highest' most general laws, must remain unexplained. (Recall Newton's "hypothesis non fingo"—his admission that while the law of gravitation explains lower-level laws such as Kepler's laws, it cannot itself be explained via more general laws.) As the principle of least action certainly belongs to the category of the most general principles employed by contemporary science, it was surprising that it could after all be explained. Moreover, the mode of explanation was, as we will see, quite unique. Second, the principle of least action has a teleological appearance that has worried philosophically minded physicists. These worries persisted even when allegations of teleology could be met (see below). The principle's reappearance in the probabilistic context of QM, a context that defies not only teleology but also strict causation gave new reasons for concern regarding its origin. Feynman's account of the principle is therefore doubly appealing.

In the form given to it by Hamilton, the principle of least action asserts, roughly, that a mechanical system moving under the influence of conservative forces takes the path for which the action has an extremal point—a minimum or maximum

[8] This lesson appears similar to Cartwright's thesis in *How the laws of Physics Lie* (Cartwright, 1983), but the differences between the two are more significant than the similarity. The causal account that Cartwrights finds superior to the law-based account does not seem to me to resolve the difficulty regarding energy-momentum conservation in GTR, nor does this difficulty point to a payoff between truth and explanatory import as Cartwright maintains; both seem to be equally jeopardized in this case.

(initially, only the first alternative was discovered, hence the principle's name).[9] More accurately, the system takes a path such that for a small variation of the path, the first-order change in the action vanishes.[10] When the principle was first formulated, the definition of the action was the subject of bitter controversy, and is indeed a thorny problem in general (more on this problem below). In classical mechanics the action is the integral over time of the difference between the kinetic and potential energies (K − P), or, since that difference is known as the Lagrangian—£—the integral of the Lagrangian over time. Thus, the action is

$$\int_{t1}^{t2} \pounds dt = \int_{t1}^{t2} (K - P)\ dt$$

and the principle says that the system moves along a path such that, for a small variation of the path, the first-order change in this integral is zero. The principle is considered the apex of classical mechanics, and has retained its centrality in subsequent physical theories, such as GTR and QM. Since it uses only a system's potential and kinetic energies, both of which are scalar functions, the principle has the advantage of being independent of the coordinates in which the Lagrangian is represented. This independence facilitates calculation, but more importantly, it has gained theoretical significance due to Einstein's postulation of general covariance as a desideratum that physical theories must satisfy. Being invariant under coordinate transformations (as scalar functions are), the action, as well as functions and equations expressed in terms of the action, automatically satisfy this requirement.

On the face of it, the principle of least action makes us wonder how a mechanical system could possibly 'choose' the right path. In terms of this teleological character, the principle stands in stark contrast to Newton's laws which are formulated in paradigmatically causal terms—forces acting on material particles and generating their acceleration. The tremendous success of Newton's system and the intuitive appeal of its underlying causal picture played a decisive role in transforming the nature of scientific explanation from teleological to causal. The principle of least action could not offer an equally intuitive causal model. Indeed, it could easily be construed as reintroducing the teleological language that Newton's system had eschewed. Eventually, the appearance of teleology was dispelled, when it was discovered that, under a wide range of conditions, Newton's differential

[9] A force is said to be conservative when it is derivable from a scalar potential, or equivalently, when the work done around a closed orbit is zero. This means that there is no friction or other dissipative forces (whose work is positive). The conservative forces restriction is not a serious limitation with regard to fundamental forces, since these forces are all taken to be conservative. Non-conservative forces are introduced when we move from the fundamental level to higher levels. The least action principle is extendable to non-conservative forces; see Goldstein (1950, ch. 2).

[10] Note that higher orders (in the Taylor expansion) of the change in the action need not, in general, be zero. The variation refers to the difference between the actual path and alternative virtual paths between the same end-points. The variation is expressed as a function of some parameter whose values correspond to each of the alternative paths.

formulation, which satisfies our causal intuitions, and the integral formulation of the least action principle, which seems to violate them, are in fact *equivalent*. For example, in the simple case of a particle moving with velocity v in a gravitational field whose potential is V, the force is dV/dx and the kinetic energy is ½ mv^2 = ½ mdx/dt. The path that satisfies the least action principle is the very path that satisfies the differential equation

$$md^2x/dt^2–dV/dx = 0,$$

which is none other than Newton's second law! Moreover, the equivalence between the differential and integral formulations of mechanics has also received a firmer grounding, upon the demonstration that the least action principle follows from the Euler-Lagrange differential equations and vice versa (Goldstein, 1950, ch. 2). The puzzle of how the particle picks the right path is resolved by noting that at every specific point along the path, it 'gets directions' as to how to proceed; it need not 'calculate' or 'take into account' the entire path or its end-point, let alone 'choose' between alternatives. The teleological picture turned out to be no more than a super-structure resting on a more satisfactory causal structure.[11] To the extent that the predictions of QM are given in terms of *probabilities*, this poses a challenge to both teleology and determinism. Particles moving in accordance with quantum mechanical laws are not expected to follow a classical path at all, let alone the privileged path determined by the principle of least action.[12] And yet the predictions of QM should converge on those of classical mechanics in the classical limit. How can we then explain the emergence of the very special path that satisfies the least action principle? Feynman drew on his path integral approach to QM to provide an answer. In his work on classical electrodynamics, Feynman had used the Lagrangian formalism rather than the Hamiltonian.[13] In moving to

[11] If the two formulations of mechanics are indeed *equivalent*, it could be argued that the causal level's being a mere 'super-structure' atop the teleological level was just as plausible as the reverse. If so, rejecting the teleological reading in favor of the causal one reflects a metaphysical or methodological preference, not the verdict of logic. Be this as it may, Feynman's approach is orthogonal to both the causal and the teleological reading.

[12] Even ignoring the indeterminacy of measurement results, the position–momentum uncertainty relations rule out the classical notion of a path, where each point represents a well-defined position and well-defined momentum.

[13] In his doctoral dissertation, Feynman had identified two problems—the infinite self-energy of the accelerating electron and the infinite degrees of freedom of the surrounding field. Feynman sought to eliminate both kinds of infinity by assuming, first, that there is no self-interaction of the electron; and second, that there is no field. The theory he proposed was purely corpuscular—particles acting on each other. In the absence of fields, it was an 'action at a distance' theory, but it satisfied relativistic locality in the sense that the mutual interaction of electrons was not instantaneous. That interaction could be expressed in terms of a field, but the field had no independent degrees of freedom other than those ensuing from particle to particle interaction. Wheeler (Feynman's dissertation advisor) and Feynman then showed that to get the correct results—i.e., no infinities—one had to assume that the absorber—the surrounding particles–responds by acting on the source by means of "advanced waves" (waves that appear to go backwards in time), a solution to

QM, however, preference for the Lagrangian formalism constituted a problem, since up to that point, the equations of QM had been formulated as a natural extension of the Hamiltonian equations of classical mechanics. The Lagrangian formalism did not seem amenable to an equally intuitive extension to the quantum domain, yet such an extension was required, Feynman believed, for creating quantum electrodynamics along the lines that guided his Lagrangian approach to classical electrodynamics. He therefore sought a quantum analogue of the classical action. The missing link was provided by Dirac, who in 1933 had published a paper entitled "The Lagrangian in Quantum Mechanics." As it appeared in the outlying *Physikalische Zeitschrift der Sowjetunion*, it had received scant attention. Like Feynman, Dirac had come to the conclusion that it was imperative to find a way to formulate QM in terms of the Lagrangian. Despite the close connection between the Hamiltonian and Lagrangian formulations of classical mechanics, there were, Dirac felt, "reasons for believing that the Lagrangian one is the more fundamental" ([1933] 2005, p. 111). In the first place the Lagrangian method allows one to collect together all the equations of motion and express them as the stationary property of a certain action function. (This action function is just the time-integral of the Lagrangian.) There is no corresponding action principle in terms of the coordinates and momenta of the Hamiltonian theory. Secondly the Lagrangian method can easily be expressed relativistically, on account of the action function being a relativistic invariant; while the Hamiltonian method is essentially non-relativistic in form, since it marks out a particular time variable as the canonical conjugate of the Hamiltonian function (ibid). Dirac proposed a "correspondence" between a certain function of the classical action and the quantum mechanical transformation matrix, which governs the wave function's transition from one instant to another.[14] Feynman seized upon this clue, developing it into a novel picture of QM that did not start out from the existing formalisms (i.e., those of Heisenberg, Schrödinger, and Dirac), and proved extremely useful in creating quantum electrodynamics. Using

Maxwell's equation generally thought to be merely mathematical, with no physical realization. Wheeler and Feynman couched their theory in the Lagrangian formalism, invoking the principle of least action. As we saw above, this formulation, in contrast to Hamilton's equations, which describe the system's development over time, describes the entire path of a particle (system) between two end-points. To Wheeler and Feynman, the advantage of this formulation seemed to be connected to the corpuscular picture they favored, for as Feynman put it later, the field variables were only "bookkeeping variables to keep track of what the particle[s] did in the past" (Feynman, 1966, p. 36). For electrons, what makes the Lagrangian approach more suitable than the Hamiltonian is that the path of an electron at a given time depends on the paths of other electrons at *other* times. Hamilton's equations, which involve positions and momenta at the same time, are therefore unsuitable. Throughout his career, Feynman's preferred formalism remained the Lagrangian. In endeavoring to construct quantum electrodynamics along the lines that worked so well in the classical case, he was seeking to define a quantum action that was the closest analogue of the classical action. Precisely at this point, he happened to learn of Dirac's virtually unknown 1933 paper.

[14] Dirac observed that exp $2\pi i/h$ multiplied by the classical action corresponds to the transformation matrix.

the Lagrangian proposed by Dirac, Feynman recast QM in terms of path integrals, and derived the Schrödinger equation therefrom. The principle of least action turned out to be the *classical limit* of this novel path-integral rendering of QM!

On Feynman's approach, QM is an algorithm for calculating probability amplitudes. Incorporating Dirac's proposal into this picture entails that probability amplitudes have a *phase* proportional to the classical action and responsible for the periodicity characteristic of quantum phenomena such as interference. The non-classical character of the theory manifests itself in the fact that probabilities interfere with each other in ways that deviate from classical expectations. Distinguishing interfering from non-interfering states is crucial for counting alternatives and calculating probabilities.[15] Feynman's idea was that the probability amplitude of an event is the sum of the amplitudes of every possible way in which the event can occur. Each possibility is represented by a path, and each path is assigned a probability amplitude. Thus a photon traveling from a source to a particular point on a screen can reach that point by infinitely many paths, each one correlated with a probability amplitude. The total probability amplitude of the event—in this case, the photon's reaching the designated point–is the sum of the probability amplitudes assigned to the possible paths, namely, the integral of the amplitude over all possible paths.

In their attempt to "demystify quantum theory," Cox and Forshaw (2011, p. 4) endorse Feynman's version of QM. Their title, *The Quantum Universe: Everything that Can Happen Does Happen*, epitomizes the path integral approach on which we integrate over all possibilities. At first sight, nothing seems farther from a lawful regime than the liberty to realize every possibility. On reflection, however, the disparity between the Cox Forshaw aphorism and the rule of law is not as striking as it appears. To see why, we need only consider the aphorism's contrapositive—everything that does not happen cannot happen. It mandates that we explain what does not happen by pointing to laws or constraints that exclude it. Pauli's principle can serve as an example. We never find multiple electrons occupying the same quantum level. There must therefore be some constraint precluding this situation, as there indeed is: Pauli's exclusion principle. The merit of Feynman's approach is that its freedom-inducing orientation facilitates the derivation of such constraints. Not only does it recover Pauli's principle, but it explains and predicts with great precision many of the constraints on elementary particle interactions.

Rather than describing the state of the system at each moment as a function of its state at the previous moment, the path integral method, like the classical principle of least action, involves looking at paths in their entirety. Feynman maintained that the path integral approach was consonant both with the time-reversal symmetry of the fundamental laws of nature (since it could be seen as countenancing particles moving backwards in time) and with Einstein-Minkowski

[15] Since, on Feynman's approach, there is no underlying equation, formalism, or intuitive model that determines quantum probabilities, such counting is far from trivial, but for our purposes here, we can set this issue aside.

four-dimensional spacetime (since it could be made Lorentz-invariant).[16] Besides, the path integral approach was beautiful: "The behavior of nature is determined by saying her whole spacetime path has a certain character" (1966, p. 35), and Feynman, as he puts it, "fell deeply in love" with this approach (1966, p. 32).

How, then, does the method of integration over paths explain the classical principle of least action? As mentioned, Feynman took the probability amplitude, represented by a complex number, to have a phase proportional to the classical action. The different paths that are integrated over thus have different phases. At the classical limit, the classical action is so large in proportion to Planck's constant h that in general, even a small variation of the action amounts to a large difference in the phase, expressed in quantum units. Paths (and probability amplitudes) that are close to one another in classical terms may thus still differ considerably in phase. These differences are reflected in the phase's rapidly changing its sign between nearby paths. Consequently, the majority of paths will be canceled out by paths with similar classical action but opposite phase. When we approach the stationary point, however, where a small variation of the path does not alter the (first order of the) action, the phase no longer oscillates, and amplitudes—paths—are not canceled out. On the contrary, here the different amplitudes add up, so that the probability amplitude is largest around the stationary point. We therefore see a large number of particles moving along this particular path—the classical path.

It is here, for the first time, that the principle of least action is given an explanation, to wit, an explanation that completely defeats the teleological interpretation that had accompanied it for more than two centuries.[17] In classical mechanics, the response to the charge of teleology was, as we saw, that the system does not 'reason' or 'calculate,' but simply follows the 'instructions' expressed by the (Newtonian, Euler-Lagrange, or Hamiltonian) differential equations. The future-oriented tendency of a system moving in accordance with the principle turned out to be an illusion, merely apparent teleology. The quantum response to the putative teleology is more radical: the system does not even get instructions; it goes whichever way it goes and no path is excluded in advance. What we, macroscopic creatures that we are, perceive as very special behavior is the result of a selection process by which many paths cancel each other out while others add up and reinforce one another. Selection processes of this kind tend to give us the impression that nature 'prefers' certain possibilities to others and 'chooses' accordingly, enticing us into teleological thinking. Moreover, and more importantly from the perspective of this paper, the emergence of the privileged classical path

[16] See Feynman (1963), I, ch. 19, Feynman (1966), and the manuscript quoted in Schweber (1994), pp. 432–433. One consequence of the path integral approach was that the continuous wave model could not be sustained.

[17] A common critique of Feynman's path integral approach is that it constitutes an ingenious mathematical technique, not a physical explanation. Even so, I would argue, the very possibility of accounting for the classical path in this way is edifying.

from the underlying multitude of quantum possibilities is a beautiful example of the emergence of a law from an underlying jungle of possibilities.[18]

3 Laws and Descriptions

The question of the relation between general laws and the individual cases to which they apply (or are supposed to apply) has occupied philosophers over the generations not only in the context of science but also in other areas such as legal and moral theory. Here is Kant's celebrated discussion of the problem.

> If understanding in general is to be viewed as the faculty of rules, judgement will be the faculty of subsuming under rules; that is distinguishing whether something does or does not stand under a given rule ... General logic contains, and can contain, no rules for judgement (Critique of Pure Reason, [Kant 1933 (1787)] A 133, B 172).

If there were rules for judgement, Kant continues, they too would require rules to tell us how to apply them, which would lead to an obvious regress. The ability to make judgements is therefore, in Kant's view, a "natural gift" (Ibid.) In the twentieth century the problem of subsuming events under laws was highlighted by Donald Davidson in a number of papers collected in his (1980b), but surfaces as well in standard texts in the philosophy of science such as Hempel's *Philosophy of Natural Science*. We can put the problem as follows. Laws are always given to us in language while the events we wish to subsume under them are not. To subsume an event (state, process, fact, action) under a law, we must describe it in terms that correspond to the language of the law (be it mathematical, physical or everyday language). This is why in Hempel's deductive nomological model of explanation, the explanandum is not the event itself, but rather a sentence describing it. And since laws assert a connection between *types* of events, we must choose, from the numerous descriptions that could fit the event in question the one that describes it by reference to the relevant type, the type that matches a type figuring in the formulation of the law. Many other descriptions may be correctly applied to the same situation (event, process and so on), but they will not make it subsumable under the said law.

In addition to their familiar roles in predicting and explaining phenomena, laws also serve to *characterize* the types of entities (states, events) that fall under them. Pauli's principle characterizes fermions and the speed (in vacuum) of 299.792458 km. per second characterizes electromagnetic radiation. As Goodman has convincingly argued in *Fact, Fiction, and Forecast* (1955), there is a close relationship between laws and the types they invoke. Laws are projectable and so

[18] For reasons noted in the introduction regarding the employment of probability theory, this jungle cannot be completely lawless. Nonetheless it is a beautiful illustration of the vision of "law without law". It would be interesting to know if Wheeler saw Feynman's construction in this light; I don't have the answer to this question.

are the predicates that designate the law-covered types. I don't need to break this particular vase in order to find out whether it is fragile. I know it is made of glass; I know that glass (in room temperature) is fragile. Because laws and the types they characterize are inseparable, I can project frangibility from one piece of glass to another, but not from glass to other types such as plastic.

Accidental generalizations also refer to types, but neither the generalizations, nor the predicates describing the types they invoke are projectable. It may happen that all the vases in my house on 1.1.2022 are made of glass, but no information can be gleaned from this generalization regarding vases in general, vases in other homes, vases in my house on 1.1.23 and so on. Goodman noted further that whereas laws are confirmed by their instances, but not by any specific instance, accidental generalizations, can only be confirmed by checking each one of their instances. From the scientific point of view, accidental generalizations are therefore inert; they do not enhance our knowledge beyond what we had already established. While Pauli's principle singles out fermions from all other entities, accidental generalizations cannot provide a physical criterion that singles out the physical state of the individuals falling under them in the same way. It is as if physics were blind to the categories grouped under accidental generalizations.

We can now turn from these two polar cases—the projectable and the unprojectable to the intermediate cases that have worried philosophers in recent decades. Consider mental events such as love, fear, expecting to pass an exam, and so on. Mental types appear to differ from the vases in my home on 1.1.22, on the one hand, and from fermions, on the other. Are they then subject to laws? Moreover are they subject to the fundamental laws of physics in the very same way that fermions are? The answer depends on one's position on reductionism. There could be weaker and stronger versions of reductionism: The weaker version asserts that all the concepts of higher level scientific theories are definable in terms of (or are at least co-extensional with) concepts of fundamental theories. Since concepts correspond to types it follows that all the types characterized by higher level laws correspond to types characterized by fundamental laws. The stronger version of reductionism asserts much more— not only concepts and types of higher-level *theories* correspond to fundamental physical types, but higher-level types *of any kind*, whether or not they figure in some scientific theory, correspond to fundamental physical types. With regard to mental types, the position that they are subject to the laws of physics clearly requires a fairly strong version of reductionism. The idea is that every mental type (say fear) is reducible to a specific neurological type, which in turn is reducible to a specific physical type. This physical type can be subsumed under physical laws that characterize it, distinguishing it from other physical types, among them (for instance) types instantiating joy. Thus, for the strong reductionist, mental types are believed to be on a par with higher level physical concepts such as pressure and temperature even if only the latter have so far been successfully reduced.[19]

[19] Interestingly, Wheeler (1983) questions the success of this reduction.

This kind of reduction of the mental has been critiqued by Davidson, Fodor, and Putnam, who subscribed to the position known as nonreductive physicalism. Generally, nonreductive physicalists do not contest reduction within science; it is the correspondence of non-scientific types to physical types that they deny. Let me follow Davidson's argument. Mental events, according to Davidson (1980a) are physical events that have mental descriptions. (The question of whether mental events are unique in exemplifying nonreductive physicalism will be addressed later.) Like other descriptions, mental descriptions assemble individual events into types, but these types, Davidson contends, are not 'held together' by laws, neither mental laws nor physical ones. And though every mental event is a physical event, the physical events that instantiate a particular mental type need not constitute a physical type and are therefore not characterized by physical laws in the way that fermions are characterizes by Pauli's principle. Reduction is therefore blocked. Mental events supervene on physical events,[20] but they are multiply realizable, that is, the individual events falling under the same mental type, though they are all physical events, can belong to various physical types. In other words, the identity between the mental and the physical is understood by Davidson as applying to token-events rather than types. And since, where there are no types, there are no laws, the argument leads to the anomalous nature of the mental. Davidson made his position seem paradoxical by showing that it enables him to combine a number of seemingly contradictory claims: every mental event is a physical event; there are causal relations between the mental and the physical; causal relations entail the existence of laws; there are no mental laws.[21] But once we understand the crucial role of types for the applicability of laws, and the crucial role of descriptions—linguistic categories—for assembling individuals into types, Davidson's solution is not as paradoxical as it at first seems.

Davidson's anomalous monism (in particular, the idea of multiple realizability on which it centers) has been accused of violating the causal closure of fundamental physics (Kim, 1993). I have argued (Ben-Menahem, 2018, chapter 7) that Kim's argument is flawed for it conflates the relation of instantiation with that of causation. Instead of repeating Kim's argument and my response, let me note that multiple realizability is quite common within physics; the temperature of a macrostate, for example, is defined as the average velocity of the molecules comprising the underlying microstate. Clearly, there are multiple ways of getting the same average, which means that the macrostate of a certain temperature is multiply realizable. Despite this multiplicity, though, in this case the higher-level concept

[20] The notion of supervenience I use here (there are others) implies a many-one relation between the mental and the physical: A mental type can be realized by events belonging to different physical types but the converse does not hold. Different types of mental events must thus be realized by different types of physical events whereas different types of physical events may instantiate the same type of mental event.

[21] Davidson's commitment to the Humean position that there is no causality without regularity is not actually essential for the main points of either "Causal Relations" (1980b) or "Mental events" (1980a) but it certainly adds to the magical appearance of his solution.

of temperature reflects a property exhibited at the fundamental level. The concept of entropy provides a more complex example of multiple realizability. By contrast with temperature, there is no property of the microstate of a system that is picked out by, or corresponds to, its entropy. What matters for the definition of entropy is only the *number of ways* (actually its measure-theoretic analogue) in which a macrostate could be realized. Entropy thus reflects a property of macrostates, not a property of the underlying level of molecules. To put it in more picturesque terms, if molecules had consciousness, they could perhaps be interested in their average velocity and thus (even though they had no concept of macrostate) have a concept that corresponds to the macro-concept of temperature. But without the notion of a macrostate they could not even entertain the multitude of a macrostate's possible realizations and would altogether miss the notion of entropy.

There are a number of lessons to draw from the case of entropy. First, higher-level concepts are essential and cannot always be eliminated in favor of lower-level ones. As we have seen, the concept of entropy does not reside entirely on the conceptual level of microstates; it requires the concept of macrostate and its 'size' to be defined. Thus, before considering mental states and their amenability to reduction, we should realize that the strong version of reductionism, which requires a type-type identity (or co-extensionality), already comes under pressure in the context of physics. Moreover, macrostates and their 'size' play an essential explanatory role. If, for instance, we ponder the stability of one particular macrostate—the equilibrium— relative to other macrostates, we need to refer to the entropy or probability of the macrostates in question and these magnitudes, we have just seen, are not completely reducible to the concepts of the fundamental level. Second, description-sensitivity is salient: macrostates do not descend from heaven with fixed identities—they are given an identity by the description we use. This does not mean macrostates are fictions—they are as real as microstates—but it means that in order to understand their behavior *as macrostates*, that is, to discover the laws that explain their behavior as macrostates, they must be characterized in a useful way. This characterization is given by us and goes beyond micro-properties of the system. Third, the correlation between microstates and macrostates is a many-one relation (or function) that manifests supervenience in the sense given to the term by Davidson. If a microstate is specified, the question of whether it realizes a certain macrostate receives a determinate answer, but the converse does not follow; the identification of a macrostate does not determine which microstate realizes it on this particular occasion.[22] Finally, the many-one function is also the formal representation of the *insensitivity* of the scope of the function to many of the features that distinguish its arguments from one another. The important characteristic of a macrostate—its probability—is relatively indifferent to many of the details characterizing the microstates that belong to this macrostate.

[22] Entropy supervenes on the microstate of the system in Botzmamm's statistical mechanics but it is not clear whether the same holds for entropy in the Gibbs formulation of statistical mechanics. If there is no supervenience reduction is in greater trouble yet.

Insensitivity of the evolution taking place on a higher level to the detailed structure of the fundamental level is not unique to the case of entropy. Another example is provided by the phenomenon known as universality: the strikingly similar behavior of very different physical systems at (or close to) specific points—*critical* points (Batterman, 2002; Morrisson, 2012, 2015).[23] Water and ferromagnetic materials have little in common in terms of their physical/chemical structure and behavior. But during phase transitions such as the water's freezing and the ferromagnet's magnetization, unexpected similarity appears not only in the overall pattern of symmetry-breaking that these transitions involve, but also in the precise values of parameters—critical exponents—that determine the characteristics of these transitions. The mechanisms are clearly distinct; electron spins, for example, play a crucial role in magnetization, but not in freezing or condensation. But the similarity between the systems manifesting universality reveals the overall pattern's insensitivity to structural and dynamic details at the fundamental level.[24] Although it is sometimes questioned whether the theory that explains universality is a physical theory, or merely a mathematical technique,[25] the situation with regard to reduction is quite similar to that of reduction in statistical mechanics. Every system exhibiting universality satisfies the requirement that higher-level patterns supervene on underlying micro-structures, but the overall patterns and the parameters that characterize them are not derived solely from the fundamental laws. The examination of the reduction of thermodynamics to statistical mechanics demonstrates that salient features of nonreductive physicalism, multiple realizability, supervenience, sensitivity to description, and insensitivity of higher levels to specifics of the lower ones, are part and parcel of physics.

Returning to mental evens, Davidson's asserts that although every individual mental event is a physical event and as such obeys physical laws, the same event *under its mental description* may not be subject to laws and may therefore be unpredictable and inexplicable qua mental event. Anomalous monism thus provides an example of events that are subject to strict laws under one description but not under another. Are mental events unique in that respect? Because of the long history of the debate on the mind-body relationship, the focus on the mental is understandable, but it tends to blind us to other applications of Davidson's insights. I

[23] Recall that universality was one of Wheeler's examples of "law without law".

[24] This insensitivity is thought to reflect the fact that at (or near) critical points there is a change in the nature of the coupling between components of the system and the range of their relevant interactions. Whereas under normal conditions long-distance coupling and correlations can be ignored, at critical points this idealization is no longer valid and all interactions must be taken into account. Calculation of these overwhelmingly complex processes is made possible by the technique known as the renormalization group, which involves iterative coarse-graining of the system, with the result that the behavior of the system on every coarse-grained level is analogous to the behavior manifested on the preceding (more fine-grained) level. In the course of this iterative process, the differences between levels within the same system, and the differences between the dynamics of different systems, are washed out.

[25] See Morrison (2012) and the references cited there.

have already noted the analogy between Davidson's account of the relation between the mental and the physical and the relation between macrostates and microstates in statistical mechanics. More generally, one need not confine the discussion to the mental (or stretch the imagination as far as Goodman's grue-some predicates), to find examples of concepts (predicates) that defy law-likeness and projectability. There are very simple physical objects whose description is irreducible to their physical characteristics. Consider a stop sign. It is certainly a physical object and belongs to the category of physical objects. It obeys the laws of physics (depending on its makeup it may obey different laws), and does not threaten to overrule any of the laws or causal relations on the fundamental level. Nonetheless, there is no physical category that corresponds to the category of stop signs. Not just because stop signs are multiply realizable (which, of course, they are), but because the concept of stop sign is open ended and non-projectable.[26] Any number of objects could become stop-signs and no physical property, or structure, or set of specific laws, distinguishes stop signs from other objects. Examples of this kind suggest amending Davidson's point about the mental. It is not a dramatic amendment, for after all, stop signs are symbols, requiring an interpreting mind to understand them. Their open-endedness thus derives from their symbolic significance and is ultimately predicated on mental activity. Such examples do suggest, however, that the crucial feature differentiating the lawful from the lawless in this context is symbolic meaning rather than mentality *per se*. Even if mental states of fear, surprise, and so on, were discovered to correspond to neurological types (which is perhaps not unreasonable) or physical types (which is far less likely), the property of being frighten*ing* or surpris*ing* would still be open-ended and lawless. Under the description of being frightening or surprising, the events and entities falling under these descriptions would be open-ended and lawless as well.

If symbolic meaning is what counts in the above examples, the idea of including the observer, or interpreter, in the reduction might suggest itself. Indeed, Meir Hemmo and Orly Shenker (2020) maintain that the category of stop signs, though not corresponding to a physical type that all and only stop signs instantiate, does correspond to a physical type that includes in addition all and only physical states of brains that perceive the objects at hand as stop signs.[27] We are back to square one, however, for Davidsonians maintain that there is no reason to accept that this type of brain state corresponds to a physical type whereas reductionists about the mental insist that it does. At this point there is no decisive argument on either side. Even at this impasse, however, we should note that reference to the observer opens a wide gap between ordinary physical types and observer-including types. Fermions and electromagnetic radiation are characterized by physicists without the mind that perceives them as such. If stop signs cannot be so described, there is a significant

[26] Multiple realizability in itself does not entail open-endedness. Universality, as we saw, is linked to multiple realizability, but it is conceivable that it is only exhibited in specific kinds of systems and is not open-ended in the way that the concept of stop sign is.

[27] See also their forthcoming paper in *Theoria* and Shenker (2015).

difference between the physical nature of the former and (the alleged) physicality of the latter.

Setting the observer dependent option aside, the lesson to draw from the extension of nonreductive physicalism beyond the mental is that the pressing question about the feasibility of reduction is not whether a multileveled structure of reality could be collapsed to its basis, but whether the various descriptions of reality could be translated in their entirety into one particular privileged description—that of fundamental physics. If this question is to be answered in the negative, as I have argued it should, then lawlessness—not only of completely accidental types but also of types that seem more similar to natural ones—is a straightforward implication. Note, further, that Davidson's position is set in the framework of a deterministic and lawful world in which all events are physical, predetermined by causes, and subject to deterministic physical laws. Lawlessness can only afflict events when grouped into different categories and referred to by different descriptions than those reducible to the language of fundamental physics. In the next section we encounter a more radical split between determinism and lawfulness.

4 Lawlessness in a Deterministic World

Determinism is usually assumed to be inseparable from lawful behavior; moreover, the connection is thought to be conceptual. To see whether there is a way of distinguishing the two concepts, it is useful to look at earlier deliberations of causation from which the more recent notion of determinism has evolved.[28] Traditionally, determinism has found its way into causal discourse through two distinct routes. First, on the most common and intuitive understanding of the causal relation, causes determine their effects in the sense that once the cause has occurred, the effect must follow (with the caveat that this is only the case provided nothing interferes and prevents the effect from happening.) What does 'must' mean is this context? It seems natural to construe 'must' as implying that recurrence of the cause guarantees recurrence of the effect. Hence the 'same cause—same effect' principle, also known as the regularity principle, as constitutive of the concept of cause. (We should actually speak of the same *type* of cause, and types, as emphasized in the previous section, involve descriptions, but here we can set this problem aside). Regularity captures one aspect of the modern sense of determinism, namely, (very roughly) that identical causes (identical initial conditions) cannot lead to diverse effects (later ones). Note that from the fact that *wherever* there is a cause it acts regularly, it does not follow that there could be no spontaneous, uncaused events. The regularity principle is conditional and is therefore consistent with the existence of chance!

[28] The first OED citation of the term 'determinism' is from 1846. Parallel terms in German and French are somewhat earlier.

Determinism has also entered into the causal discourse by another route—the unconditional principle, also known as the universality principle, on which *every* event is determined by causes. This construal of determinism rules out the spontaneous occurrence of events, i.e., it rules out chance. The two principles are therefore quite distinct. On the one hand, it is possible that whenever the same (type of) cause recurs, the same (type of) effect recurs, but that uncaused events also occur. On the other hand, it could be the case that every event has a cause, but the same (type of) cause does not invariably lead to the same (type of) effect. If, however, we *define* causation (as Hume did) in terms of the regularity requirement, then a world obeying universality, that is, a world in which every event has a cause is, *ipso facto*, a world in which the same (type of) cause leads to the same (type of) effect. The converse does not follow; causation can be defined via the regularity requirement, but a world obeying regularity may still allow for chance. It is tempting to link regularity to the existence of laws that provide bridges between causes and their effects, ensuring the invariable relation between them. As we will see, this temptation should be resisted.

Contemporary science does not define determinism in the terminology of cause and effect. Instead, determinism is taken to require that two copies of a closed system that agree in their fundamental physical parameters at some time *t*, agree on the values of these parameters at all other times (or at least future times).[29] This definition captures both of the above principles. If any two systems that are in the same physical state at one point in time continue to be in identical states at all other (future) times, deviation from either universality or regularity is excluded.[30] Note that on this definition, determinism does not guarantee predictability, for it is conceivable (and does indeed happen) that a system behaves deterministically but, for various reasons, resists complete knowledge and predictability. My question, however, is not whether determinism implies predictability, but whether determinism implies the existence of laws.[31]

The answer to this question was provided (independently) by Gregory J. Chaitin, Ray Solomonoff and Andrei N. Kolmogorov, the founding fathers of the subject

[29] The concept of a closed system is an idealization that cannot be fully actualized. Furthermore, the notion of the value of a parameter at a specific time also needs refinement, since some physical magnitudes, e.g., velocity, involve change over time, convergence to a limit, and so on. It may also be useful to introduce quantitative measures—degrees of correlation–rather than make do with the binary division between determinism and indeterminism.

[30] In many contexts, for example, that of the free will problem, the focus of philosophical interest is the truth of determinism 'in reality,' but it is more accurate to think of determinism (indeterminism) as a characteristic of *theories*. A theory is deterministic if it entails the above-stated condition, that is, it entails that two systems that are in identical states at one point are identical throughout. Ascribing determinism (indeterminism) to the world is, then, simply shorthand for ascribing this property to our best theory of the world.

[31] The connection between determinism and predictability is made by Laplace in describing the omniscient demon, and is sometimes still assumed in the literature, e.g., the *Encyclopedia Britannica* entry on determinism. On the difference between the two concepts, see, e.g., Pitowsky (1996).

called Algorithmic Complexity, Kolmogorov Complexity, or Algorithmic Information Theory (AIT).[32] Very schematically, the theory introduces a new definition of the randomness of a series of numbers (digits of a number, results represented by numbers) that is independent of the process by which these numbers are produced. Flipping a fair coin n times could produce any one of the possible 2^n series of heads and tails with equal probability.[33] A very small number of these series will exhibit recognizable patterns (say, 10 consecutive heads followed by 10 consecutive tails) whereas most others will not. Thus, while there is no difference between the 2^n series in terms of their origin in a chancy process, and thus no difference in their probability, there is a huge difference between them in terms of the kind of orderliness they exhibit. The aim of the new approach was to capture this intuitively perceived difference in structure in precise mathematical terms.

The crux of the Chaitin, Kolmogorov, and Solomonoff definition is a distinction between sequences that can be computed by means of a general law, a mathematical formula or an algorithm directly connecting every member of the sequence with every other member, and sequences where no such law exists. The latter are the random ones. When a law exists—say, the law characterizing the sequence of even numbers, $X_n = 2n$—it supplies a shortcut that enables the calculation of every member (in principle, an infinite numbers of members). Of course, for finite sequences, we could always reproduce a sequence by *listing* its members. The question is therefore whether there is a law that is more economical than such a list, a law that 'compresses' the sequence and enables calculation of its members rather than listing them one by one. These ideas have been further refined by quantifying information so that we think of the amount of information contained in a sequence in terms of the number of its bits (each one of which represents a choice or a decision between two equally probable possibilities) and the complexity of a sequence as the number of bits required to reproduce it. Randomness is then defined in terms of complexity: a sequence is random if the minimal algorithm that reproduces it is of (approximately) the same complexity as the sequence. It has been shown that random sequences exist; indeed for sequences of a (not too small) length n, the random ones by far outnumber nonrandom ones. And yet, while it can be shown that a sequence is not random (by finding an algorithm that compresses it), randomness cannot in general be demonstrated.[34] This impossibility—a new limit theorem—is

[32] There are earlier works on mathematical randomness, but the above definition is certainly a major landmark in the history of the subject. Antony Eagle's entry "Chance versus Randomness" in the *Stanford Encyclopedia of Philosophy* (Eagle, 2018) provides a wonderfully clear and detailed philosophical and historical account of the development of complexity theory.

[33] The results of coin tossing are standardly taken to be produced by chance, but if one insists they are after all subject to deterministic laws, one can think of other processes, say certain results of a quantum measurement.

[34] The randomness of finite sequences can be proved when they are of lower complexity than the complexity of the system used for the proof.

closely related to Gödel's incompleteness theorem and other epoch making works on computability by Turing and Church.[35] Chaitin therefore says:

> In a nutshell, Gödel discovered incompleteness, Turing discovered uncomputability, and I discovered randomness—that's the amazing fact that some mathematical statements are true for no reason, they're true by accident. (Chaitin, 1999, p. v)

The definition of randomness of a sequence, we have seen, is independent of the process that generates the sequence, allowing for nonrandom sequences generated by chance. It is thus natural to raise the question of whether the opposite situation also obtains: Could random sequences be generated by a deterministic process? Surprisingly, the answer is positive. Examples of such systems are supplied by cellular automata, studied in great detail by Wolfram (1994). A fascinating example known as the *Game of Life* was devised by John H. Conway in 1970 (Berlekamp et al., 1982; Gardner, 1970). In this game, every step is uniquely determined by the game's update function together with the configuration reached in the previous step. The game is run on a two-dimensional configuration of cells that have two possible states, 'dead' and 'alive.' As an example of an update function consider: A living cell at step n remains alive at step $n + 1$ if and only if, at step n, two or three of its neighbors were alive. A dead cell at step n is revived at step $n + 1$ if and only if, at step n, it had three living neighbors. A live cell at step n with more than three live neighbors dies at step $n + 1$. Obviously, the game does not manifest chance in the sense that flipping a coin does; there is no open, undecided future. Rather, the game is deterministic both in the sense that every step is uniquely determined by the previous one, and in the sense that starting from the same initial configuration (and using the same update function) the series of steps will repeat itself indefinitely. The system thus obeys universality and regularity, the above-mentioned principles that characterize determinism. At the same time, since for many initial configurations there are no shortcut laws, the game involves randomness in the sense of Chaitin, Kolmogorov and Solomonoff. For random runs, the only way of finding out what configuration the game would reach at a certain step is to actually run the game (a computer simulation of it) up to that point. These examples mandate a conceptual distinction between determinism and lawfulness. Even though the possibilities opened by this distinction are still mathematical rather than physical, their very existence should warn us against the traditional conflation of these diverse concepts.

To conclude: Having briefly reviewed the motivation underlying Wheeler's position of "law without law", I have addressed three more limited questions,

[35] In his non-technical (1975), Chaitin explains the impossibility of demonstrating randomness via an analogy with Berry's paradox— finding the smallest positive integer whose definition is longer than this sentence. He shows why the impossibility of finding such a number (actually proving its existence) follows from Gödel's incompleteness theorem together with the result that it is impossible to prove that a particular series of digits is of a higher complexity than that of the program that produces it. The latter impossibility amounts to a proof of the impossibility of proving randomness (See the previous footnote for exceptions).

each one of which suggests a distinct kind of (restricted) lawlessness. First, I have presented Feynman's derivation of the principle of least action as an illustration of the possibility of a law emerging from (nearly) lawless origins. Second, I have argued that lawfulness is description-sensitive. The concept of entropy—certainly a physical concept—served as an example of a concept that is transparent or invisible to the fundamental level of physics and is therefore not entirely reducible to it. Davidson's anomalous monism (asserting the lawlessness of the mental) and non-reductive physicalism in general can draw encouragement from the example of entropy. Finally, I have urged distinguishing determinism from lawfulness. Although our best theory of the world may well be *in*deterministic, the fact that lawlessness is compatible with determinism is, I believe, of great philosophical value.

References

Batterman, R. W. (2002). *The devil in the details*. Oxford University Press.

Ben-Menahem, Y. (2018). *Causation in science*. Princeton University Press.

Berlekamp, E. R., Conway, J. H., & Guy, R. K. (1982). *Winning ways for your mathematical plays II*. Academic.

Blum, A., & Furlan, S. (this volume). How John Wheeler lost his faith in the law

Cartwright, N. (1983). *How the laws of physics lie*. Clarendon.

Chaitin, G. J. (1975). Randomness and mathematical proof. *Scientific American, 232*, 47–52.

Chaitin, G. J. (1999). *The unkowable*. Singapore: Springer.

Cox, B. & Forshaw, J. (2011). *The quantum universe: Everything that can happen does happen*. London: Allen Lane

Davidson, D. (1980a). [1970] "Mental events". In *Essays on actions and events* (pp. 207–224). Clarendon Press.

Davidson, D. (1980b). [1967] "Causal relations". In *Essays on actions and events* (pp. 149–162). Clarendon Press.

Deutsch, D. (1986). On Wheeler's notion of "Law without law" in physics. *Foundations of Physics, 16*(6), 565–572.

Dirac, P. A. M. (1933). The Lagrangian in quantum mechanics. *Physikalische Zeitschrift der Sowjetunion, 3*, 64–72. Reprinted in L.M. Brown (ed.) *Feynman's Thesis* (Singapore: World Scientific) 111–119.

Duerr, P. M. (2020). *Gravitational energy and energy conservation in general relativity and other theories of gravity*. DPhil Thesis, Oriel College, Oxford University.

Eagle, A. (2018). Chance versus randomness. In E. N. Zalta (Ed.), *Stanford Encyclopedia of Philosophy*. Spring 2021 Edition. URL: https://plato.stanford.edu/archives/spr2021/entries/chance-randomness/

Feynman, R. P. (1963). *The Feynman lectures on physics I*. Addison-Wesley.

Feynman, R. P. (1966). "The development of the space-time view of quantum electrodynamics" (Nobel lecture). *Physics Today, 19*(8), 31–44.

Gardner, M. (1970). The fantastic combinations of John Conway's new solitaire game 'life'. *Scientific American, 223*(4), 120–123.

Goldstein, H. (1950). *Classical mechanics*. Reading, MA: Addison-Wesley

Goldstein, H. (1973). [1950] *Classical mechanics*. Addison-Wesley.

Goodman, N. (1955). *Fact, fiction, and forecast*. Harvard University Press.

Hemmo, M., & Shenker, O. (2012). Measures over initial conditions. In Y. Ben-Menahem & M. Hemmo (Eds.), *Probability in physics* (pp. 87–98). Springer.
Hemmo, M., & Shenker, O. (2020). A dilemma for Davidson's anomalous monism. http://philsci-archive.pitt.edu/19563.
Hemmo, M., & Shenker, O. (n.d.). "Flat Physicalism" forthcoming in *Theoria.*
Hoefer, C. (2000). Energy conservation in GTR. *Studies in History and Philosophy of Modern Physics, 31*, 187–199.
Kant, I. (1933). [1787] *Critique of pure reason* (N. Kemp Smith, Trans.). London: Macmillan
Kim, J. (1993). *Supervenience and mind.* Cambridge University Press.
Morrison, M. (2012). Beyond calculation: Extracting physical information from mathematical methods. *Iyyun, 61*, 149–166.
Morrison, M. (2015). *Reconstructing reality: Models, mathematics, and simulations.* Oxford University Press.
Pitowsky, I. (1996). Laplace's demon consults an oracle: The computational complexity of prediction. *Studies in History and Philosophy of Modern Physics, 27*, 161–180.
Pitowsky, I. (2012). Typicality and the role of the Lebesgue measure in statistical mechanics. In Y. Ben-Menahem & M. Hemmo (Eds.), *Probability in physics* (pp. 41–58). Springer.
Pitts, J. B. (this volume). On two slights to Noether's first theorem: Mental causation and general relativity
Poincare, H. (1952). [1902] *Science and hypothesis.* Dover
Schweber, S. S. (1994). *QED and the men who made it.* Princeton University Press.
Shenker, O. (2015). (Hebrew) "Davidson on descriptions: A principle theory". *Iyyun, 64*, 171–190.
Steiner, M. (1998). *The applicability of mathematics as a philosophical problem.* Harvard University Press.
Weyl, H. (1952). *Symmetry.* Princeton University Press.
Wheeler, J. A. (1983). On recognizing 'law without law. *American Journal of Physics, 51*, 398–404.
Wolfram, S. (1994). *Cellular automata and complexity.* Addison-Wesley.

Revaluing Laws of Nature in Secularized Science

Eli I. Lichtenstein

Abstract Discovering laws of nature was a way to worship a law-giving God, during the Scientific Revolution. So why should we consider it worthwhile now, in our own more secularized science? For historical perspective, I examine two competing early modern theological traditions that related laws of nature to different divine attributes, and their secular legacy in views ranging from Kant and Nietzsche to Humean and 'governing' accounts in recent analytic metaphysics. Tracing these branching offshoots of ethically charged God-concepts sheds light on how our ethical ideals and ideas of natural order can still be valuably integrated. Early modern intellectualists valued the law-governed order of nature as a sign of divine Reason. In turn, Reason traditionally ascribed to God has now been partly reclaimed for humans, reframing the value of natural order anthropocentrically, in terms of the value of our own intelligence. Alternatively, Reason may be reclaimed for nature itself, as in an 'objective' idealism or metaphysical rationalism. However, beyond divine Reason, an influential voluntarist tradition in theology stressed a connection between laws of nature and God's Power or free Will. Tracking how divine Power has been reinvested in human beings provides a broader context for instrumentalism and related lineages of empiricism. But secularization can also transfer Power from God to the impersonal natural world. In this light, current scientific interest in lawlike order may also reflect the inherent value of brute necessity or inhuman causal power in nature: this is a deeper way to reject anthropocentrism and to show our respect for the environment.

Keywords Secularization · Science and religion · Anthropocentrism · Environmental value · Intellectualism · Voluntarism

I am grateful to Gordon Belot, Yemima Ben-Menahem, Filipa Melo Lopes, Tad Schmaltz, and an audience at Fordham University for helpful feedback on earlier versions of this material.

E. I. Lichtenstein (✉)
University of Michigan, Ann Arbor, MI, USA
e-mail: elicht@umich.edu

Y. Ben-Menahem (ed.), *Rethinking the Concept of Law of Nature*,
Jerusalem Studies in Philosophy and History of Science,
https://doi.org/10.1007/978-3-030-96775-8_13

1 Introduction

As the idea of laws of nature rose to a new, more central status in science during the seventeenth century, it was framed by Christian visions of a law-giving God.[1] But if searching for lawlike order in nature was closely tied to religious glorification during the Scientific Revolution, its value in our more secularized science is less clear. It may seem obvious that any remaining value in a non-religious concept of laws of nature is epistemic, if not pragmatic. But reflecting on how competing early modern theologies linked law to God, and on their varied paths to secularization, can reveal a wider, more dynamic landscape.

Intellectualists in theology stressed connections between laws of nature and divine intellect or reason. In turn, Reason once ascribed to God has now been partly reclaimed for humans, recasting the value of natural order anthropocentrically, in terms of the value of our own rationality or ability to impose intelligible form onto the raw matter of sensations or disjointed facts. While overt in Kant, who views the idea of a law-governed system of nature as 'required by reason', this shift extends even to recent Humeans who see laws as simple summaries of the total set of basic physical facts. Or Reason can be read back into nature itself, in the spirit of an impersonal rationalism that sees intelligible structure as an ultimate ground of events, like Plato's 'most real' Forms. This unites some recent views in which laws of nature 'govern' physical reality with many older idealisms.

But a distinct voluntarist tradition in theology, which heavily influenced Newton and Boyle along with many other scientific pioneers, instead emphasized a link between law-governed natural order and God's power or free will. Noting how divine Power has been reinvested into human beings sheds light on other secularized accounts of science, like instrumentalism and related lineages of empiricism, as we will see. But some free exercises of power imputed to God can also be transferred to nature. It also pays, then, to consider how interest in natural order may reflect the value of impersonal causal power, rather than the value of pragmatic utility, human intellect, or the intelligibility of reality. This is a way to more deeply reject anthropocentrism and embrace environmental value.

Tracing these theological legacies illuminates the current range of opinion about what laws of nature are, and why they matter. It brings into view certain secularized ideas of natural order and its value which, though marginalized or untested, merit more focus. And it suggests promising ways of thinking about how natural science connects to ethics.

Minimizing the religious genealogy of scientific ideas and methods is therefore less benign than it may seem. Positing a clear divide between religious and secular

---

[1] Many historians locate a rapid rise to prominence of the idea of laws of nature in the seventeenth century, often identifying Descartes as its first major figure. See e.g. Oakley, 1961a, 433; Henry, 2004, 87–88; Garber, 2013, 45; Crombie, 1996, 86–87; Daston & Stolleis, 2008, 2. Joseph Needham (1951a, 29) points to the sixteenth century, too. On pre-sixteenth-century ideas of laws of nature, see also Ruby, 1986; Kedar & Hon, 2017.

views of science, with the secular taken to be largely insulated from skepticism of the religious, obscures how science relates to other forms of understanding, and makes it all too easy to ignore ideological assumptions that influence current science. Even when they seem to be purely descriptive, secularized views of laws often rely on implicit value judgments for motivation. And accounts of laws of nature that implicitly value 'reason' over 'sense', or stress human control over nature, often look worse after taking stock of their conceptual ties to contentious historical ideas of God. At a minimum, this context shows how metaphysical analysis of order in nature can be as much an expression of values as it is a dispassionate inquiry into the basic structure of reality. But even where value-neutral metaphysical analysis is possible, it may still be undesirable. In this light, the role of ethically charged God-concepts in the history of the idea of laws of nature offers insight into how secularized ethical ideals and ideas of natural order can be valuably integrated.

2 Laws of Nature Beyond the 'Death of God'

The influence of Christianity on the Scientific Revolution is widely recognized, but the idea that this religious context bears deeply on current science is just as widely contested. We can easily find that Descartes asked his friend Mersenne, in 1630, to "not hesitate to assert and proclaim everywhere that it is God who has laid down these laws in nature just as a king lays down laws in his kingdom."[2] In *Principles of Philosophy*, Descartes even tries to show that specific 'laws of nature', like the straight-line motion of unimpeded bodies, follow directly from God's immutability and simplicity.[3] Newton, too, suggests openly in his treatise on *Opticks* that it was "by the Counsel of an Intelligent Agent," with the "help" of "active Principles" or "Laws of Nature" like gravity, that "all material Things seem to have been composed" of "Particles[. . .]associated in the first Creation."[4] To be sure, he was often more circumspect about theology. Still, for Newton just as much as Descartes and countless other early modern natural philosophers, the law-governed cosmos that they had come to reveal was a clear and awe-inspiring sign of God—again, this much is largely uncontested.[5] Yet many who accept it are still extremely reluctant to identify any strong trace of early modern theology in present-day science.

Even historians of science and religion tend to downplay the current significance of this religious backdrop. Joseph Needham claims, for instance, that "[i]n the

---

[2] Letter to Mersenne, 15 April 1630 (Descartes, 1991, 23).

[3] Descartes, 1982, 59–62.

[4] Query 31 (Newton, 1730, 377–378). See also Newton to Bentley, 10 Dec. 1692 (Cohen, 1958, 280).

[5] On historians' view of Newton on God, see Henry, 1994, 123; Snobelen, 2001, 198. See also note 37.

outlook of modern science there is, of course, no residue of the notions of command and duty in the 'laws' of Nature."[6] In a similar vein, Richard S. Westfall suggests that natural philosophy "does not[...]have to concern itself with God at all," although it once did—and even asserts that "[w]ith us it does not."[7] It is no great surprise, then, that one is even harder-pressed to find influential accounts of laws of nature in recent philosophy of science or metaphysics which ascribe any deep importance to the idea's religious genealogy.[8]

But if this situation is to be expected in one sense, in another it is still strange: ahistorical and an artificial limit on the analysis of science, at least, if not just untenable. After all, why would there not be 'residues' of divine command in the idea of laws of nature, given its origins? Where even historians draw this sort of stark divide between past and present, it is fair to surmise that we may not want to know. Seeing residues of theology in science might make us uncomfortable, of course, as it can seem tantamount to rejecting scientific naturalism or endorsing epistemological relativism. But early modern ideas of God have left an imprint on current views of lawlike order in nature, regardless.

This is not to deny that science has undergone a process of secularization. It is not to fall prey to the genetic fallacy. And it is certainly not to say that scientists do or ought to believe in God or gods. But it is to say that we would do well to more frankly examine deep currents of structural continuity, as well as discontinuity, between the religious metaphysics underlying early modern science and the secularized metaphysics of current science. This can productively recontextualize debates about how to analyze the scientific concept of law. Confronting the early religious framing of laws of nature may even call into doubt the ongoing viability of the idea.[9] But this approach also illuminates a basic question that is even more often overlooked: why think that finding laws of nature would be valuable, in the first place? Distinct early modern theological traditions valued laws in relation to different attributes of God. Some of these traits are more commonly or more easily secularized. And secularization has shifted divine traits into many different things.

In particular, it will prove helpful to contrast divine Reason or Intellect against divine Power or free Will. And it will prove helpful to distinguish two areas into which these traits can be displaced through processes of secularization:

---

[6] Needham, 1951b, 229. His elaboration that laws are "now thought of" as "descriptions not prescription" (ibid., 229) is belied by ongoing interest in 'governing' accounts of laws. See also Needham, 1951a.

[7] Westfall, 1986, 234.

[8] Compare e.g. Lewis, 1983, 1994; Ramsey, 1978; Loewer, 1996 ('best system' accounts); Armstrong, 1983, Armstrong, 1993; Dretske, 1983; Tooley, 1977, 1987 ('necessitarian' accounts); Carroll, 1994; Lange, 2000, 2009; Maudlin, 2007 ('antireductionist' or 'primitivist' accounts). One partial exception is 'antirealist' accounts—e.g. Giere (1999, 23) argues that "the original view of science as discovering universal laws of nature had little basis in the actual practice of science, but was imported largely from theology"; so, for Giere, it is natural that 'laws' should be purged from secularized science.

[9] E.g. see Ben-Menahem & Ben-Menahem, 2020, 53. See also Giere, 1999.

anthropocentric theories reinvest divine traits into humanity, while *ecocentric* theories reinvest divine traits into impersonal nature. Taken together, these two divides suggest four basic secularized views about the value of laws of nature: anthropocentric and ecocentric forms of *rationalism*, which ground the value of laws in the value of human intellect or rational intelligibility; but also anthropocentric and ecocentric forms of *dynamism*, which ground the value of laws in the value of power—whether it be human control over nature, impersonal causal forces, or brute constraints imposed by arational patterns in nature that we cannot control.

Many of these positions are compatible, since we can value several things at once. But they often come apart in practice, particularly when tied to distinct views about what laws of nature are. And while some paths to secularization have been historically favored, others have been largely overlooked—especially forms of dynamism that find basic value in arational order or impersonal force, focusing less on reason or human causal power.

3 Intellectualist and Voluntarist Theology at the Rise of Modern Science

A voluntarist tradition in theology made a deep impact on early modern thinking about laws of nature. Voluntarists placed particular emphasis on divine power or will, celebrating "an omnipotent God from whom the world did not proceed by any necessary emanation, but who called it into being by the autonomous *fiat* of his will."[10] And they came to see laws of nature as a paradigmatic expression of this free Will or Power. As Francis Oakley has shown, although the "principal source" of this conception of a "legislating God whose most striking attribute was his irresistible power" was the Old Testament, it only achieved a "wide currency" in the sixteenth and seventeenth centuries, in a form "directly descended" from nominalist positions worked out after the Condemnation of 1277, by theologians like William of Ockham.[11] A competing intellectualist tradition in theology instead placed more central emphasis on God's intellect, wisdom, or reason.

In the medieval context in which it came to the fore, this voluntarist-intellectualist rivalry derived from divergent views about the status of neo-Aristotelian philosophy vis-à-vis Christian theology.[12] Voluntarists worried about Aristotelian theses that seemed to deny or stand in tension with God's power or autonomy. As the historian of science John Henry explains, for example: "[i]f Aristotle said creation of a vacuum was impossible then it did not undermine God's omnipotence to say God could not create a vacuum—for these [intellectualist] thinkers it was as though creation of a vacuum was not simply a physical impossibility, but a *logical*

[10] Oakley, 1961a, 450–451.

[11] Oakley, 1961a, 449 *et passim*; see also Oakley, 1961b.

[12] Henry, 2009, 80.

impossibility."[13] Voluntarists were skeptical, stressing that God acts with free Will. In so doing, their point was "not to insist that God did things in an irrational way"; it was simply "to ensure that supposedly logically-driven conclusions did not get out of hand," by denying God's power or implying that "he was not fully in control of his own actions, because he always had to follow fully the dictates of reason, and acted, therefore, out of necessity."[14] Voluntarists could even allow some genuine contradictions that God could not circumvent. They just denied that everything deemed 'contradictory' by intellectualists was truly contradictory or impossible for God.

In turn, throughout the past century, a range of historians of science, religion, and philosophy have argued that voluntarist theology played a pivotal role in the development of modern science. This impact was perhaps most notable in relation to Newton, Boyle, and others in England,[15] but voluntarism also influenced continental natural philosophers like Gassendi and (some argue) Descartes.[16] In a study of seventeenth-century science, Eugene M. Klaaren thus finds that "hypotheses, attention, the search for particulars, and laws of nature were all part and parcel of a new empirical way of knowing indebted to a critical (and carefully constructive) voluntarist theology of creation."[17] John Henry qualifies that the sudden seventeenth-century rise of the idea of laws of nature is "best seen as the exploitation of a ready made theological tradition *in support* of innovations," rather than a genuine "*explanation for* those innovations."[18] But, with this caveat, Henry still judges that "there can be no doubt of the relevance of providentialist disputes, and theological voluntarism in the efforts of Descartes, Boyle, Newton and Leibniz to justify their use of laws of nature."[19] Certain historians do remain more deeply skeptical of any 'voluntarism and science' hypothesis.[20] But the real force of the conceptual connections that its proponents adduce, together with the fact that many historians have ably defended stronger claims of influence, more than suffices for my purposes, which are largely conceptual and ethical.

The most widely discussed aspect of this link between voluntarism and science goes beyond laws of nature, to broader connections between voluntarism and empiricist orientations toward knowledge. Correspondingly, there is said to be a

---

[13] Henry, 2009, 80.

[14] Henry, 2009, 80.

[15] On Newton's voluntarism, see Force, 1990, 89; Shapin, 1981; Snobelen, 2001, 176–177. Cf. Harrison, 2004. On voluntarism's influence on Newton's teacher, Isaac Barrow, see Malet, 1997. On Boyle vis-à-vis voluntarist views of creation in seventeenth-century England, see Klaaren, 1977. See also Burtt, 1925, 290.

[16] On Gassendi's voluntarism, see Osler, 1991, 157 *et passim*. See also Osler, 1994. On scholarly debate regarding Descartes, vis-à-vis voluntarism and intellectualism, see note 22 and references therein.

[17] Klaaren, 1977, 123.

[18] Henry, 2004, 91.

[19] Henry, 2004, 91.

[20] See especially Harrison, 2002, 2004, 2009.

regular (albeit not inviolable)[21] link between intellectualist theology and rationalist views of knowledge. Figures like Newton and Boyle are taken to be paradigmatic voluntarist-empiricists, while philosophers like Leibniz—and perhaps Descartes, although again here there is more debate[22]—are paradigms of the intellectualist-rationalist camp.[23] One locus classicus for this thesis is a series of papers by Michael B. Foster, published in *Mind* in the 1930s.[24]

Foster surveys several helpful points of contrast between empiricism and 'Greek' rationalism.[25] First, the modern empiricist "describes" natural substances, whereas the ancient Greek rationalist (or medieval Scholastic) "defines" them. Here to 'define' is to specify an object's essence in a way that allows for a priori species-genus classifications going beyond the empiricist's a posteriori "comparison of similar sensible qualities."[26] For an object to be definable, Foster suggests, its form must be both intelligible and its real (not merely nominal) essence.[27] This is the case for objects produced by a Greek-style Artificer, who purposively forms pre-existing matter toward some prior end. But the empiricist's Creator-God is less a technician and more an artist, with no end in mind before creation. For this Creator, the goal of creation is just the created world itself, such that a natural object has no intelligible form or purpose as a prior ground.[28] God creates nature's matter and form simultaneously and bound inextricably. Foster takes the analogy to art even further: the meaning of a painting cannot be abstracted from the work, as it is tied to its sensuous qualities and not to any pre-conceived plan; so the meaning of a painting is "not intelligible in the sense in which the purpose of a wheelbarrow is."[29] (One might add that a beautiful work of art exceeds its generically-describable features, if not even any fixed 'meaning'.) Likewise, then, there is "something more" in the

[21] See note 22.

[22] Harrison (2002) argues that Descartes was both a rationalist and a voluntarist, and uses this example to "break[] the inexorability of the logic of a connection between voluntarism and empiricism" (Harrison, 2002, 66). Note that others have argued that Descartes is better described as an intellectualist (e.g. Davis, 1991; Osler, 1994). Here I would also echo John Henry (2009, 84): positions like voluntarism and empiricism can be deeply linked, historically and also conceptually, even if there is no "inexorable logic" binding them together—as the case of Descartes arguably illustrates. Cf. Harrison, 2004, 2009.

[23] In England, Henry More and his notion of "divine reason" is a plausible example (Klaaren, 1977, 122).

[24] Foster, 1934, 1935, 1936. Foster's focus is on voluntarism. Henry (2009, 79) also invokes Lovejoy's *The Great Chain of Being* (Lovejoy, 1936) as a canonical study of the opposing (now 'intellectualist') approach.

[25] Foster, 1934, 454.

[26] Foster, 1934, 461.

[27] Foster, 1934, 461.

[28] Foster, 1934, 462.

[29] Foster, 1934, 462.

world made by a Creator, which is "*not* capable of being conceived in distinction from the sensible material in which it is expressed."[30]

The empiricist thus discovers things' properties "by observation and experiment," whereas the rationalist does so "by 'intuitive induction' and demonstration."[31] And the empiricist seeks relations of cause and effect, not the rationalist's ground and consequent. Here think of Newton invoking impressed force as the cause of acceleration; by contrast, think of Descartes claiming that extension is the essence of matter, and arguing that this entails the impossibility of a vacuum. Experience can be crucial to rationalists, too, but in a distinct way: rationalism "implies that, however sensuous experience may be necessary as a step preliminary to the discovery of essence, every sensuous element is discarded from the act in which the essence is known"[32]—as in the case of Aristotelian induction.

Lastly, and most to the point here, Foster connects empiricism to voluntarism. The rationalist construes materiality, or the sensuous aspect of things, as a "defect" or deficit of being.[33] Again, that is, for the rationalist, intelligible form or universal quality is the essence of sensible objects, such that the concrete particular is reduced to a mere appearance or incomplete realization of the universal—'less real' than it and 'due to' it as a formal cause, if not a final end toward which an Artificer formed matter.[34] Intelligible form thus grounds the reality of concrete things, exhausting their essential traits. For the modern empiricist, by contrast, materiality is a more positive aspect of being: that aspect of reality exceeding intellectual intelligibility or rational determination, that which is fully real and yet not fixed by universal form or necessitated by reason. Foster relates this excess of rational necessity to God's freedom. And this ties voluntarism to empiricism: "*voluntary* activity of the Creator[. . .]terminates on the *contingent* being of the creature"; but if "the contingent" is thus "essential to nature" while also being "knowable only by sensuous experience," then "experience must be indispensable to the science of nature."[35]

Rather than investigating the logical (im)possibility of vacuums, then, early modern voluntarists emphasized that the only way to tell whether God had allowed empty space was to see if one could in fact find or make it. Why? Because, they said, God had a free choice in the matter. God's power is thus not rationally necessitated, but 'arbitrary'.[36]

---

[30] Foster, 1934, 462.

[31] Foster, 1934, 454.

[32] Foster, 1936, 2.

[33] Foster, 1934, 455 and 463–464.

[34] Here I am elaborating more freely on Foster's account. Compare Plato's view that Forms are 'really real Being' (*Phaedrus* 247e), 'more real' (*Republic* 515d) than the sensible particulars that participate in them. Note that this does not mean really or more *existent*—see Vlastos, 1965, 1966; Guthrie, 1975.

[35] Foster, 1934, 464.

[36] Peter Harrison (2013, 147), a skeptic of 'voluntarism and science' hypotheses, stresses that 'arbitrary' just meant that "God's will is not bound by considerations external to him"; still, he

Likewise, for voluntarists, the only way to tell whether matter could be active was to empirically investigate bodies. Again, their assumption was that there is nothing in the essence of matter which makes it impossible for God to choose to make it active. Here there are two things that God might be doing 'freely' or 'arbitrarily'. God could directly order matter, imposing law by physically arranging material objects at every moment. Or God could do something more indirect, like adding activity to inherently inert matter.[37] (Compare Newton's claim that "[i]t is inconceivable, that inanimate brute Matter should, without the Mediation of something else, which is not material, operate upon, and affect other Matter without mutual Contact." This is "one Reason why I desired you would not ascribe innate Gravity to me." Indeed, "[t]hat Gravity should be innate, inherent and essential to Matter" is "so great an Absurdity" that "no man who has in philosophical Matters a competent Faculty of thinking, can ever fall into it."[38]) Either way, however, law-governed order manifests God's total power or free will, more so than God's reason.

In contrast to current philosophers of science or metaphysicians who defend so-called 'governing' accounts of laws of nature, then, seventeenth-century voluntarists like Newton and Boyle thought that abstract patterns are the wrong kind of thing to govern, constrain, or produce physical effects, in themselves. Rather, an early modern voluntarist might see abstract patterns simply as summarizing the impact of laws taken as acts of Will: laws govern nature only insofar as they are free exercises of power, not just intelligible rules. To the extent that they can be viewed as intelligible rules, laws are like the abstracted 'meaning' of a painting, not like the pre-conceived purpose of a tool or the geometric essence that a visible line imperfectly copies. The artwork is not a copy of its meaning—the artwork grounds its meaning, not vice versa. Just so, nature grounds its intelligible form. Viewed as intelligible patterns, laws describe but do not govern nature. But viewed instead as acts of Will that give rise to a world displaying these patterns, laws do govern nature. God governs nature in virtue of the act of Will through which God creates nature.

There is no prior idea of, or rationale for, specific laws of nature, on this view—just God's free choice in the act of creation. And brute matter does not obey laws by its own nature. So it takes Will or Power, not just Reason, to produce a law-governed world. This connects to voluntarists' emphasis on God's transcendence, as opposed to immanent presence in nature: voluntarists stressed that law is imposed on nature from without. By contrast, rationalists might say that divine Reason permeates

allows that the Newtonians here differed from Leibniz, who thought God had to create the best of all possible worlds.

[37] Newton scholars disagree on this issue. E.g. Henry (1994), arguing against A. Rupert Hall's, Alexandre Koyré's, and I. Bernard Cohen's view that "Newton relied more or less directly upon God to explain the force of gravity" (Henry, 1994, 125), suggests that Newton did believe in the possibility of gravity as action at a distance, grounded in a *non-essential* power of attraction *superadded* to matter by God. Cf. Janiak, 2008, 2013; for responses to Janiak, see Henry, 2011, 2014. See also Koyré, 1965, 149.

[38] Newton to Bentley, 25 Feb 1692/3 (Cohen, 1958, 302–303).

nature, that matter conforms to law by its own nature, or that intelligible form is the inner essence of natural objects.[39]

This voluntarist emphasis on divine transcendence is on display in Newton's view that "[t]his Being governs all things, not as the soul of the world, but as a Lord over all; and on account of his dominion he is wont to be called Lord God[...]or Universal Ruler."[40] Newton's insistence that God is 'not the soul of nature' is a repudiation of the rationalist conceit that divine reason is immanent in nature, or that intelligible form is the essence of material substance. (For Boyle, too, it is "the *Creator* who, and not the *World*, nor the *Soul* of It, is the True God."[41]) Newton's emphasis on 'dominion' in turn relates to voluntarists' focus on divine Power. And God displays this Power by imposing law on nature, since (the idea is) order can be imposed on nature only if not already present in it:

> Matter does not move[...]in accord with mathematically precise laws of nature [...]because of any Neoplatonic overflow of God's being into the world or because of any Hobbist, Cartesian, or Leibnizian notion of necessary rational order intrinsically immanent within matter or imposed once and for all long ago by a deity who long since has absented himself from the daily operations of creation. Such metaphysical views dilute the total subordination of matter to the will of God and are the metaphysical equivalent of theological idolatry. Rather, matter exists and ordinarily operates in accord with natural law for one reason: God wills it so by divine *fiat*.[42]

God's dominion may thus be the "first metaphysical principle" of Newtonian science.[43]

More broadly, early modern voluntarists—following Calvin, as well as Ockham[44]—saw the idea that God "Pervades all things" as a blasphemous way of "making a shadow deity to drive away the true God":[45] an all-powerful Lord who imposes law on nature from without. Intellectualists valued laws vis-à-vis God's Reason, if not nature's rational soul. But voluntarists valued natural law more so as a sign of God's free and total Power.

4 Ecocentric Rationalism: Valuing Nature-As-Intelligible

It can seem self-evident that the value of a drive to find lawlike order in nature is epistemic, whether one takes the relevant epistemic ideal to be truth, empirical adequacy, knowledge, or basic explanation. But now-intuitive epistemic views

---

39 On transcendence and immanence, see also Dupré, 1976 (theology), Crosby, 2003 ('religion of nature').

40 Quoted in Burtt, 1925, 291.

41 Boyle, 1686, 120.

42 Force, 1990, 84. See also Jacob & Jacob, 1980, 265; Shapin, 1981, 195; Oakley, 1961a, 436.

43 Force, 1990, 84.

44 On the influence of Calvinism on seventeenth-century scientific thought, see Klaaren, 1977.

45 *Calvin, Institutes of the Christian Religion I:5:5* (Calvin, 1960, *58).*

of laws of nature can obscure secularized offshoots of the voluntarist theology that guided many central figures in the Scientific Revolution, notably in England. Narrowly epistemic views of laws of nature partly reflect value-laden decisions to emphasize human reason or impersonal intelligibility at the expense of other basic ideals, just as the theological divide between intellectualists and voluntarists was—more obviously—partly rooted in divergent values. It is not a genetic fallacy, or conflation of the current and past meaning of science, to acknowledge these ongoing value-laden decisions and their similarities to past theology.

Of course, this does not mean that anyone who focuses on connections between laws of nature and ideals like intelligibility must be a hopeless 'rationalist', or that those who instead stress the pragmatic utility or contingency of laws 'do not value rationality'. Early modern voluntarists still praised God's reason, after all, and intellectualists still respected God's power. Still, divergent views about how to weigh shared values can be real disagreements, with substantial impact on scientific decision-making.[46] In secularized societies, it may also be easier to focus on epistemic values in science even while attending appropriately to other kinds of value in other areas of life. But this sort of compartmentalization is not necessarily to our credit. And since it is hard to take a broad perspective on ourselves, we often behave in ways that reflect and reinforce contentious or culturally-specific hierarchies of value without seeing that we are. Adopting a certain view of laws of nature may well be ethically loaded, then, even when this is not intended.

In this light, and as an inroad to understanding how voluntarist ideas about laws of nature have been secularized, we may first consider the secularization of intellectualist focus on God's reason. A rationalist position intermediate between traditional theism and full atheism might look like Einstein's view: he saw the universe as a "single significant whole,"[47] whose "harmony of natural law[. . .]reveals an intelligence of such superiority that, compared with it, all the systematic thinking and acting of human beings is an utterly insignificant reflection."[48] Einstein meant this 'intelligence' to be impersonal—not benevolent or purposive, for example—such that he also believed that "there is no anthropomorphic conception of God" underwriting the "cosmic religious feeling" he approvingly described.[49] Still, Einstein's appeal to an inhuman intelligence clearly evokes earlier intellectualist currents in natural theology, and especially calls to mind his claim to "believe in Spinoza's God, who reveals himself in the lawful harmony of being."[50]

To many, of course, Einstein's account of this amoral cosmic intelligence will still seem too anthropomorphic. This suggests two more thoroughly secularized versions of intellectualism: first, a retreat from claims about cosmic intelligence to

---

[46] See e.g. Kuhn, 1977; compare Longino, 1995, Lichtenstein, 2021a.

[47] Einstein, 1949, 26.

[48] Einstein, 1949, 29.

[49] Einstein, 1949, 26.

[50] Letter to Rabbi H. S. Goldstein, 25 April 1929 (The Albert Einstein Archives at The Hebrew University of Jerusalem, Reel 33, Item 272).

more modest ones about the *intelligibility of nature*; second, a focus on the *intelligence of humans* engaged in scientific inquiry. These are not mutually exclusive. For instance, scientists may focus on nature's intelligibility, while valuing it derivatively upon the value of our intellect. Focus on intelligibility in nature can therefore belie a kind of indirectly *anthropocentric* approach. Still, at least in principle, a more robustly *ecocentric* version of rationalism is possible, on which the value of laws of nature stems from the inherent value of impersonal intelligibility, not human intelligence. In practice, however, most forms of secularized intellectualism are plausibly anthropocentric, whether directly or indirectly.

Focusing on the intelligibility of nature involves recovering common tropes from ancient idealism, early modern rationalism, or even more 'objective' strands of post-Kantian idealism. Explicitly reverting to older idealisms may also be preferable, in several ways, to the views of laws of nature recently advanced by 'antireductionists' like Tim Maudlin or advocates of related 'governing' accounts.[51] On Maudlin's view, laws of nature "operate to generate or produce" later states of the universe from its initial state.[52] But Boyle and other early modern voluntarists had perfectly good reason to deny that laws of nature can generate physical effects. Boyle proposes what seems like a stronger explanatory scheme, setting aside his appeal to God, in *The Christian Virtuoso* (1690):

> I look upon a law as a moral, not a physical cause, as being indeed but a notional thing, according to which, an intelligent and free agent is bound to regulate its actions. But inanimate bodies are utterly incapable of understanding what a law is, or what it enjoins, or when they act conformably or unconformably to it; and therefore the actions of inanimate bodies, which cannot incite or moderate their own actions, are produced by real power, not by laws; though the agents, if intelligent, may regulate the exertions of their power by settled rules.[53]

As Boyle rightly observes, unthinking objects are not able to understand laws of nature.[54] So, while a causally powerful God might 'produce' physical effects in inanimate objects, abstract rules or ideas seem incapable of doing so. Boyle's picture may make better sense than Maudlin's, then: God produces law-governed order in a display of Power; insofar as laws of nature are abstract Rules and not displays of Power, they cannot affect unthinking matter; and matter does not obey laws of its own accord, since it is inherently passive. Maudlin does qualify elsewhere that when

[51] On "governing" vs. "descriptive" or "Humean" accounts of laws, see e.g. Loewer, 1996; Beebee, 2000; Schneider, 2007; Bhogal, 2020. For a recent survey of non-Humean views, see Hildebrand, 2020. For an account looking to combine aspects of governing and Humean views, see e.g. Roberts, 2008.

[52] Maudlin, 2007, 174.

[53] Quoted in Crombie, 1996, 85

[54] Compare Suárez's similar view that "acceptation of law" in nature is "metaphorical, since things which lack reason are not, strictly speaking, susceptible to law, just as they are not capable of obedience" (Suárez, 1944, 22; quoted in Ben-Menahem & Ben-Menahem, 2020, 50). Spinoza also suggests that this use of 'law' is only metaphorical—see Spinoza, 2007, 58; Ben-Menahem & Ben-Menahem, 2020, 51.

he says laws 'generate' or 'govern' the universe's evolution, these are "metaphors" that can "fire the imagination," but which are "not offered as analyses."[55] In clarifying his view's non-metaphorical content, however, he then says that "[l]aws are the patterns that nature respects; to say what is physically possible is to say what the constraint of those patterns allows."[56] But why 'respect' and 'constraint' are here any less metaphorical than 'produce' or 'govern' remains unclear.[57]

Suppose that laws of nature are 'the patterns that nature respects'. Are these patterns abstract or concrete? It is not clear what Maudlin thinks. But claiming that *abstract* patterns generate physical states is extravagant, or else metaphorical to a fault. More restrained is something like an old-fashioned rationalist view: natural phenomena are not caused or temporally *produced* by the intelligible patterns they instantiate; rather, they are non-causally and atemporally *grounded* in these patterns. Indeed, atemporal grounding is what many advocates of governing accounts of law seem to have in mind.[58] Maudlin's picture, in which laws of nature take the initial state of the universe as input and produce later states as output, is a more unusual sort of reification of deductive-nomological explanation—although this may be what certain others mean by 'govern'.

As one example of an older rationalism invoking something like grounding, note that for post-Kantian absolute idealists, "*everything in reality conforms to reason, which consists in the forms, ideas, or purpose of things.*"[59] For Friedrich Schlegel and other Early German Romantics, this view had a "Platonic provenance" and aesthetic dimension: for them as for Plato, "reason is the intelligible structure of things, what we perceive in beauty through the power of intellectual intuition."[60] Despite widespread concern about the speculative character of post-Kantian ide-

[55] Maudlin, 2007, 15.

[56] Maudlin, 2007, 15.

[57] 'Respect' is ambiguous: for a state to respect laws could just mean that it *conforms* to laws; but it could also mean that it *considers* laws, or *refrains from interfering with* them. The first, weaker sense is innocuous, but fails to distinguish Maudlin's view from Humean accounts, which he rejects. So he must mean something stronger, related to 'ontological dependence' (Maudlin, 2007, 174–178). But this is also ambiguous: Humeans can agree that facts would be different if the laws were very different, e.g.—since facts would have to be different in order for laws that just summarize facts to be very different—so even Humeans can admit a weak sense in which facts 'depend on' laws. So Maudlin seems to need a stronger notion of causal influence or acausal grounding, to insist that laws cause or ground events, and not vice versa. But laws are ill-suited to causally affect inanimate objects—despite Maudlin seeming to rely on 'production' being quasi-causal, or not being able to run backwards in time—see Lichtenstein, 2021d. Or, if it is grounding that is in play, then causal terms like 'produce' are misleading, even as metaphors.

[58] The relevant kind of 'grounding' is often associated with 'ontological dependence', and has to be more than "purely logical" determination, since a 'descriptive' Humean about laws could agree that "you can *deduce* future facts from current facts plus the laws" (Beebee, 2000, 578). For a Humean view on which laws "generate descriptions of particular cases" in something like a purely logical way, see Ward, 2007.

[59] Beiser, 2003a, 69.

[60] Beiser, 2003a, 70. *See also* Beiser, 2003b.

alism or the apparent mysticism of Romanticism, this appeal to the essential 'conformity' of reality to reason may well be more defensible than Maudlin's appeal to the 'productive' power of laws. Or perhaps Maudlin could say that laws are *concrete* patterns. But then they would be tasked with generating the patterned phenomena they comprise—that is, laws would have to 'produce' and then 'respect' themselves. Again, many traditional idealisms may be better. Maudlin's picture of laws producing later states of the universe from its initial state, if not also some other recent governing views of law, conflates old idealistic tropes with modern ideas of causation, to the detriment of both. Metaphor or not, the idea that laws of nature generate physical states is a relatively implausible residue of older visions of God performing double duty as both a physical Cause and a moral-rational Ground.

The more overt ethical content of early modern natural theology is also to its comparative advantage. For Boyle and Newton, God's ability to command nature through a freely 'ordained' rule of law was related to the nature and value of freedom or creative power. For intellectualists, law-governed natural order displayed reason and its value. For Maudlin and other contemporary metaphysicians discussing laws of nature, however, it seems that far less is ever at stake. From one standpoint, of course, this is to respect a real fact-value distinction. But from another standpoint, it is to lose a sense of how modern science and ethics can and should be interrelated. If one were to integrate the most plausible version of a governing account of laws of nature into a broader worldview, in this spirit, the result might well be a more frank metaphysical idealism and exaltation of 'ideal' form or rational intelligibility across ethics, politics, and art, as well as in science.

More generally, now-intuitive ties between laws of nature and epistemic ideals like truth, knowledge, or understanding partly reflect the legacy of intellectualist currents in the theology that underwrote the development of the concept of laws of nature. And the most promising version of this intellectualist program, in a more secular context, may well be a fairly traditional philosophical idealism. Distinguishing subjective and objective forms of idealism is crucial here. As a scholar of post-Kantian idealism puts it, the basic idealistic thesis that "reality depends upon the ideal or the rational" includes subjective variants in which the ideal or rational is "the subjective, mental, or spiritual," but also objective variants in which the ideal is just "the archetypical, intelligible, or structural."[61]

Ecocentric rationalism, in the present sense, simply adds two twists to objective idealism: an emphasis on *law-like order* as the kind of 'ideal' or 'rational' structure that is at issue; and an evaluative claim to the effect that the rational is not just the highest reality or ultimate ground, but also *inherently valuable*. This shift in focus, from an older idealistic view of Form as genus or kind to a view of Laws of Nature as the quintessential 'ideas', might also be connected to wider intellectual historical dynamics. Informed by his reading of major developments in early twentieth-century physics, for instance, the neo-Kantian philosopher and historian of ideas Ernst Cassirer described a broad shift from old concepts of objects

[61] Beiser, 2002, 11.

as reified 'substances' to modern ideas of objects as mere principles of 'functional' interconnection among measurable properties.[62] Differences may remain between those who take laws to govern nature from without, as with a transcendent God, and those who see law as nature's inner essence. Either way, the idea that laws govern nature is, in part, a residue of the idea that divine Reason is the ground or soul of nature.

5 Anthropocentric Rationalism: Valuing Human Intellect

A second kind of secularized rationalism is more overtly anthropocentric.[63] Here consider Kant's view that a "systematic unity" of knowledge is "required by reason"—albeit as a purely "regulative" ideal, not a "constitutive" part of our experience like space, time, or causality.[64] Kant ties this unity to laws of nature, claiming that reason transforms cognition from "merely a contingent aggregate" to a "system" connected "in accordance with necessary laws."[65] So, Kantians may say that laws of nature are "statements that play a particular role in the system that would emerge from an ideally extended inquiry."[66] Kant arguably takes particular laws' necessity to stem from their role in systematizing experience.[67] But he at least views it as an a priori constraint on all laws that they figure prospectively in a total system of nature, posited by reason itself as a purely ideal limit.

Kant's functional description (if not analysis) of laws of nature vis-à-vis their role in idealized rational inquiry is similar, in important ways, to David Lewis's

[62] Cassirer, 1910. For analysis of Cassirer's view and its intellectual historical context, see Friedman, 2000.

[63] Even 'objective' idealism can be indirectly anthropocentric, insofar as it values things as possible objects of the kind of rational or intellectual capacity that many take to elevate us above less intelligent animals.

[64] *Critique of Pure Reason,* A648/B676 (Kant, 1998).

[65] *Critique of Pure Reason* A645/B673 (Kant, 1998). Compare Kitcher, 1986, 206.

[66] Kitcher, 1986, 215.

[67] Kitcher, 1986. Messina (2017) labels Kitcher's view of Kantian laws, as deriving their necessity directly from systematic unification, a 'Best System' interpretation, in contrast to accounts according to which Kant thinks that (i) knowing particular laws to be necessary requires deriving them from underlying synthetic a priori principles, along with empirical data (e.g. Friedman, 1992, 2013); or (ii) empirical laws are necessary rules that obtain by virtue of specific objects' natures (e.g. Messina, 2017; Watkins, 2005). See also Breitenbach, 2018. I take no real stand on this debate—I am mainly concerned to survey secularized views of the value of laws of nature, and my basic point about Kant is just that he takes the idea of nature as a law-governed unity to be a purely regulative ideal rooted in 'reason' rather than inherent in nature as sensibly given to us. Kitcher's account is the most superficially compatible with the analogy I draw between Kant and Lewis. Watkins-style necessitation accounts draw Kant a bit closer to 'bottom-up' dispositional essentialists like Ellis (Messina, 2017, 137; see Sect. 7, below). And Friedman's view draws Kant a bit closer to rationalists like Descartes. But all parties to this debate can still agree that Kant thinks that particular laws must conform to the pure idea of a system of nature.

'best system' account. This approach, anticipated by Mill and Ramsey, analyzes laws of nature as truths that summarize the 'Humean mosaic' of basic natural properties at each spacetime point in whichever way best balances theoretical virtues of simplicity and strength.[68] Lewis's view is Humean in two senses: laws summarize the world without 'governing' it, and modal properties like causal dispositions are excluded from the mosaic that the laws summarize. Other best-system accounts of laws admit fundamental powers into nature.[69]

Lewis has been diagnosed as a "transcendental [i.e. Kantian] idealist" by Gordon Belot, tracing a chain of his theses: "the causal relations of our world are determined by which counterfactual statements are true (Lewis, 1986a); which counterfactual statements are true depends on the laws of nature (Lewis, 1986b); and the laws of nature depend on our cognitive capacities (Lewis, 1986c, Appendix C)."[70] Lewis does admit that his view "makes lawhood depend on us"—"a feature of the approach that I [Lewis] do not at all welcome!"—since the relevant standards of simplicity, strength, and their balance are just "to be those that guide us in assessing the credibility of rival hypotheses as to what the laws are."[71] This does not entail certain egregious kinds of relativism: since we can "take our actual standards as fixed" when asking what the laws would be under counterfactual circumstances, a best system analysis does not entail that eras or cultures with different standards have different laws of nature.[72] Still, Lewis is here closer to subjective idealism about laws and causes (albeit not the Humean mosaic) than to Kantian idealism, since our present cognitive standards of simplicity and strength are evidently more contingent than Kant intends for his 'ideas of reason'. Kant's transcendental subject is meant to be more generically rational than a subject of 'actual standards' of present-day scientific practice.

Kant's approach is still anthropocentric, in that he sees space and time as nothing more than forms of our sensible intuition—which is not shared by intuitive intellects like God, and maybe not even by other finite rational beings, like intelligent aliens.[73] Even so, it might be that all rational beings are beholden to an idea of systematic unity. God might thus intuit reality as a non-spatiotemporal but still law-governed whole. Or perhaps God's direct grasp of reality as a whole, for Kant, precludes cognition via laws of nature uniting its (derivative) parts.[74] Either way, laws' teleological connection to an idea of systematic unity shared at least by all rational beings like us, if not just by all rational beings, is less subjective than laws' dependence on 'actual standards' that vary across time and cultures.

[68] See Lewis, 1983, 1994. Compare Mill, 1843; Ramsey, 1978.

[69] E.g. see Demarest, 2017.

[70] Belot, 2010, 430. See also Belot (2022).

[71] Lewis, 1986c, 122.

[72] Lewis, 1986c, 122.

[73] For analysis of Kant's views on aliens and alternate forms of sensible intuition, see Belot (2022).

[74] Compare Kant, 2000, 5:406–408. I take no stand here on which of these options is closer to Kant's view.

Lewis himself later came to doubt that he could escape the "ratbag idealist" view that "we can change the laws[. . .]just by changing the way we think" simply by focusing on "our actual and present standards." Rather, we must "hope" if not "presuppose" that "nature is kind," in the sense that "the best system will be *robustly* best—so far ahead of its rivals that it will come out first under any standards of simplicity and strength and balance."[75] While it seems doubtful that this aim is satisfiable, given the wide scope of 'any' standards and the fact that even reasonable scientists often disagree, it is at least somewhat closer to Kant.[76] At any rate, both views raise a similar prospect: valuing the pursuit of laws of nature may be a way of valuing our own theoretical rationality. After all, for Lewis or Kant it is we rational beings who evidently 'govern' nature by imposing lawlike order on lawless mosaics or sensations. Of course, lawlike patterns are really 'out there' in the Humean mosaic, on Lewis's view. But so, too, are all the other patterns that are not laws. The only extra ontological weight laws have is their fit with standards of simplicity etc. that are framed as generically rational. So Lewis is right: his laws really do depend on us—if not in a ratbag idealist way, then in a loosely Kantian way. In this light, it seems fair to view Lewis, like Kant, as valuing human reason via ideas of natural law.[77]

Kant's path toward secularization suggests an analogy: the constructive role of the transcendental subject vis-à-vis the phenomenal world, for Kant, is broadly like God's creative role vis-à-vis nature, for earlier natural philosophers. Newton's view of space as the 'sensorium' of God even seems to have directly influenced Kant's account of space and time as pure 'forms' of human sensible intuition.[78] Likewise, Kant's claim that a total system of nature is required by reason seems to be a secularized legacy of intellectualist claims about natural order reflecting God's Reason. Kant's view of systematic order in nature as a merely "projected" unity or "regulative" ideal—"to be regarded not as given in itself, but as a problem only"[79]— is also productively viewed as a residue of religious views of a transcendent God, beyond nature. The constructive power of Kant's cognizing subject is a secularized recapitulation of Christian ideas of divine Creation. And Kant's creation myth has a rationalist tenor, despite his overt aim to analyze the limits of reason. Despite this

---

[75] Lewis, 1994, 479.

[76] This 'hope' that nature is 'kind' also evokes Duhem's appeal to an "act of faith" from which we are "powerless to rid [our] reason," in assuming that the "logical order" into which physical theory arranges experimental laws reflects "real relations among things" (Duhem, 1991, 26–27).

[77] Lewis's appeal to simplicity and strength arguably does less to establish that laws or causal relations have species-wide intersubjective validity than do Kantian tactics like the 'metaphysical deduction' of categories from forms of judgment or the 'transcendental deduction' meant to show that any object given to us via sense must be categorically determined. That said, standards like simplicity may inform Kantian science aimed at law-governed unity, as scientists judge which theories are most 'systematic'. Kant and Lewis may hope that reasonable people will agree, but this is a value-laden article of faith.

[78] Insole, 2011.

[79] *Critique of Pure Reason* A647/B675 (Kant, 1998).

critique, Kant treats 'rational' ideas of God, freedom, and nature's law-governed unity as regulative aims that give ultimate purpose to all our action and inquiry.

While Kant's metaphysics may be abstruse, the core idea of anthropocentric rationalism is intuitive: the value of the scientific drive to disclose law-governed order in nature derives from the value of our own basic power to impose rational order on the phenomena we encounter in experience. Pursuing laws of nature is thus valued as a kind of collective self-expression on the part of humanity. That is, it is valued as the joint expression of a basic human capacity—and, in particular, a power of reason or intellect.

Rationalistic accounts of the value of the scientific pursuit of law-governed order may thus be tied to contentious theories of human nature as comprising a basic power of 'reason' or 'intellect', often systematically opposed to 'sense', which supposedly makes us better than less intelligent animals, plants, and inanimate objects. But this hierarchical view of humans' bipartite nature as 'rational animals', 'embodied minds', or 'conditioned and yet free beings' is far from clearly true or basic. So, accounts of laws of nature and their value arguably should not be predicated on it. Of course, current analyses of laws of nature typically do not involve explicitly endorsing a view of human nature as two-sided and asymmetrically valuable. Still, intellectualist currents in secularized accounts of laws of nature are plausibly tied to rationalistic conceptions of human nature more implicitly.

Dynamism—here meaning any position that grounds the value of laws of nature in the inherent value of power—could be viewed, in this light, as a 'sensuous' correlate to intellectualist focus on 'rational' aspects of lawlike order. Together, these two approaches may suggest a more balanced picture of humans and their value. Or, more hierarchically, dynamists may insist that lawlike order is grounded in causal relations, not vice versa.[80] This inverts intellectualist views, if order represents the ideal or rational and causal power represents the natural or arational. This inversion could be valuable, in turn, if one takes an affirmation of physical reality to be better than the kind of deference to ideal Form or mathematical structure that one finds in Plato or newer idealisms. In this vein, Nietzsche criticizes the metaphysical drive to dismiss the sensuous world as a mere appearance of some higher reality, arguing it reflects an unhealthy "instinct of slander, detraction, and suspicion against life."[81] By contrast, he claims that "[t]oday we possess science precisely to the extent to which we have decided to *accept* the testimony of the senses"[82]—evoking earlier links between empiricism and non-intellectualist theology. Or dynamism about laws of nature could move further past contrasts like reason-sense and subject-object. For instance, the relevant idea of legislative power

[80] Compare Nancy Cartwright's view that "the source of order in nature is not laws but powers and mechanisms" (Cartwright, 2016, 56). She denies that there are "universal laws, laws that hold everywhere and everywhen" (ibid.; see also Cartwright, 1983, 1999; Cartwright & Ward, 2016; cf. Sklar, 2003). My 'dynamism' is less concerned with universality: causal powers can ground universal laws.

[81] *Twilight of the Idols* "'Reason' in Philosophy" §6 (Nietzsche, 1974b, 481).

[82] *Twilight of the Idols* "'Reason' in Philosophy" §3 (Nietzsche, 1974b, 484).

may be one that subsumes both arational causal power in inanimate objects and the intellectual power of cognizing humans.[83]

One might protest that to view *God* as a disembodied Mind, or *law* as a rational Ground, is not to worship *ourselves* as disembodied minds or just rational. But this is to misunderstand the charge of projection. Of course, as Feuerbach puts it, "God is not what man is" and "man is not what God is": humans are imperfect, finite, temporal, weak, and sinful, whereas God is perfect, infinite, eternal, almighty, and holy.[84] Still, as Feuerbach goes on to suggest, it could be that God's traits are ultimately just "our positive, essential qualities, our realities," projected beyond actual human limitations by our understanding in the fullest expression of its abstractive power.[85] God is humanity "purified" of all traits that the believer sees as evils or imperfections.[86] A disembodied God (not as Christ)[87] is a vision, not of how humans are in all respects, but of how it is best for us to be—insofar as we are rational or free 'spiritual' beings, not just animals or sensuous beings. And secular rationalism can carry the same ethical baggage, whether intended or not. In fact, the risk of one-sided intellectualism may be greater for us now, as narrow focus on the epistemic payoff of laws of nature comes to substitute for praise of a more multidimensional God.

6 Anthropocentric Dynamism: Valuing Human Power

Given that the secular legacy of intellectualism remains ethically loaded in ways that we might not endorse, it is important to consider how focus on God's power or free will can also be secularized. A broad tradition of dynamism, which grounds the value of laws in the value of power, again includes both more and less anthropocentric positions.

A first way to secularize voluntarist focus on God's freedom is by translating it into focus on human freedom. And one way to do this is to reground scientific drives to lawlike order in our pragmatic interests rather than our theoretical rationality. Even if law-governed order is not 'required by reason', as Kant claims, finding

[83] Compare my suggestions about 'cognitive dominion' vis-à-vis 'ecological twists' on Nietzsche, below.

[84] Feuerbach, 1957, 33.

[85] Feuerbach, 1957, 38–39. This influenced Marx's similar view of religion—compare e.g. Marx, 1978, 53.

[86] Feuerbach, 1957, 181.

[87] Feuerbach (1957, 155) suggests that Christianity distinctively views "the individual by himself" as "a perfect being." Its divinity is the perfection of the human species made present under the alienated guise of God's self-standing individuality, reflecting man's desire to individually achieve a human mode of being that is in fact realizable only by a social whole (82–83, 150–160, 182–184). This individualism is in turn given its "most unequivocal expression" in the doctrine of God's incarnation as Christ (154).

lawlike patterns in nature can clearly still be useful. By the same token, even if laws of nature do not give insight into approximate truths about unobservable reality, as scientific realists may hope, it can still be pragmatically valuable to find lawlike patterns in observable phenomena.

One main legacy of early modern voluntarist theology in secularized accounts of laws of nature, then, is a pragmatic streak typical of neo-empiricist views. God's Power has been partly reinvested in our own power to order experience in ways that have extra-epistemic value. Nor is this surprising, given voluntarism's historical ties to empiricism.

In this light, we may reconsider Lewis's best-system view, and in particular his appeal to simplicity as a criterion for lawhood. For contemporary empiricists like Bas van Fraassen, super-empirical criteria like simplicity are "*pragmatic* virtues," which "do not concern the relation between the theory and the world, but rather the use and usefulness of the theory." For van Fraassen, simplicity is therefore just "a function of our pleasures and interests."[88] It is possible, then, to reframe a Lewisian best-system account of laws of nature—although Lewis would presumably not do this—within a dynamist rather than rationalist framework, by stressing that simpler systems of law tend to be more useful.[89]

Van Fraassen's account has historical roots in earlier twentieth-century empiricism. Ernst Mach, for instance, also took the scientific preference for systematic organization of experimental laws and data to be largely pragmatic.[90] Machian instrumentalists point out that scientific interest in simplicity ensures that empirical information is ordered into cognitively manageable and readily communicable forms. Deference to simpler theories is thus justified pragmatically, with an eye toward an "exquisite economy of thought."[91]

This association between natural order and human practical control over empirical information also sheds light on prominent late modern critiques of science. For example, Heidegger associates pervasive technical order with the modern triumph of an ethos of Nietzschean will to power.[92] In other words, trading partly on intuitive ties between science and technology, critics in this vein argue that modern science is at root concerned with controlling things—at least cognitively, if not also practically.[93] Law-governed order may thus be viewed as one product of a broader drive toward calculability, which critics like Heidegger see as a perverse impulse to flatten all things into just 'resources' standing by for efficient use or technocratic

---

[88] van Fraassen, 1980, 87–88. Compare Rowbottom, 2010, 247.

[89] For recent pragmatically-oriented best system accounts, see Dorst, 2019; Jaag & Loew, 2020—both are framed as fleshing out alternatives to 'simplicity and strength', however, not as developing pragmatist interpretations of these standards. See also Cohen & Callender, 2009.

[90] Duhem is often grouped with Mach, but this ignores his notion of 'natural classification' (see note 76).

[91] Mach, 1895, 198.

[92] See Zimmerman, 1990, 34; Thomson, 2019, 179. Compare Jünger, 1993. See also Lichtenstein, 2021b.

[93] Compare e.g. Heidegger, 1977, 23. See also Young, 2002, Thomson, 2019.

management. So, even if Kant is right that a lawful system of nature is required by reason, this may simply go to show that reason requires exerting a kind of cognitive dominion or intellectual mastery over its objects. In this more abstract way, secularized science may again recapitulate aspects of voluntarist theology.

However, one does not have to worship a 'will to power' in order to find the drive to predict and manipulate the world, partly by way of finding lawlike patterns in it, to be generally good. Being useful is no crime, after all. And even if the view that *knowledge is power* can reflect an unhealthy control-fetish, it need not. Also, while Nietzsche's basic vision of science as the "transformation of nature into concepts for the purpose of mastering nature" may be approving, he still harshly criticizes scientistic drives to render all things 'calculable'.[94] It may even be to Nietzsche's credit that his critique of scientism is less sentimental than Heidegger's critique of 'Nietzschean' scientism: concerns about human control should be met with more respect for power in nature, not fear of ourselves. Even if science masters its objects, that is, this may just make it like all forces in nature.[95] A deeper worry is that faith in science can lead us to conflate mastery with rationalization and so to devalue both arational natural forces and non-scientific kinds of understanding.

A second basic way to secularize voluntarism is to translate God's freedom into natural contingency, without invoking any anthropomorphic will. In a human-centered outlook, this could involve viewing laws of nature as partial constraints on events rather than as comprehensively determining them. This tracks a wider divide between 'freedom-excluding' and 'freedom-granting' ideas of legality, as Yemima Ben-Menahem notes:

> On the first, usually deemed applicable to state officials, everything that they are not mandated by law to do is prohibited. On the second, usually deemed applicable to citizens, everything that is not prohibited by law is permitted. Similarly, we can think of the constraints imposed on natural processes either as necessitating everything that happens, or as excluding certain occurrences, but leaving a considerable amount of freedom: whatever is not excluded may happen.[96]

Adopting one versus the other of these two legal models may affect empirical scientific methodology—the freedom-excluding model may, for instance, lead physicists to try to derive the values of physical constants, like particles' masses, from underlying principles instead of accepting them as 'brute facts'.[97] But the choice can also have important ethical implications. For example, Ben-Menahem points out that different views of contingency and inevitability in history can impact divergent explanations of, and attitudes toward, issues like gender-based

---

[94] *The Will to Power* §610 (Nietzsche 1968, 328); *The Gay Science* §373 (Nietzsche 1974a, 335). For discussion, see Lichtenstein, 2021b.

[95] Not all interaction is or should be hierarchical, but many asymmetrical relations of influence in nature are unproblematic—and humans exerting control over various parts of nature is perfectly acceptable.

[96] Ben-Menahem, 2018, 18.

[97] Ben-Menahem, 2018, 19.

sociopolitical inequality.[98] In this light, there may be deep resonances between value-laden emphasis on contingency in nature and adopting a freedom-including model of law that secularizes some aspects of voluntarist theology.

In a distinct variation on this theme, Nancy Cartwright advocates for abandoning the idea of "universal rule of law," replacing laws with "powers and mechanisms" as "the source of order in nature."[99] One benefit of this, she argues, is ethical: "[t]he contingency of events that can be allowed once the universal rule of law is rejected can leave space for us to play an active role in what the world will come to be like."[100] Of course, others will insist that agency is compatible with exhaustive determination by universal natural laws.

None of this is to say that ethical or political commitment to specific ideals of freedom can or should settle the answers to questions like *'Will we be able to find deterministic laws of history?'* or *'Is there an underlying reason why the Higgs boson has a mass of 125 GeV, rather than some other value?'*. These are empirical questions, in large part. Still, ideas about freedom can affect appraisals of the prior probability of different answers to questions like these. Ethical or political values can also affect extra-epistemic aspects of scientific method, like assessing which questions are most worth addressing. And, for better or worse, social values may even affect our basic standards of explanatory value or theoretical virtue. But also vice versa: scientific metaphysics and methods can influence ethics, just as ethical ideals can influence approaches to science.

7 Ecocentric Dynamism: Valuing Impersonal Power

Not all paths to secularizing voluntarist theology involve focusing on human freedom or pragmatic interests. According to a more ecocentric version of dynamism, the value of pursuing lawlike order in science directly reflects the value of impersonal causal power or arational patterns in nature. For those who accept that humans are not the center of reality but who also do not posit a personal God, this approach to laws of nature has clear advantages, despite being commonly overlooked. It also avoids the claim that abstract form ultimately grounds physical reality—which is contentious metaphysics, but also, in many cases, a way of indirectly praising our own rationality or intelligence.

98 Ben-Menahem, 2018, 46–47.

99 Cartwright, 2016, 56.

100 Cartwright, 2016, 60. Cartwright qualifies that her focus on contingency is "not motivated by the free will problem," but rather stems from an aspiration to metaphysical modesty (ibid., 60). Still, conscious 'motivation' by ethical views is not the only way that metaphysics can reflect or reinforce ethical views.

One less anthropocentric way of secularizing emphasis on God's free will is to stress that laws of nature are contingent.[101] This could just mean that the specific content of laws of nature is not logically necessary or otherwise 'required by reason'. Intuitively, then, the actual laws of nature could have been different, whereas 1 + 1 is 2 in all possible worlds. For example, certain basic physical constants, like the elementary charge or the fine structure constant, may just be brute facts despite figuring in fundamental laws. It also may well be impossible to know a priori whether phenomena like action at a distance are too 'spooky' or 'unintelligible' to be real, or whether phenomena like straight-line motion are so 'natural' that they must be the physical default. Or, if laws are necessary insofar as they track necessary connections between causal powers and their effects, still, the existence of these causal powers is presumably contingent.[102] Laws of nature are also arational, if we reject personified gods, in the sense that no intelligent being has set them in place. Even if laws of nature provide reasons for (law-governed) phenomena, there are no reasons for basic laws themselves. Or, if causal relations are reasons for basic laws, still, most causal powers are wholly impersonal and arational. So, while intelligent beings like humans can use laws of nature to understand phenomena, there is a sense in which we ultimately cannot understand laws themselves: they are arational and so empirical.

For Newton, God's 'absolute' power was tied to the possibility of miracles, or violations of the laws that reflect God's regularly 'ordained' power.[103] But secular reinterpretations of divine Power may forgo miracles, insisting just that laws of nature are contingent (even if inviolable), and rejecting putative links between causal and rational order, or Spinozistic parallels between physical necessity and conceptual entailment.

In this light, consider ongoing efforts to analyze laws of nature in terms of a relation of "non-logical or contingent necessitation" between universals, as in views developed by Dretske, Tooley, and Armstrong.[104] In response to challenges to elaborate what this necessity relation comes to, Armstrong later identified it with "the causal relation[...]now hypothesized to relate types not tokens."[105] So, even if laws of nature do not govern the world themselves, merely descriptive laws may still track concrete powers which govern nature by causing regular physical effects.[106] One might value the scientific pursuit of laws of nature, then, at least

[101] Certain metaphysicians do argue that (some or all) laws of nature are *necessary* truths—e.g. see Bird, 2005; Ellis, 2001. Cf. Sidelle, 2002; Vetter, 2012. I address Bird-style dispositional essentialism, below.

[102] Compare Bird's claim that his 'dispositional essentialism' "entails that laws are necessary at least in the weak sense that they hold in all possible worlds where the grounding universal exists" (Bird, 2007, 169).

[103] See Harrison, 1995, 2002; Henry, 2009.

[104] Armstrong, 1983, 85. See also Dretske, 1983; Tooley, 1977, 1987.

[105] Armstrong, 1993, 422.

[106] Views of this sort may even have a Humean dimension. E.g. Demarest (2017) defends a "Potency-Best System Account of Laws," pairing a Humean view of *laws* (as systematizing, not

partly insofar as one values causal power itself—if not especially the causal power exerted by general kinds of concrete things, like magnetic fields generically. This is one basic path to secularizing voluntarist focus on God's freedom, by way of teasing impersonal Power further apart from anthropomorphic Will.

But the most de-anthropomorphized, and in this sense secularized, analyses of causality may be Humean ones that divorce it from power, too. Robust natural necessity or causation can be tied to actions and hence to agents, after all, whereas a thin notion of regular association (as in Humean supervenience) can be understood without agents.[107] On the other hand, to the extent that this Humean landscape is a more disenchanted field of inert patterns—just one static little fact after another,[108] with no active agents or even impersonal 'doers' to identify with—it may be harder to find deep value in it. This is one reason why Humean views fit well with focus on a quasi-transcendent human subject of practical or theoretical rationality: they present fundamental physical reality as devoid of immanent reason or power, in a way that can make human agency seem more valuable, by contrast.[109] Nature becomes a blank slate that has deeper meaning only in relation to us. This image of a world stripped of its own reason and power can be tied to Heidegger's critique of a modern drive to reduce all things to just resources standing by for use. Or, in an ecological twist on Nietzsche, one might embrace this objectifying aspect of Humean views, while also valuing how impersonal forces in nature equally work to objectify us.

Why would impersonal causal power or arational necessity be valuable, in itself? The idea is less strange than it may seem; in fact, it is intuitive. Think of being awestruck in the face of overwhelming natural forces, like hurricane winds or the sheer mass of a cliff face. Or consider how often polytheistic religions posit gods that personify powers like the sun and ocean, or phenomena like life and death or time and necessity. Maybe natural order is an image of primal forces broadly like these, then, and less so a mark of cosmic rationality or human interests. To pursue lawlike natural order out of basic respect for causal power or physical necessity is no more odd, in itself, than are attitudes of awe or respect before the causal power of divine forces. Nor does this have to be the only source of value in revealing laws of nature, for it to be one important source of this value.

Rationalists like Kant work to explain aesthetic experiences of the sublime, like those elicited by forces of nature, in terms of a feeling of the deeper superiority

---

governing, the world) with an anti-Humean *fundamental ontology* (including basic 'potencies' in nature). Compare Bird, 2007. Demarest (2017, 44) takes Drestke, Tooley, and Armstrong all to claim that laws govern nature. Note, then, that when I say laws describe causes without governing, I do not ascribe this view to Armstrong.

107 Thanks to Yemima Ben-Menahem for this suggestion.

108 Compare Lewis's description of Humean Supervenience as the view that "all there is to the world is a vast mosaic of local matters of particular fact, just one little thing and then another" (Lewis, 1986d, ix).

109 A similar critique applies to Kantians vis-à-vis (what they see as) inscrutable 'things in themselves'.

of reason over sense[110]—just as he works to relate law-governed order to a pure 'idea of reason'. But Kant may be wrong, in both cases. It may seem absurd to respect impersonal forces. But from another standpoint it seems equally absurd to treat natural science or aesthetic response to nature as just an occasion to celebrate our own intellect or power over nature. This is a basic failure of objectivity: a lack of receptivity to value in the broader world.[111]

Dynamists may further distinguish internal and external physical necessity, in the spirit of Aristotle's distinction between nature (*physis*) as "a cause of movement in the thing itself" and force (*dunamis*) as "a cause in something else, or in the thing itself regarded as something else."[112] Stones move downwards by nature, then, but upwards only "on compulsion and by force."[113] As the religious idea of an omnipotent Law-Giver is secularized, it can thus be replaced either by the idea of matter conforming to lawlike patterns by its own nature, or else by the idea of patterns in observable phenomena which are caused by underlying forces.[114] In this light, one might value lawlike order vis-à-vis either 'natural' or 'enforced' order. That is, building on Nietzsche's claim that the idea of laws of nature can be analyzed either so as to "comfortably accommodate the democratic instincts of the modern soul"—by reading "equality before the law" into nature[115]—or else in terms of "a tyrannically ruthless and pitiless execution of power claims," the value of laws of nature could be rooted in the value of either obedience to internally-given laws or compulsion by superior external force.[116] This difference may have ethical and political implications, as well as effects on scientific imagination. Inner necessity is intuitively tied to freedom, for instance, in a way that external force is not. Freedom, not to do otherwise, but in a sense closer to autonomy: acting in accord with one's own nature. We arguably fail to be free in this sense when we are weak-willed, or ruled by alien desires. But even a stone may be 'self-determined' insofar as its own mass makes it fall or press down, and 'unfree' insofar as it is cast up by force. So, ecocentric dynamism could be tied to seeing freedom as an inner necessity ubiquitous in nature. Or it could be tied to shifts away from moralized

110 Compare Kant, 2000, 5: 251–260.

111 For elaboration of this ideal of objectivity in application to art, see Lichtenstein, 2021c. On how Kant's aesthetics encodes a rationalistic view of humanity, and a distinct approach, see also Lichtenstein, 2019.

112 Aristotle, *De Caelo* 310b. See also Machamer, 1978.

113 Aristotle, *Eudemian Ethics* 1224a15.

114 Compare T. C. Hammond's claim that natural processes are "no longer conceived as the arbitrary determinations of creative power," but rather as "the necessary expressions of an inner reality developing itself in and by these very processes"—which he links to the idea of an immanent God (Hammond, 1911, 198). To others, 'arbitrary determinations of creative power' presumably sounds apt.

115 Compare Jacob & Jacob, 1980, 255–256; Christianson, 1984, 307 (cf. Force, 1990, 91–92).

116 Nietzsche, 2002, 30; see also Ben-Menahem & Ben-Menahem, 2020, 53. Compare Nietzsche, 1996, 216.

focus on freedom, toward frank recognition of external conditions of and limits on agency. After all, a stone falls to the ground 'by its nature' only given Earth's gravity.

Finally, lawlike order can be valued as constant being. By contrast, as Water Stace notes, the "unreality of the world" has often been associated with "its character as *flux*":

> In Plato's doctrine it is always on the becoming of the world that the predicate of unreality is fastened. Being is real, becoming, because it participates both in being and in non-being is the half-real. In Indian thought, too, it is especially the fleetingness of things which makes them unreal. And the unreality of time has always been a favorite doctrine of idealistic philosophers.[. . .]The passage of time means to most men primarily the loss of dear ones, the onset of old age, death. Therefore the flux of the world is, in the imagination of men, deeply identified with the evil of the world, the inner worthlessness of things. Hence if unreality means worthlessness, it is above all time and change which are felt to be unreal.[117]

Stace's focus is mysticism in religion. But it is no more inherently religious to value laws of nature as a sign of fixed order than it is to value them as a sign of cosmic intelligibility or the dignity of human reason. And if laws are analyzed as concrete patterns or in terms of causal powers, one avoids most of what can make Plato's view seem too otherworldly. Impermanence does not have to be worthless, for timeless being to merit our respect. But laws of nature can also help us to understand how change and stable form are interrelated.

Regardless of its exact elaboration, ecocentric dynamism about laws of nature can reflect a rejection of anthropocentrism and a basic embrace of environmental value. Whether one shows respect for causally-central natural kinds, or arational necessity, or brute patterns, or natural motion, or enforced order, or fixed being, one emphasizes traits that are fully expressed within impersonal nature at least as much as within humans. Physical necessity and causal power are not bound to intellect or reason in the way that laws of nature are, for religious or secular rationalists. But if lawlike order reflects the impact of inhuman causal powers, it is also less directly tied to human freedom or control over nature than it is for anthropocentric offshoots of voluntarism. By shifting focus away from distinctively human virtues and interests, views that connect lawlike order to more impersonal forces may help us to recover a deeper feeling for value in the natural world.

References

Armstrong, D. M. (1983). *What is a law of nature?* Cambridge University Press.

Armstrong, D. M. (1993). The identification problem and the inference problem. *Philosophy and Phenomenological Research, 53*, 421–422.

Beebee, H. (2000). The non-governing conception of Laws of nature. *Philosophy and Phenomenological Research, 61*(3), 571–594.

[117] Stace, 1952, 129–130. To be clear, Stace is here discussing religious mysticism and not laws of nature.

Beiser, F. C. (2002). *German idealism: The struggle against subjectivism, 1781–1801*. Harvard University Press.
Beiser, F. C. (2003a). *The romantic imperative: The concept of early German romanticism*. Harvard University Press.
Beiser, F. C. (2003b). Hegel and *Naturphilosophie*. *Studies in History and Philosophy of Science, Part A, 34*, 135–147.
Belot, G. (2010). Transcendental idealism among the Jersey metaphysicians. *Philosophical Studies, 150*, 429–438.
Belot, G. (2022). Ratbag idealism. In Y. Ben-Menahem (Ed.), *Rethinking the concept of Laws of nature*. Springer.
Ben-Menahem, Y. (2018). *Causation in science*. Princeton University Press.
Ben-Menahem, H., & Ben-Menahem, Y. (2020). The rule of law: Natural, human, and divine. *Studies in History and Philosophy of Science Part A, 81*, 46–54.
Bhogal, H. (2020). Humeanism about laws of nature. *Philosophy Compass, 15*(8), e12696.
Bird, A. (2005). The Dispositionalist conception of laws. *Foundations of Science, 10*, 353–370.
Bird, A. (2007). *Nature's metaphysics: Laws and properties*. Oxford University Press.
Boyle, R. (1686). A free enquiry into the vulgarly Receiv'd notion of nature; made in an essay, *Address'd to a Friend*. Printed by H. Clark for John Taylor. (Original work published 1686).
Breitenbach, A. (2018). Laws and ideal unity. In W. Ott & L. Patton (Eds.), *Laws of nature* (pp. 108–121). Oxford University Press.
Burtt, E. A. (1925). *The metaphysical foundations of modern science*. Kegan Paul/Trench/Trubner & Co.
Calvin, J. (1960). *Institutes of the Christian religion* (Vol. 1) (J. T. McNeill, Ed., and F. L. Battles, Trans.). Westminster John Knox Press. (Original work published 1536).
Carroll, J. (1994). *Laws of nature*. Cambridge University Press.
Cartwright, N. (1983). *How the Laws of physics lie*. Oxford University Press.
Cartwright, N. (1999). *The dappled world: A study of the boundaries of science*. Cambridge University Press.
Cartwright, N. (2016). Contingency and the order of nature. *Studies in History and Philosophy of Science Part C, 58*, 56–63.
Cartwright, N., & Ward, K. (Eds.). (2016). *Rethinking order: After the Laws of nature*. Bloomsbury.
Cassirer, E. (1910). *Substanzbegriff und Funktionsbegriff: Untersuchungen über die Grundfragen der Erkenntniskritik*. Bruno Cassirer.
Christianson, G. E. (1984). *In the presence of the creator: Isaac Newton and his times*. The Free Press.
Cohen, I. B. (Ed.). (1958). *Isaac Newton's papers and letters on natural philosophy and related documents*. Harvard University Press..
Cohen, J., & Callender, C. (2009). A better best system account of lawhood. *Philosophical Studies, 145*(1), 1–34.
Crombie, A. C. (1996). Infinite power and the Laws of nature: A medieval speculation. In *Science art, and nature in medieval and modern thought* (pp. 67–88). Hambledon Press.
Crosby, D. A. (2003). Transcendence and immanence in a religion of nature. *American Journal of Theology & Philosophy, 24*(3), 245–259.
Daston, L., & Stolleis, M. (Eds.). (2008). *Natural law and laws of nature in early modern Europe: Jurisprudence, theology, moral and natural philosophy*. Ashgate.
Davis, E. B. (1991). God, man, and nature: The problem of creation in Cartesian thought. *Scottish Journal of Theology, 44*, 325–348.
Demarest, H. (2017). Powerful properties, powerless Laws. In J. D. Jacobs (Ed.), *Causal powers* (pp. 38–53). Oxford University Press.
Descartes, R. (1982). *Principles of philosophy* (V. R. Miller, & R. P. Miller, Trans.). Kluwer. (Original work published 1644).
Descartes, R. (1991). *The philosophical writings of descartes, volume III: The correspondence* (J. Cottingham, R. Stoothoff, D. Murdoch, & A. Kenny, Trans.). Cambridge University Press.

Dorst, C. (2019). Towards a best predictive system account of Laws of nature. *British Journal for the Philosophy of Science, 70*(3), 877–900.
Dretske, F. I. (1983). Laws of nature. *Philosophy of Science, 44*(2), 248–268.
Duhem, P. (1991). *The aim and structure of physical theory* (2nd ed.) (P. P. Wiener, Trans.). Princeton University Press. (Original work published 1914).
Dupré, L. (1976). Transcendence and immanence as theological categories. *Proceedings of the Catholic Theological Society of America, 31*, 1–10.
Einstein, A. (1949). *The world as i see it* (A. Harris, Trans.). Watts & Co. (Original worked published 1934).
Ellis, B. (2001). *Scientific Essentialism*. Cambridge University Press.
Feuerbach, L. (1957). *The essence of Christianity* (G. Eliot, Trans.). Harper. (Original work published 1841).
Force, J. E. (1990). Newton's god of dominion: The Unity of Newton's theological, scientific, and political thought. In J. E. Force & R. H. Popkin (Eds.), *Essays on the context, nature, and influence of Isaac Newton's theology* (pp. 75–102). Kluwer.
Foster, M. B. (1934). The Christian doctrine of creation and the rise of modern natural science. *Mind, 43*(172), 446–468.
Foster, M. B. (1935). Christian theology and modern science of nature (I.). *Mind, 44*(176), 439–466.
Foster, M. B. (1936). Christian theology and modern science of nature (II.). *Mind, 45*(177), 1–27.
Friedman, M. (1992). Causal Laws and the foundations of natural science. In P. Guyer (Ed.), *The Cambridge companion to Kant* (pp. 161–199). Cambridge University Press.
Friedman, M. (2000). *A parting of the ways: Carnap, Cassirer, and Heidegger*. Open Court.
Friedman, M. (2013). *Kant's construction of nature: A Reading of the metaphysical foundations of natural science*. Cambridge University Press.
Garber, D. (2013). God, Laws, and the order of nature: Descartes and Leibniz, Hobbes, and Spinoza. In E. Watkins (Ed.), *The divine order, the human order, and the order of nature* (pp. 45–66). Oxford University Press.
Giere, R. N. (1999). *Science without Laws*. The University of Chicago Press.
Guthrie, W. K. C. (1975). *A history of Greek philosophy, v. IV: Plato: The man and his dialogues, early period*. Cambridge University Press.
Hammond, T. C. (1911). Immanence and transcendence. *The Irish Church Quarterly, 4*, 198–215.
Harrison, P. (1995). Newtonian science, miracles, and the laws of nature. *Journal of the History of Ideas, 56*, 531–553.
Harrison, P. (2002). Voluntarism and early modern science. *History of Science, 40*, 63–89.
Harrison, P. (2004). Was Newton a voluntarist? In J. E. Force & S. Hutton (Eds.), *Newton and Newtonianism: New studies, international archives of the history of ideas 188* (pp. 39–63). Kluwer.
Harrison, P. (2009). Voluntarism and the origins of modern science: A reply to John Henry. *History of Science, 47*(2), 223–231.
Harrison, P. (2013). Laws of nature in seventeenth-century England: From Cambridge Platonism to Newtonianism. In E. Watkins (Ed.), *The divine order, the human order, and the order of nature: Historical perspectives* (pp. 127–148). Oxford University Press.
Heidegger, M. (1977). *The question concerning technology and other essays* (W. Lovitt, Trans.). Garland.
Henry, J. (1994). 'Pray do not ascribe that notion to me': God and Newton's gravity. In J. E. Force & R. H. Popkins (Eds.), *The books of nature and scripture: Recent essays on natural philosophy, theology and biblical criticism in the Netherlands of Spinoza's time and the British Isles of Newton's time* (pp. 123–147). Springer.
Henry, J. (2004). Metaphysics and the origins of modern science: Descartes and the importance of Laws of nature. *Early Science and Medicine, 9*(2), 73–114.
Henry, J. (2009). Voluntarist theology at the origins of modern science: A response to Peter Harrison. *History of Science, 47*(1), 79–113.

Henry, J. (2011). Gravity and *De gravitatione*: The development of Newton's ideas on action at a distance. *Studies in History and Philosophy of Science Part A, 42*, 11–27.
Henry, J. (2014). Newton and action at a distance between bodies—A response to Andrew Janiak's 'Three Concepts of Causation in Newton. *Studies in History and Philosophy of Science Part A, 47*, 91–97.
Hildebrand, T. (2020). Non-Humean theories of natural necessity. *Philosophy Compass, 15*, e12662.
Insole, C. (2011). Kant's transcendental idealism and Newton's divine sensorium. *Journal of the History of Ideas, 72*(3), 413–436.
Jaag, S., & Loew, C. (2020). Making best systems *best for us*. *Synthese, 197*, 2525–2550.
Jacob, J. R., & Jacob, M. C. (1980). The Anglican origins of modern science: The metaphysical foundations of the Whig constitution. *Isis, 71*(2), 251–267.
Janiak, A. (2008). *Newton as philosopher*. Cambridge University Press.
Janiak, A. (2013). Three concepts of causation in Newton. *Studies in History and Philosophy of Science Part A, 44*, 396–407.
Jünger, E. (1993). Total mobilization. In R. Wolin (Ed.), *The Heidegger controversy: A critical reader* (pp. 119–139). The MIT Press. (Original work published 1930).
Kant, I. (1998). *The critique of pure reason* (P. Guyer, & A. W. Wood, Trans.). Cambridge University Press. (Original work published 1781 (1st ed.), 1787 (2nd ed.))
Kant, I. (2000). *Critique of the power of judgment* (P. Guyer, & E. Matthews, Trans.). Cambridge University Press. (Original work published 1790).
Kedar, Y., & Hon, G. (2017). 'Natures' and 'Laws': The making of the concept of law of nature—Robert Grosseteste (c. 1168–1253) and Roger Bacon (1214/1220–1292). *Studies in the History and Philosophy of Science, 61*, 21–31.
Kitcher, P. (1986). Projecting the order of nature. In R. E. Butts (Ed.), *Kant's philosophy of physical science* (pp. 201–235). Springer.
Klaaren, E. M. (1977). *Religious origins of modern science: Belief in creation in seventeenth-century thought*. Eerdmans.
Koyré, A. (1965). *Newtonian Studies*. Chapman & Hall.
Kuhn, T. S. (1977). Objectivity, value judgment, and theory choice. In *The essential tension: Selected studies in scientific tradition and change* (pp. 320–339). The University of Chicago Press.
Lange, M. (2000). *Natural Laws in scientific practice*. Oxford University Press.
Lange, M. (2009). *Laws and Lawmakers*. Oxford University Press.
Lewis, D. (1983). New work for a theory of universals. *Australasian Journal of Philosophy, 61*, 343–377.
Lewis, D. (1986a). Causation. In *Philosophical papers* (Vol. 2, pp. 159–213). Oxford University Press.
Lewis, D. (1986b). Counterfactual dependence and time's arrow. In *Philosophical papers* (Vol. 2, pp. 32–66). Oxford University Press.
Lewis, D. (1986c). A subjectivist's guide to objective chance. In *Philosophical papers* (Vol. 2, pp. 83–132). Oxford University Press.
Lewis, D. (1986d). Introduction. In *Philosophical papers* (Vol. 2, pp. ix–xvii). Oxford University Press.
Lewis, D. (1994). Humean Supervenience Debugged. *Mind, 103*, 473–490.
Lichtenstein, E. I. (2019). Sensory Force, sublime impact, and beautiful form. *British Journal of Aesthetics, 59*(4), 449–464.
Lichtenstein, E. I. (2021a). (Mis)Understanding scientific disagreement: Success versus pursuit-worthiness in theory choice. *Studies in History and Philosophy of Science Part A, 85*, 166–175.
Lichtenstein, E. I. (2021b). Classical form or modern scientific rationalization? Nietzsche on the drive to ordered thought as apollonian power and Socratic pathology. *Journal of Nietzsche Studies, 52*(1), 105–134.
Lichtenstein, E. I. (2021c). Artistic objectivity: From Ruskin's pathetic fallacy to creative receptivity. *British Journal of Aesthetics, 61*(4), 505–526.

Lichtenstein, E. I. (2021d). How anti-humeans can embrace a thermodynamic reduction of Time's causal arrow. *Philosophy of Science, 88*(5), 1161–1171.
Loewer, B. (1996). Humean supervenience. *Philosophical Topics, 24*, 101–127.
Longino, H. E. (1995). Gender, politics, and the theoretical virtues. *Synthese, 104*(3), 383–397.
Lovejoy, A. O. (1936). *The great chain of being: A study of the history of an idea*. Harvard University Press.
Mach, E. (1895). On the economical nature of physical inquiry (T. J. McCormack, Trans.). In *Popular scientific lectures* (pp. 186–213). Open court.
Machamer, P. K. (1978). Aristotle on natural place and natural motion. *Isis, 69*(3), 377–387.
Malet, A. (1997). Isaac Barrow on the Mathematization of nature: Theological voluntarism and the rise of geometrical optics. *Journal of the History of Ideas, 58*(2), 265–287.
Marx, K. (1978). Contribution to the critique of Hegel's *philosophy of right*: Introduction. In R. C. Tucker (Ed.), *The Marx-Engels reader* (2nd ed., pp. 53–65). Princeton University Press.
Maudlin, T. (2007). *The metaphysics within physics*. Oxford University Press.
Messina, J. (2017). Kant's necessitation account of Laws and the nature of natures. In A. Breitenbach & M. Massimi (Eds.), *Kant and the Laws of nature* (pp. 131–149). Cambridge University Press.
Mill, J. S. (1843). *A system of logic, rationcinative and inductive*. Cambridge University Press.
Needham, J. (1951a). Human Laws and Laws of nature in China and the West (I). *Journal of the History of Ideas, 12*(1), 3–30.
Needham, J. (1951b). Human Laws and Laws of nature in China and the West (II): Chinese civilization and the Laws of nature. *Journal of the History of Ideas, 12*(2), 194–230.
Newton, I. (1730). *Opticks: Or, a treatise of the reflections, refractions, inflections, and Colours of light* (4th ed.). William Innys.
Nietzsche, F. (1968). *The will to power* (W. Kaufmann, & R. J. Hollingdale, Trans.). Vintage.
Nietzsche, F. (1974a). *The gay science: With a prelude in rhymes and an appendix in songs* (W. Kaufmann, Trans.). Vintage. (Original work published 1882).
Nietzsche, F. (1974b). Twilight of the idols. In W. Kaufmann (Ed. & Trans.). *The portable Nietzsche* (pp. 463–564). Penguin. (Original work published 1889).
Nietzsche, F. (1996). *Human, all too human* (R. J. Hollingdale, Trans.). Cambridge University Press. (Original work published 1878).
Nietzsche, F. (2002). *Beyond good and evil* (R. Horstmann, & J. Norman, Eds., and J. Norman, Trans.). Cambridge University Press. (Original work published 1886).
Oakley, F. (1961a). Christian theology and the Newtonian science: The rise of the concept of the Laws of nature. *Church History, 30*(4), 433–457.
Oakley, F. (1961b). Medieval theories of natural law: William of Ockham and the significance of the voluntarist tradition. *American Journal of Jurisprudence, 6*, 65–83.
Osler, M. J. (1991). Fortune, fate, and divination: Gassendi's voluntarist theology and the baptism of epicureanism. In M. J. Osler (Ed.), *Atoms, pneuma, and tranquillity* (pp. 155–174). Cambridge University Press.
Osler, M. J. (1994). *Divine will and the mechanical philosophy: Gassendi and Descartes on contingency and necessity in the created world*. Cambridge University Press.
Ramsey, F. (1978). *Foundations*. Routledge and Kegan Paul.
Roberts, J. T. (2008). *The law-governed universe*. Oxford University Press.
Rowbottom, D. P. (2010). What scientific progress is not: Against Bird's epistemic view. *International Studies in the Philosophy of Science, 24*(3), 241–255.
Ruby, J. E. (1986). The origins of scientific 'law'. *Journal of the History of Ideas, 47*, 341–359.
Schneider, S. (2007). What is the significance of the intuition that Laws of nature govern? *Australasian Journal of Philosophy, 85*, 307–324.
Shapin, S. (1981). Of gods and kings: Natural philosophy and politics in the Leibniz-Clarke disputes. *Isis, 72*(2), 187–215.
Sidelle, A. (2002). On the metaphysical contingency of laws of nature. In T. S. Gendler & J. Hawthorne (Eds.), *Conceivability and possibility* (pp. 309–336). Clarendon Press.
Sklar, L. (2003). Dappled theories in a uniform world. *Philosophy of Science, 70*(2), 424–441.

Snobelen, S. D. (2001). 'God of Gods, and Lord of lords': The theology of Isaac Newton's general scholium to the principia. *Osiris, 16*, 169–208.

Spinoza, B. (2007). *Theological-political treatise* (J. Israel, Ed., & M. Silverthorne, and J. Israel, Trans.). Cambridge University Press. (Original work published 1670).

Stace, W. T. (1952). *Time and eternity: An essay in the philosophy of religion*. Princeton University Press.

Suárez, F. (1944). *Selections from three works* (G. L. Williams, A. Brown, & J. Waldron, Trans.). Clarendon Press.

Thomson, I. (2019). Technology, ontotheology, education. In A. J. Wendland, C. Merwin, & C. Hadjioannou (Eds.), *Heidegger on technology* (pp. 174–193). Routledge.

Tooley, M. (1977). The Laws of nature. *Canadian Journal of Philosophy, 7*, 667–698.

Tooley, M. (1987). *Causation: A realist approach*. Clarendon Press.

van Fraassen, B. C. (1980). *The scientific image*. Oxford University Press.

Vetter, B. (2012). Dispositional essentialism and the Laws of nature. In A. Bird, B. Ellis, & H. Sankey (Eds.), *Properties, powers and structures: Issues in the metaphysics of realism* (pp. 201–215). Routledge.

Vlastos, G. (1965). Degrees of reality in Plato. In R. Bambrough (Ed.), *New essays on Plato and Aristotle* (pp. 1–20). Routledge.

Vlastos, G. (1966). A metaphysical paradox. *Proceedings and Addresses of the American Philosophical Association, 39*, 5–19.

Ward, B. (2007). Laws, explanation, governing, and generation. *Australasian Journal of Philosophy, 85*(4), 537–552.

Watkins, E. (2005). *Kant and the metaphysics of causality*. Cambridge University Press.

Westfall, R. S. (1986). The rise of science and the decline of orthodox Christianity: A study of Descartes, Kepler, and Newton. In D. C. Lindberg & R. L. Numbers (Eds.), *God and nature* (pp. 218–237). University of California Press.

Young, J. (2002). *Heidegger's Later Philosophy*. Cambridge University Press.

Zimmerman, M. E. (1990). *Heidegger's confrontation with modernity: Technology, politics, and art*. Indiana University Press.

Index

Y. Ben-Menahem (ed.), *Rethinking the Concept of Law of Nature*,
Jerusalem Studies in Philosophy and History of Science,
https://doi.org/10.1007/978-3-030-96775-8

MIX
Papier aus verantwortungsvollen Quellen
Paper from responsible sources
FSC® C105338

If you have any concerns about our products,
you can contact us on
ProductSafety@springernature.com

In case Publisher is established outside the EU,
the EU authorized representative is:
Springer Nature Customer Service Center GmbH
Europaplatz 3, 69115 Heidelberg, Germany

Printed by Libri Plureos GmbH
in Hamburg, Germany